Cilia and Nervous System Development and Function

Kerry L. Tucker • Tamara Caspary
Editors

Cilia and Nervous System Development and Function

Springer

Editors
Kerry L. Tucker
Anatomy and Cell Biology
University of Heidelberg
Heidelberg, Germany

Tamara Caspary
Human Genetics
Emory University School
of Medicine
Atlanta, Georgia, USA

ISBN 978-94-007-5807-0 ISBN 978-94-007-5808-7 (eBook)
DOI 10.1007/978-94-007-5808-7
Springer Dordrecht Heidelberg New York London

Library of Congress Control Number: 2013930057

Printed on acid-free paper

Springer is part of Springer Science+Business Media (www.springer.com)

Acknowledgements

K.L.T. and T.C. would like to thank Martijn Roelandse at Springer-Verlag for his patient support of the whole project. K.L.T is especially grateful to Frau Sara Doll at the Institute of Anatomy at the University of Heidelberg for enthusiastic assistance with research into the history of cilia investigations in the last several centuries.

Contents

Cilia and Nervous System Development and Function

Introduction

Investigation of the cellular organelle called the cilium has an august history dating back to no less a personage than Antonie Philips van Leeuwenhoek at the end of the seventeenth century. Using his handmade, palm-sized microscope, Leeuwenhoek observed protozoa propelled through water by the coordinated beating of sheets of motile cilia. Ever since, biologists have been fascinated by this tiny organelle. Cilia are 1–3 μm long, 250 nm wide, microtubule-based organelles projecting from the plasma membrane of unicellular organisms and of many cell types in the animal, protist, and to a very limited extent, plant and fungi kingdoms. The term "cilium" is Latin for eyelid, and by association an eyelash, thus referring not only to the hair-like structures visible at high magnification, but also calling up images of the wave-like movement made by the lashes in the blink of an eye. The Danish naturalist Otto Friedrich Müller coined the term in his book *Animalcula Infusoria; Fluvia Tilia et Marina* (1786) to describe certain members of the protozoan phylum Ciliophora. Despite their fascinating rhythmic beating patterns, clarifying the ultrastructural basis of ciliary motility took another 250 years and was only made possible thanks to advances in electron microscopy and biochemistry.

The second half of the nineteenth century brought the discovery of a variant form of cilia, which is now called the primary cilium, although who first recorded its existence is somewhat controversial. Some authors refer to a two-page description of the sea lamprey inner ear epithelia published by Alexander Ecker ("Flimmerbewegung im Gehörorgan von *Petromyzon marinus*." *Archiv fur Anatomie, Physiologie und Wissenschaftliche Medicin*, 1844, pp. 520–521). Though Ecker did see an epithelium, in which the individual cells bore a single cilium, his article very clearly identifies the cilia as *motile*. The first true description of primary cilia probably comes from an analysis of lancelet development, in which the epithelium of the surface ectoderm could be shown to bear individual cilia (Aleksandr Kovalevskij, "Entwicklungsgeschichte des *Amphioxus lanceolatus*." Mémoires de l'Académie impériale des sciences de St. Pétersbourg VII Série Tome XI, 1867, pp. 1–17).

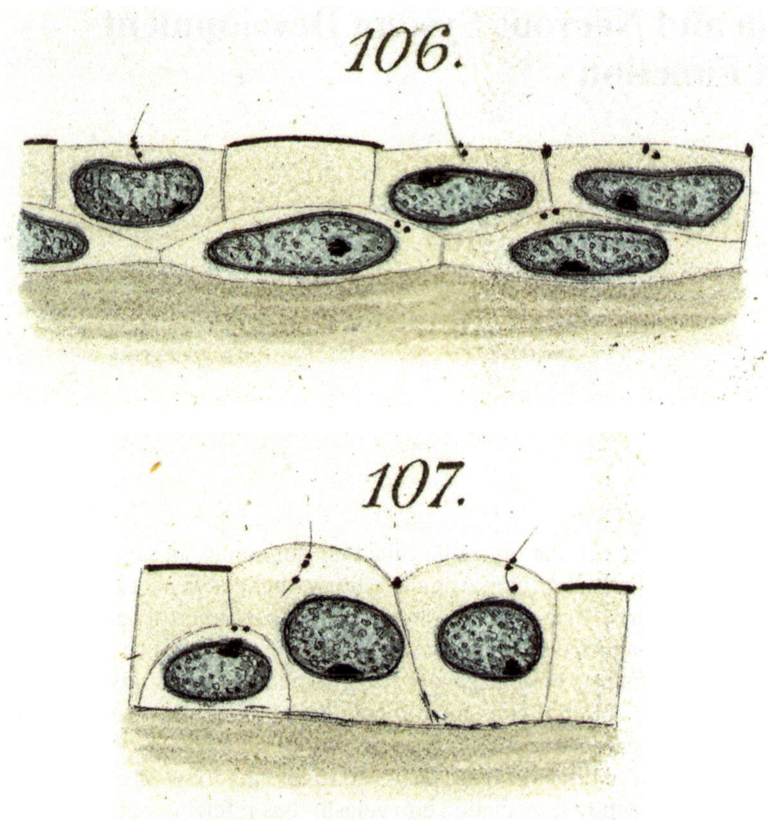

Fig. 1 First published image of primary cilia in human tissue, epithelial cells of the seminal vesicle. The primary cilium can be seen protruding from the apical edge of the cells, described in the original figure legend as a "Central-Flagellum in the superficial cells." K. W. Zimmermann, "Beiträge zur Kenntniss einiger Drüsen und Epithelien." *Archiv für Mikroskopische Anatomie und Entwicklungsgeschichte* (1898) V. 52(3), pp. 552–706

Several reports followed sporadically over the next three decades, but they did not arouse much interest, until the German anatomist Karl Wilhelm Zimmermann became the first person to describe primary cilia in mammals (Fig. 1, "Beiträge zur Kenntniss einiger Drüsen und Epithelien." *Archiv für Mikroskopische Anatomie und Entwicklungsgeschichte* (1898) V. 52(3), pp. 552–706).

Describing a histological preparation of the seminal vesicle of an executed criminal (Fig. 1), he wrote:

> *Wenn auch nicht in allen Fällen, so doch in vielen, sah ich auch über das oberflächliche Körperchen hinaus den Verbindungsfaden sich fortsetzen und frei in das Lumen hineinragen. Wir haben also in den genannten Fällen Einrichtungen ("Centralgeissel"), wie wir sie in gewissen Abschnitten der Niere durchaus als Regel kennen gelernt haben. Ob in den anderen Fällen der überaus feine "Aussenfaden" prinzipiell fehlte, ob er abgerissen war, oder ob sein Fehlen nur vorübergehend, also periodisch ist, oder schliesslich ob er zwar*

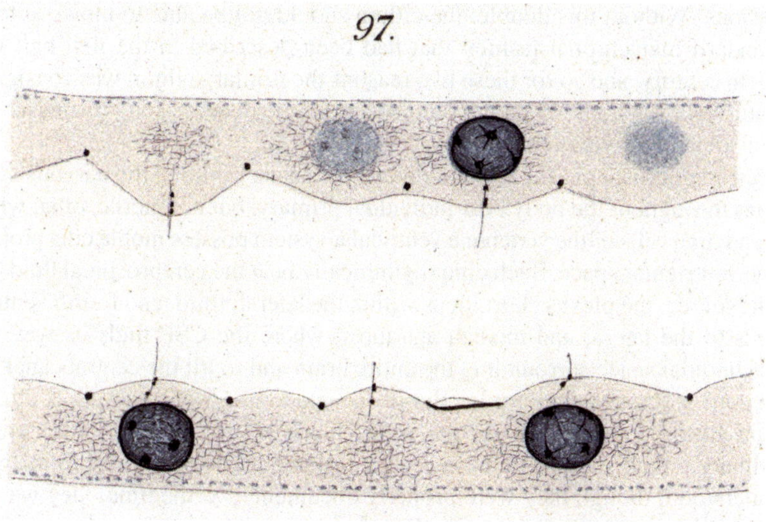

Fig. 2 First published image of primary cilia in mammalian tissue, epithelial cells of the distal convoluted tubule of the rabbit kidney. The primary cilium can be seen protruding from the apical edge of the cells, described in the original figure legend as a "Central-Flagellum". K. W. Zimmermann, "Beiträge zur Kenntniss einiger Drüsen und Epithelien." *Archiv für Mikroskopische Anatomie und Entwicklungsgeschichte* (1898) V. 52(3), pp. 552–706

vorhanden, aber an die Zelloberfläche angelegt und mit ihr künstlich verklebt war, so dass ich ihn nicht sehen konnte, das sind Fragen, in welche weitere Untersuchungen Licht bringen müssen, sowie in die Bedeutung der Einrichtung überhaupt.

Though not in every case, but indeed in most, I saw this "connecting thread" continuing above the superficially-located [basal] body and extending freely into the lumen. We have then, in the cases mentioned here, structures ("Central-Flagellum") similar to those that we regularly observed in certain segments of the kidney.[1] Whether, in those other cases [in which cilia were not found], this extremely fine "outer thread" was missing altogether; whether it was torn off; or whether it was missing temporarily, appearing only periodically in the tissue; or whether it was actually present but lying close to the cell surface and artificially glued to it so that I could not distinguish it, these are questions which further investigations must illuminate, including the significance of the whole structure itself.[2]

Zimmermann was able to distinguish primary from motile cilia, to catalog their presence in a number of organs in both rabbit (Fig. 2) and man (Fig. 1), and to hypothesize a sensory function for them; perhaps for these reasons he is consistently credited in the literature with the discovery of this organelle.

In contrast to motile cilia, only one copy of the so-called "primary" cilium is found in each cell, and over half a century after Zimmermann's discovery, ultrastructural analysis revealed that the central microtubule doublet was missing from these

[1] Earlier in the article, Zimmermann describes primary cilia jutting into the lumen from the epithelia of various tubules of the rabbit kidney (Fig. 2).

[2] Translation courtesy of C. Lulu Bradford.

protrusions. Without this doublet the cilium should not be able to move, at least in the standard metachronal fashion that had been described in the first half of the twentieth century, and so for these two reasons the primary cilium was consigned to the status of a vestigial organelle, reminiscent of the decades-long dismissal of the appendix and the thymus as "vestigial" organs.

Electron microscopic studies first described primary cilia in the nervous system. Neurons throughout the body bear individual primary, but not motile, cilia, whereas the ependymal cells of the vertebrate ventricular system possess motile cilia projecting into the ventricular space. Such cilia rhythmically beat the cerebrospinal fluid (CSF) from its source, the plexus choroideus within the lateral, third, and fourth ventricles, onwards to the lateral and median apertures, where the CSF finds its way to the subarachnoidal space surrounding the entire brain and to fill the central canal of the spinal cord. Although the motile cilia of the ependyma were viewed as clinically relevant almost immediately for the ventricular swelling seen in hydrocephalus, the primary cilia displayed by neurons and the stem cells giving rise to them were misunderstood; though they were properly documented at the time, they were also written off as vestigial and promptly forgotten.

In the past 10 years, this long-neglected organelle has emerged to take its rightful place in the spotlight, as a torrent of research points to a crucial role for primary cilia in the development and function of the central nervous system. A common theme of these studies is the critical dependency upon cilia of signal transduction of the Sonic hedgehog and (more recently) the Wnt signaling pathways to regulate fate decisions and morphogenesis. It is also becoming increasingly clear that both primary and motile cilia are important in the activity of the nervous system, including postnatal neurogenesis, the primary processing of sensory information, the control of body mass, and higher faculties, such as behavior and cognition, where they serve as "antennae" for neurons to sense and process their environment. In this book we describe the structure and behavior of cilia and the various tissues throughout the brain and spinal cord that depend on cilia for their proper development and operation. In doing so, we explore the many responsibilities of cilia in neurogenesis and neural activity.

The book is divided into chapters that cover the following topics. We start with a detailed ultrastructural and molecular description of primary and motile cilia (Chap. 1: Primary and Motile Cilia: Their Ultrastructure and Ciliogenesis). The manifold roles that cilia play in the development of the nervous system are explained in the next two chapters (Chap. 2: Primary Cilia, Sonic Hedgehog Signaling, and Spinal Cord Development; Chap. 3: Primary Cilia and Brain Development). This is followed by several chapters focusing on the regulatory control primary cilia exert in neuronal activity in the postnatal mammalian body, which is just coming to light (Chap. 4: Primary Cilia in Cerebral Cortex: Growth and Functions on Neuronal and Non-Neuronal Cells; Chap. 5: Primary Cilia and Inner Ear Sensory Epithelia; Chap. 6: Neuronal Cilia and Obesity). There follows a chapter examining the motile cilia found in the ventricles of the brain (Chap. 7: Motile Cilia and Brain Function: Ependymal Motile Cilia Development, Organization, Function, and Their Associated Pathologies). Finally, in the last two chapters we discuss the pathological consequences that arise when cilia disappear or do not work properly (Chap. 8: Primary Cilia and Brain Cancer; Chap. 9: Abnormalities of the Central Nervous System Across the Ciliopathy Spectrum). In this book we can of course only graze the surface of the deep pool of knowledge gathered about cilia and their control of virtually every aspect of organogenesis, and increasingly, their more recently discovered involvement in function/dysfunction in many organs of the mammalian body. By restricting our attention to the nervous system, we hope to provide both a generous introduction for the interested layperson, as well as a detailed exposition for those curious neuroscientists who may wish to delve more deeply into the field.

Kerry L. Tucker and Tamara Caspary

Chapter 1
Primary and Motile Cilia: Their Ultrastructure and Ciliogenesis

Sigrid Hoyer-Fender

Contents

Abstract Cilia are microtubule-based organelles surrounded by a specialized ciliary membrane. They emerge from basal bodies situated underneath the cell membrane. Basal bodies not only anchor cilia inside the cytoplasm but also they are essential for ciliary assembly. The basal body either originates from the oldest centriole of a pair of centrioles found at the centrosome, the main microtubule organizing center of the cell, to give rise to one primary cilium, or is generated in large numbers prior to the formation of multiple cilia. Large numbers of motile cilia are generally present on the cell surface of epithelia specialized for fluid movement. Motile cilia have a 9+2 axonemal structure with nine outer microtubule

S. Hoyer-Fender (✉)
Johann-Friedrich-Blumenbach-Institute of Zoology and
Anthropology-Developmental Biology, Georg-August-University of Göttingen,
GZMB, Ernst-Caspari-Haus, Justus-von-Liebig-Weg 11, Göttingen, Germany
e-mail: shoyer@gwdg.de

K.L. Tucker and T. Caspary (eds.), *Cilia and Nervous System Development and Function*,
DOI 10.1007/978-94-007-5808-7_1, © Springer Science+Business Media Dordrecht 2013

doublets surrounding two centrally-located singlet microtubules, and additional accessory structures. By contrast, primary cilia are found as single entities on most cells of the body. Their axoneme is of a 9+0 organization i.e. the centrally located singlet microtubules as well as the accessory structures are missing. Primary cilia are generally immotile. However, they are essential for sensing and transducing environmental signals. Primary cilia are therefore critical for embryonic and postnatal development, as well as for tissue homeostasis in adulthood. Impaired ciliary function causes a tremendous number of severe and diverse human diseases. Thus, unravelling the molecular mechanisms of cilium generation and maintenance is of crucial importance not only for medical research. In the past decade, pioneering studies have provided important insights into the underlying molecular mechanisms and the different players involved. This chapter addresses the differences and similarities in structure and function of motile and primary cilia as well as their assembly. Electron microscope data of ciliary structure and assembly are combined with molecular data to present an overview of the mechanisms involved in centriole and basal body generation, cilium formation and maintenance, cilium disassembly, and the diverse functions that are ascribed to cilia.

Keywords Acentriolar pathway • Alar sheets • Axoneme • Basal body • Basal foot • Cartwheel • Centriole disengagement • Centriolar pathway • Centriole • Centriole duplication • Centrosome • Centrosome splitting • Centrosome cycle • Ciliary membrane • Ciliary pocket • Ciliary pore • Ciliogenesis • Ciliopathies • Cilium disassembly • Daughter centriole • Distal appendages • Intraflagellar transport • Mother centriole • Motile cilia • MTOC • Necklace • PCM • Procentriole • Primary cilia • Rootlet • Subdistal appendages • Transition fibers • Transition zone

1.1 Introduction

Cilia are membrane-bound, microtubule based organelles that in most cases project from the cell surface into the cellular environment. Each cilium is anchored at the cell membrane by its basal body. Based on their ultrastructural organization cilia can be classified into a few subtypes. However, they observe a wide variety in overall appearance and execute a lot of diverse functions. Additionally, besides common cilia highly modified cilia exist that do not project from the cell surface. Instead, these modified cilia are found in the cytoplasm mostly of sensory cells and contribute to sensation (e.g. the connecting cilium of the rod cells of the mammalian retina; De Robertis 1956). Cilia are evolutionary conserved organelles present throughout the animal kingdom. In invertebrates, however, they are restricted to sensory neurons. Higher plants and fungi do not possess cilia. In mammals, cilia are present on almost all cells of the body. According to their ubiquitous occurrence cilia are essential organelles, and it is becoming more and more obvious that ciliary malfunction causes a large number of severe diseases.

1.2 Overview of Ciliary Structure and Function

The core structure of all eukaryotic cilia and flagella is the axoneme bounded by the ciliary membrane. Although continuous with the plasma membrane, the ciliary membrane is a specialized compartment to which molecules gain restricted access. The axoneme consists of a cylindrical array of nine microtubule (MT) doublets. Each MT doublet is composed of one complete MT consisting of 13 protofilaments, the A tubule, to which an incomplete B tubule consisting of 10 protofilaments only is attached. Protofilaments itself are assembled by the polymerization of αβ-tubulin dimers. Motile cilia, generally, in addition harbour two centrally-located singlet microtubules and accessory structures. Motile cilia are therefore mostly of a 9+2 type (Fig. 1.1a). Commonly, they concentrate in large numbers on the surface of specific mammalian cells. Their orchestrated beating is essential for fluid and cell movement. The long flagellum of the sperm can be viewed as a single motile cilium and shares its underlying 9+2 structure. However, most cells in mammals bear one single immotile cilium on their surface called the primary cilium or monocilium (Wheatley et al. 1996). Primary cilia are mostly of a 9+0 organization i.e. they lack the central singlet microtubules (Fig. 1.1b). Additionally, accessory structures are also absent. Primary cilia are often adapted to serve specialized sensory functions. Moreover, there are also examples of 9+2 immotile cilia (e.g. olfactory receptor cilia, Fig. 1.1c; Moran et al. 1982) and 9+0 motile cilia (e.g. cilia of node cells; Fig. 1.1d; Sulik et al. 1994).

Motile cilia are found on the surface of specialized epithelia usually in large numbers. Their coordinated beating propels fluid and cells over the epithelial surface. Multiciliated epithelia are found in mammals in the respiratory tract, the paranasal sinuses, the oviduct, and the ventricular system of the brain (Fawcett and Porter 1954; Sorokin 1968). In the human airway, motile cilia are important to propel inhaled material out of the lung. Moreover, it has been found that these motile cilia

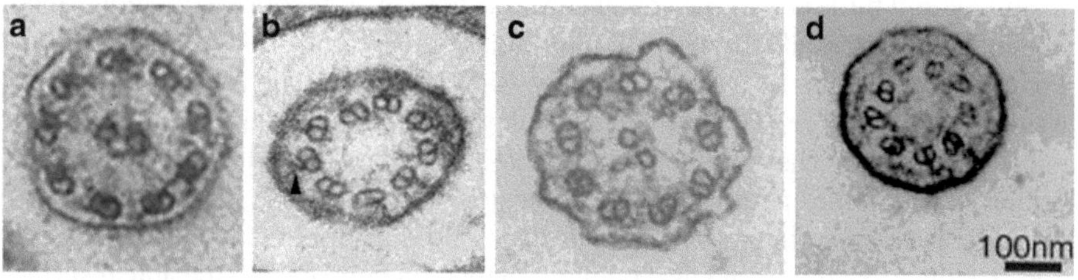

Fig. 1.1 Transmission electron micrographs of ciliary cross-sections. (**a**) A motile cilium from rat brain ependymal cells with typical 9+2 organization. ×67,000 (With permission from Brightman and Palay 1963). (**b**) Cross-section through a primary cilium of 3T6 fibroblasts with typical 9+0 organization. Y-shaped bridges are seen that tether the MT doublets to the membrane (*arrowhead*). ×77,000 (With kind permission from Wheatley 1972) (**c**) Cross-section through an olfactory receptor cilium of man with 9+2 structure but missing dynein arms. × 83,100 (With permission from Moran et al. 1982). (**d**) Cross section through a nodal cilium of mice (With permission from Nonaka et al. 1998)

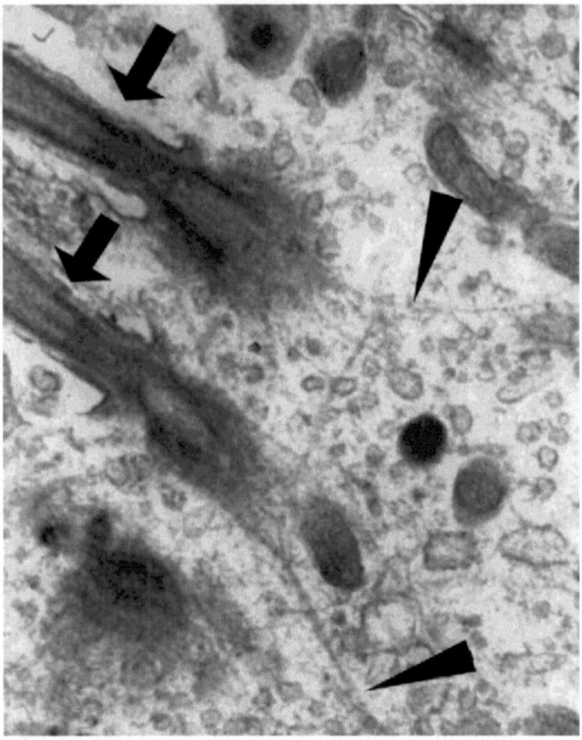

Fig. 1.2 TEM showing two cilia (*arrows*) protruding from the surface of an ependymal cell of the lateral ventricle of the cat. Ciliary rootlets emanate from the basal bodies (*arrowheads*). ×28,000 (With permission from Klinkerfuss 1964)

express bitter taste receptors. Binding of bitter compounds increased the intracellular Ca^{2+} concentration and stimulated ciliary beat frequency. Therefore, motile cilia, at least in the lung airway, are also sensory organelles (Shah et al. 2009).

The ventricles of the brain and the central canal of the spinal cord are lined by a continuous epithelium, the ependyma. The ependyma of lateral ventricles in mammals and birds and of the spinal cord is a ciliated epithelium with cilia extending into the ventricular lumen (Sotelo and Trujillo-Cenoz 1958; Tennyson and Pappas 1962; Klinkerfuss 1964). Usually 40–60 cilia protrude from the surface of a single ependymal cell (Brightman and Palay 1963; Chamberlain 1973; Nakayama and Kohno 1974). These kinocilia have a 9+2 organization and are far longer than primary cilia. In the ependymal cells of the golden hamster (*Mesocricetus auratus*) kinocilia have a length of ~10 μm and are of 150–175 nm in diameter (Blinzinger 1962). Cilia grow out from basal bodies situated directly beneath the apical surface of ependymal cells (first observed by Purkinje 1836). The basal bodies of ciliated ependymal cells from the lateral ventricle and the third ventricle emanate cross-striated rootlets and one basal foot (Fig. 1.2; Brightman and Palay 1963; Klinkerfuss 1964). Rootlets are filamentous appendages that emerge from the lateral wall and the proximal tip of the basal body to extend into the cytoplasm. They are approximately 0.4 μm in length with frayed ends. The basal foot is a conical structure laterally associated with the basal body. Kinocilia of brain ependyma are important for circulation of the cerebrospinal fluid.

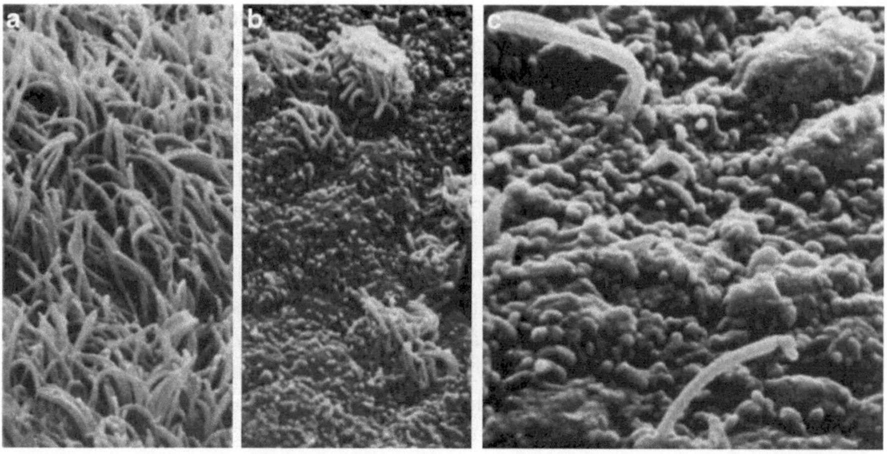

Fig. 1.3 Scanning electron micrographs of the ependymal surface of the third ventricle of the rabbit. Ciliary tufts protruding into the ventricular lumen in the upper two-third of the ventricle (**a**; ×2,800). In the lower one-third of the third ventricle ciliated cells occur less frequently (**b**; ×2,800) and eventually single cilia are occasionally seen rather than ciliary bundles (**c**; ×7,000) (With permission from Bruni et al. 1972)

Based on the morphology of the ependymal surface of the third ventricle, three regions can be identified. The upper two-thirds (in rabbit, mouse, rat and man) are lined by ependymal cells that protrude clusters of cilia from their surface. But in the lower one-third of the third ventricular wall ciliary bundles on the luminal ependymal surface become less frequent and eventually ependymal cells occasionally carry only single cilia (Bruni et al. 1972) (Fig. 1.3). Tufts of cilia on ependymal cells of the rat brain are chronologically generated during embryonic and fetal development. First, ependymal cells emerge only one single cilium but consecutively grow further cilia to possess later on tufts of them (Chamberlain 1973). Along the immature neural tube ciliated ependymal cells are not present. Gradual covering of the central canal and ventricles by ciliated ependyma takes place concordantly with regression of the ventricular zone and radial glial cells (in: Del Bigio 2010). However, radial glial cells themselves bear primary cilia (Cohen et al. 1988; Willaredt et al. 2008; Arellano et al. 2012). In adulthood, ependymal cells in the central canal of rats regularly protrude two cilia whereas those in the rabbit have single as well as multiple cilia. Cilia observe a polarity towards the caudal direction along the central canal suggesting that ciliary activity moves the cerebrospinal fluid caudalwards (Nakayama and Kohno 1974).

As in the brain ependyma, the close vicinity of cells bearing multiple motile cilia with those projecting only one cilium was also found in the oviduct epithelium (Chamberlain 1973; Odor and Blandau 1985; Hagiwara et al. 2002). But cells either possess 9+2 cilia or a solitary cilium. Ciliated cells with 9+2 cilia never seem to possess primary cilia (Wheatley et al. 1996).

Solitary primary cilia have been identified in the central nervous system, not only on ependymal cells but also on granular neurons and on astroglial cells of the fascia dentata of the hippocampal region in the rat (Dahl 1963). Moreover, using antibodies that selectively detect neuronal cilia, such as antibodies against somatostatin receptor subtype 3 (Sstr3) or type III adenylyl cyclase, the presence of a primary cilium could be demonstrated on most neurons in the mammalian brain (Handel et al. 1999; Fuchs and Schwark 2004; Bishop et al. 2007; Arellano et al. 2012). The neuronal cilium is of 9+0 structure and has a two-centriole basal organization. However, primary cilia are found on most cells of the body (a complete list of cells and tissues harbouring a primary cilium can be found at: http://www.bowserlab.org/primary-cilia/cilialist.html).

By the observation of ciliated secretory cells in the pars distalis of the mouse hypophysis and the structural similarity of these cilia to sensory cilia, Barnes suggested a sensory function of monocilia. Moreover, she noticed that motile cilia are of 9+2 structure and are associated with a single centriole or basal body whereas cilia that lack the central microtubules (and are therefore of 9+0 structure) are mostly associated with two centrioles (Barnes 1961).

Primary cilia are thought to function as mechanosensors and/or chemosensors. A mechanosensory function, possibly in addition to a chemosensory function, has been postulated for primary cilia of bone and cartilage cells and odontoblasts (Magloire et al. 2004; Malone et al. 2007; Anderson et al. 2008; Whitfield 2008). The presence of primary cilia in chondrocytes of mouse embryos has first been observed by Scherft and Daems in 1967. A role in the perception of extrinsic stimuli e.g. mechanical loading from the extracellular matrix and the transduction to the cell body has been ascribed to primary cilia in cartilage chondrocytes (Poole et al. 1985). The close association of the primary cilium with the Golgi apparatus suggested a functional feedback mechanism to facilitate directed secretion of extracellular matrix components in response to biomechanical load (Poole et al. 1997, 2001). The recent identification of extracellular matrix receptors on chondrocyte primary cilia strengthens the view that primary cilia act as mechanosensors (McGlashan et al. 2006). More convincingly, their function as mechanotransducers in chondrocytes has been demonstrated by studying chondrocytes that lack primary cilia. However, this study also showed that primary cilia sense mechanical loading via perception of ATP release in compressed cartilage (Wann et al. 2012).

Primary cilia, additionally, act as mechanosensors in a variety of other cell types, including cholangiocytes, smooth muscle cells, embryonic node cells, renal epithelial cells, and endothelial cells (Praetorius and Spring 2001, 2002; McGrath et al. 2003; Masyuk et al. 2006; Lu et al. 2008; Nauli et al. 2008). Endothelial primary cilia might be involved in blood pressure control (reviewed in: Nauli et al. 2011). In endothelial and renal epithelial cells the primary cilium senses fluid shear and transmits the signal via polycystin-1/polycystin-2 and increase of Ca^{2+} to affect gene expression (Praetorius and Spring 2001, 2002; Hierck et al. 2008; Nauli et al. 2008; AbouAlaiwi et al. 2009). Polycystin-1 (PC1) and polycystin-2 (PC2) are integral membrane proteins encoded by *PKD1* and *PKD2*, respectively. Mutations in *PKD1* or *PKD2* cause autosomal dominant polycystic kidney disease (ADPKD), a lethal disorder characterized by progressive development and enlargement of fluid-filled cysts in the kidney and eventually renal failure. PC1 activates a G-protein signaling pathway, whereas PC2 functions as a Ca^{2+}-permeable cation channel. A subfraction of both proteins heterodimerizes and colocalizes to the primary cilium in renal epithelial cells and some other cells. Bending of the primary cilium in renal epithelial cells results in PC1/PC2 mediated Ca^{2+} increase (Nauli et al. 2003; reviewed in: Chapin and Caplan 2010). Moreover, primary cilia are essential for proper regulation of Ca^{2+}-signaling (Siroky et al. 2006). Renal cilia, in addition, seem to be chemosensory since the type 2 vasopressin receptor (V2R) has been localized to them (Raychowdhury et al. 2009). The discovery that ADPKD is caused by mutations in genes encoding cilia-associated proteins in turn created a new class of genetic disorders, the "ciliopathies".

Primary cilia of the corneal endothelial cells that protrude into the anterior chamber seem to have an osmoregulatory and/or chemosensory function (Svedbergh and Bill 1972; Gallagher 1980; Collin and Collin 2004). In the olfactory receptor cells cilia have been suggested to be involved in the detection of olfactory stimuli by binding of odorant molecules and sense transduction (Moran et al. 1982). This view was confirmed later by the identification of odorant receptors in cilia of olfactory receptor cells (Menco et al. 1997; reviewed in: Jenkins et al. 2009). The olfactory receptor cell bears 10–30 cilia at its apical side projecting into the nasal cavity. Olfactory cilia emerge from basal bodies underneath the cell membrane and are of the 9+2 structure but lack dynein arms (Fig. 1.1c).

Furthermore, primary cilia are implicated in key signaling pathways like platelet-derived growth factor (PDGF), Wnt, and Hedgehog signaling (reviewed in: Christensen et al. 2007). The platelet-derived growth factor receptor alpha (PDGFRα) localizes to the primary cilium in quiescent fibroblasts. Ligand binding activates PDGFRα followed by activation of Akt and the Mek1/2-Erk1/2 pathways but signaling depends on the presence of the primary cilium (Schneider et al. 2005). PDGFRα signaling via the primary cilium monitors directional migration during wound healing (Schneider et al. 2010). Intriguingly, already in 1977 Albrecht-Buehler observed that primary cilia on migrating fibroblasts orient in the direction of cell migration. A role of primary cilia in Wnt signaling has been suggested by the finding that the ciliary protein Inversin (also named nephrocystin-2 or NPHP2) acts as a molecular switch between canonical versus non-canonical Wnt pathways. Inversin interacts with Dishevelled (Dvl) and targets cytoplasmic Dvl for ubiquitin-dependent degradation (Simons et al. 2005; reviewed in: Lienkamp et al. 2012). Degradation is controlled by the anaphase-promoting complex/cyclosome (APC/C) (Ganner et al. 2009; Miyamoto et al. 2011). Even though the involvement of primary cilia in Wnt signaling during development is under debate (Huang and Schier 2009; Ocbina et al. 2009), evidence for a direct role of the primary cilium in restraining the canonical Wnt/β-catenin pathway has recently been presented. Jouberin (Jbn), the protein product of a gene mutated in the ciliopathy Joubert syndrome, is sequestered in cilia. Activation of the canonical Wnt pathway releases Jbn from cilia, which in turn interacts with β-catenin to facilitate translocation to the nucleus and transcription of β-catenin-dependent target genes (Lancaster et al. 2011). In contrast, the essential role of primary cilia in Hedgehog (Hh) signaling is very well established (Huangfu and Anderson

2005). Hh signaling is mediated by the Hh receptor patched (Ptc) and the transmembrane protein smoothened (Smo). Ptc localizes to primary cilia in the absence of Hh stimulation, thus preventing accumulation of Smo in the cilium. Upon Hh activation, Ptc leaves the cilium that in turn allows Smo accumulation (Corbit et al. 2005; Rohatgi et al. 2007). Smo then activates Hh signaling at the ciliary tip by promoting the formation of Gli2 activator and inhibition of the formation of the Gli3 repressor isoform, a proteolytic product of the full length Gli3 protein (Haycraft et al. 2005; Tasouri and Tucker 2011). Notably, formation of Gli repressors in the absence of Hh signaling also requires primary cilia (Liu et al. 2005; May et al. 2005). Furthermore, in epidermal cells Notch-receptors and Notch-processing enzymes colocalize to primary cilia, and Notch signaling is transduced via primary cilia (Ezratty et al. 2011). Finally, the essential requirement of primary cilia in skin and hair follicles for hair morphogenesis is well documented (Lehman et al. 2009; Croyle et al. 2011).

The first described defect in ciliary motility leading to primary cilia dyskinesia (PCD) was observed by Afzelius 1976. This immotile cilia syndrome was caused by the absence of dynein arms of the axoneme and resulted in altered left-right axis patterning (situs inversus) and chronic bronchitis/sinusitis. That severe diseases are also caused by impaired function of non-motile primary cilia was first observed by Cole and co-workers in 2000 (Pazour et al. 2000), and led to a boost in ciliary research interest. Now it is widely accepted that primary cilia sense, transduce and coordinate signaling pathways and are therefore critical in embryonic and postnatal development, as well as in tissue homeostasis in adulthood. Primary cilia are important organelles for human health and development and their malfunction has been implicated in a wide range of human diseases including kidney diseases, vision and hearing loss, anosmia, obesity, polydactyly, neural tube defects, and left-right asymmetry collectively referred as ciliopathies (reviewed in: Badano et al. 2006; Fliegauf et al. 2007; Gerdes et al. 2009; Goetz and Anderson 2010; Pazour and Rosenbaum 2002; Veland et al. 2009).

1.3 Of Centrosomes, Centrioles and Basal Bodies

Each cilium is anchored inside the cell by its basal body that is located apically beneath the cell membrane. Formation of primary cilia starts from the distal end of the mature or mother centriole that has been transformed into the basal body. Centrioles and basal bodies share the same basic structure. Both are barrel-shaped organelles consisting of a symmetrical array of microtubules (MT) arranged in mammals in a ninefold triplet configuration (Fig. 1.4). Centrioles are usually found in a pairwise configuration at the cell center adjacent to the nucleus, and, together with their associated pericentriolar material (PCM), constitute the centrosome (Fig. 1.5). Basal bodies, on the other hand, function in cilia formation and are often located at the cell surface. Basal bodies and centrioles, although exerting different functions, can be transformed into each other (Sorokin 1962, 1968; reviewed in: Dawe et al. 2007a; Hoyer-Fender 2010).

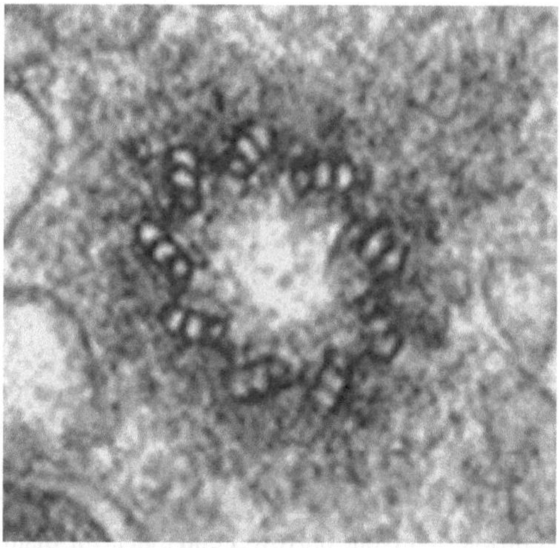

Fig. 1.4 Symmetrical array of nine triplet microtubules in centrioles of mouse L929 cells. ×72,500 (With permission from Rattner and Phillips 1973)

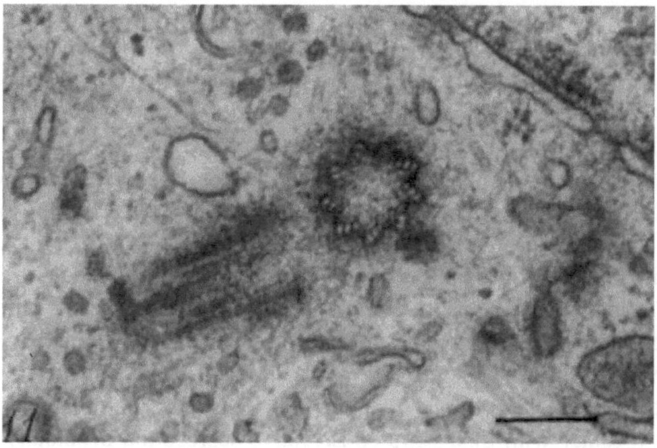

Fig. 1.5 A pair of centrioles (a diplosome) at the beginning of G1 phase. ×60,000 (With permission from Vorobjev and Chentsov 1982)

1.3.1 Ultrastructure of Centrioles and Basal Bodies

The centrosome is the main microtubule organizing center (MTOC) in animal cells. It consists of a pair of centrally-positioned centrioles surrounded by a somewhat amorphous zone of pericentriolar material (PCM). A main protein component of the PCM is γ-tubulin, present in multi-protein complexes constituting γ-tubulin ring complexes (γ-TuRCs). The γ-TuRC anchors microtubule minus ends to the centrosome (Moritz et al. 1995). Hence, the centrosome is viewed as the organizing center of the cytoskeleton in interphase cells. In mitotic cells, the centrosome forms the poles of the mitotic spindle to allow for correct chromosome segregation (reviewed in: Kellogg et al. 1994; Doxsey 2001; Bornens 2002). However, centrioles seem not to be essential for all somatic cell divisions. Early cleavage stage embryos of the mouse are devoid of centrioles. Electron microscopy has not identified centrioles in the cells of embryos/morulae before the 16-cell stage. Then, from the 16-cell stage on, centrioles are clearly present (Szöllösi et al. 1972; Gueth-Hallonet et al. 1993; reviewed in: Hoyer-Fender 2011). Since all eukaryotic organisms that form cilia or flagella at any time during their life cycle possess centrioles their main function therefore might be to provide the basal body to extend a cilium or flagellum (Pickett-Heaps 1971; reviewed in: Carvalho-Santos et al. 2011).

Centrioles are tiny organelles of about 0.5 μm in length and 0.2 μm in diameter. Nine microtubule triplets are arranged to form a cylindrical structure (Vorobjev and Chentsov 1980). Each triplet is composed of three different microtubules, a complete A (inner) tubule, and two incomplete tubules, termed the B (middle) and C (outer) tubules. Triplet microtubules are interlinked throughout the length of the centriole, and run in parallel, slightly twisted bundles. The C microtubule, however, is often shorter, giving rise to centrioles with one narrower end that is the distal end. Centrioles have therefore a defined proximo-distal orientation. At the proximal end, inner (A) and outer (C) microtubules of neighbouring triplets are connected by links, whereas in the middle and distal regions the links form a dense ring-like structure in the lumen of the centriole (Alvey 1986). When transformed into a basal body a cilium grows out from the distal end. At the transition zone the basal body then passes into the axoneme of the cilium that basically consists of nine microtubule doublets that are extensions of A and B MT.

Centrioles are generally present pairwise in an approximately orthogonal orientation and are interlinked by fibrous structures (Paintrand et al. 1992). In actively dividing cells, duplication of centrioles, therefore, has to be synchronized with the cell cycle (Robbins et al. 1968; Vorobjev and Chentsov 1982; Alvey 1985) (Fig. 1.6). In S phase, a procentriole forms at the proximal end of each centriole that elongates to full-length daughter centrioles throughout S to G2 phases (Fig. 1.7). At the beginning of mitosis each cell therefore possesses two pairs of centrioles with their associated PCM constituting two centrosomes that generate the spindle poles. During cell division, each centrosome is then transmitted to one daughter cell. This templated pathway of centriole duplication, together with their semi-conservative segregation eventually give rise to centrosomes harbouring a pair of centrioles that are both of different age. Whereas the younger centriole (or daughter centriole) was generated in the last cell cycle, the older or mother centriole originated at least two cell divisions before. Moreover, both centrioles not only differ in age but they are structurally and functionally not equivalent. Specifically, it is the mother centriole only that initiates assembly of the primary cilium.

Maturation of the centriole takes 1.5 cell cycles and is accomplished by the formation of appendages and the capacity to generate a cilium. During the second half of mitosis, distal appendages are generated at the former daughter centriole and are thus found on one centriole of a centriole pair at each mitotic spindle pole. In the G1 phase of the next cell cycle subdistal appendages, which have disappeared in mitosis at the mother centriole, are generated anew on the oldest centriole in every cell. As a result, a pair of centrioles is present in each cell comprising one daughter centriole generated in the last cell cycle and one older centriole now matured to a mother centriole (Vorobjev and Chentsov 1982). The mother or mature centriole is characterized by the presence of subdistal and distal appendages (Figs. 1.8 and 1.9).

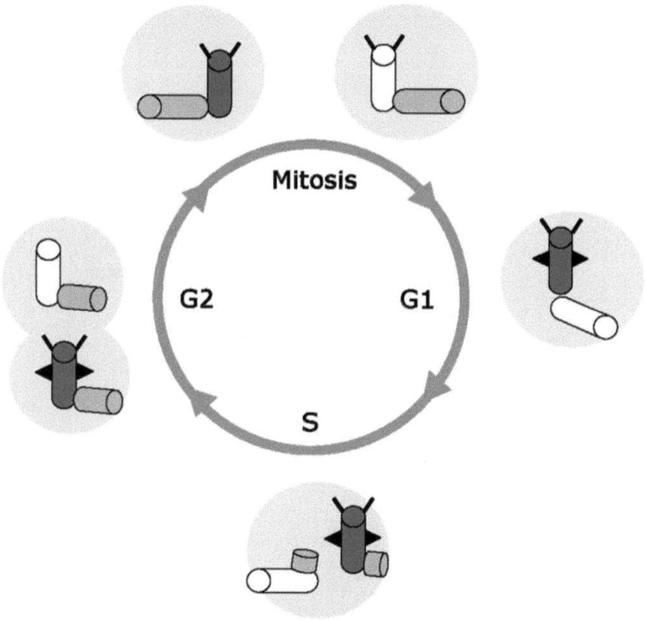

Fig. 1.6 Conservative centriole duplication and semi-conservative segregation during the cell cycle. The mature centriole (*dark grey cylinder*) is characterized by subdistal (*arrow heads*) and distal (*black lines*) appendages. Each centriole pair is surrounded by its associated PCM (*grey sphere*) to constitute the centrosome. Procentriole (*light grey cylinders*) formation starts in S phase at the proximal ends of both mother and daughter (*white cylinder*) centrioles. Procentrioles elongate to full length from S to G2 phase. In mitosis centrosomes generate the spindle poles

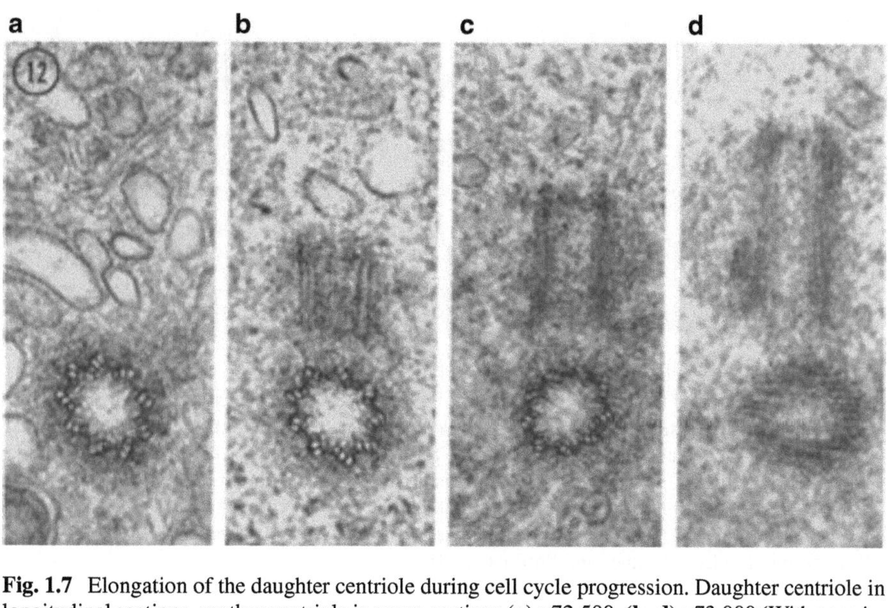

Fig. 1.7 Elongation of the daughter centriole during cell cycle progression. Daughter centriole in longitudinal sections, mother centriole in cross-sections (**a**) ×72,500, (**b–d**) ×73,000 (With permission from Rattner and Phillips 1973)

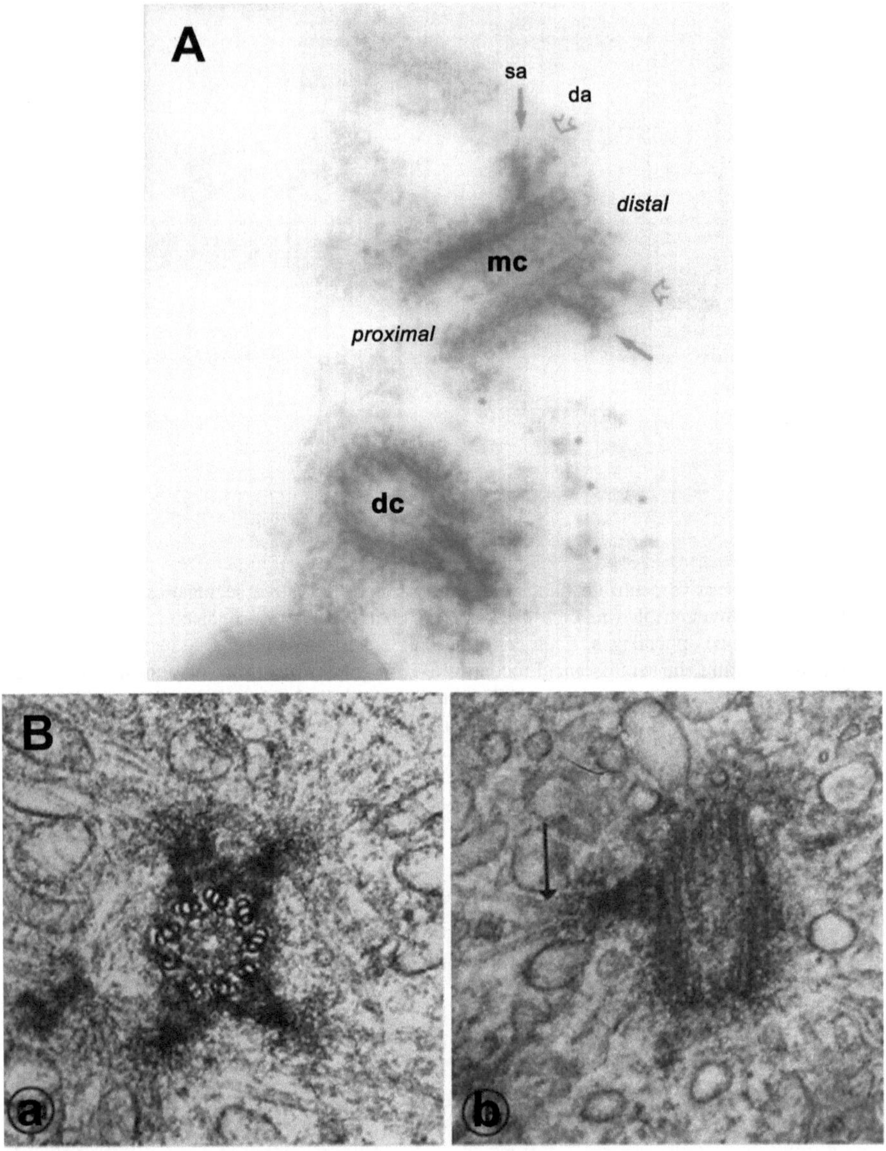

Fig. 1.8 (**A**) Isolated centrosomes of human lymphoblasts consisting of a pair of centrioles. Longitudinal section of the mother centriole (*mc*) showing distal appendages (*da*, *open arrow*) and subdistal appendages (*sa*, *filled arrow*). Cross-section of the daughter centriole (*dc*) (With permission from Paintrand et al. 1992. Modified). (**B**) Cross-section through the distal region of a mother centriole (*a*). Four wedge shaped striated subdistal appendages extend from the centriolar wall. ×92,250. (*b*) Longitudinal section through a mother centriole comprising one wedge shaped striated subdistal appendage. Microtubules impinge on the head of the appendage (*arrow*). ×91,250 (With permission from Rattner and Phillips 1973)

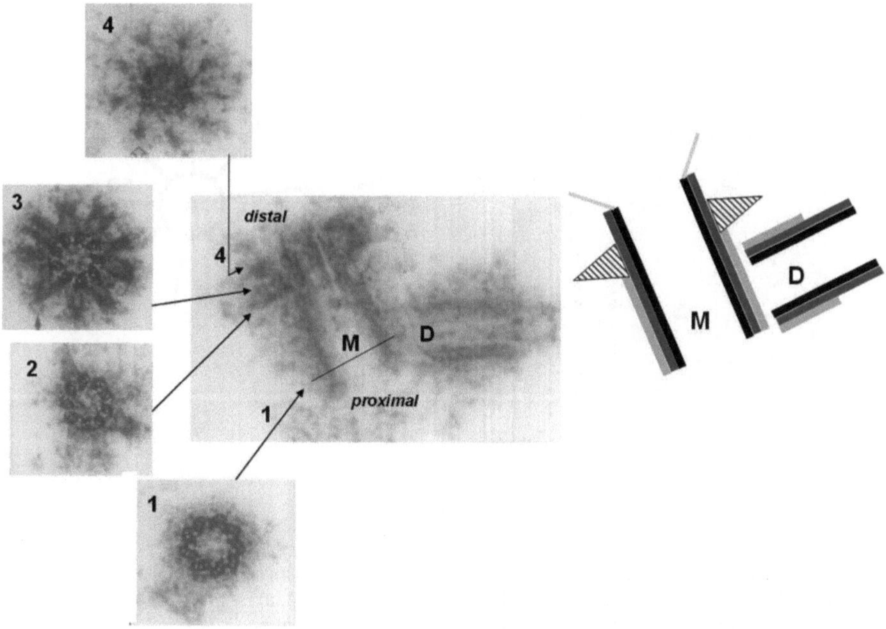

Fig. 1.9 A pair of centrioles consisting of a mother (mature, *M*) centriole and a daughter centriole (*D*) in longitudinal section. The mature centriole harbours subdistal and distal appendages. Cross-sections of the mature centriole shown on the left site from proximal (*1*) to distal (*4*). The approximate positions of cross-sections are marked in the longitudinal section. The centriolar cylinder consists of nine sets of microtubule triplets at its proximal end, and nine sets of microtubule doublets at its distal end. Tubules A and C of adjacent triplets at the proximal end seem to be connected. Subdistal appendages are conical structures that insert approximately at that region of the centriolar cylinder where the C MT get lost. Subdistal appendages are sites of MT anchoring. They are variable in number, location and thickness and usually less than nine are present. Distal appendages however feature a strict ninefold symmetry. They seem to be associated with the edge of each B tubule and protrude as blades of a turbine (Electron micrographs with permission from Paintrand et al. 1992)

Distal appendages correspond to transition fibers or alar sheets of basal bodies that participate in the association of the basal body to the plasma membrane. It is not known whether distal appendages and transition fibers differ in protein composition. Subdistal appendages most likely correspond to basal feet of basal bodies (and the pericentriolar satellites described by Vorobjev and Chentsov 1982). Both function by anchoring microtubules at their tips (Fig. 1.8B) (Piel et al. 2000). Subdistal appendages consist of a conical striated stem and a small round head and therefore resemble basal feet in this regard, too. The stem is attached to two adjacent MT triplets of the centriole/basal body wall. Subdistal appendages vary significantly in number, thickness and distribution along the proximo-distal axis of the mother centriole. At least two or three, but usually less than nine subdistal appendages are found at the distal part of the mother centriole (Fig. 1.8B). In contrast, distal appendages

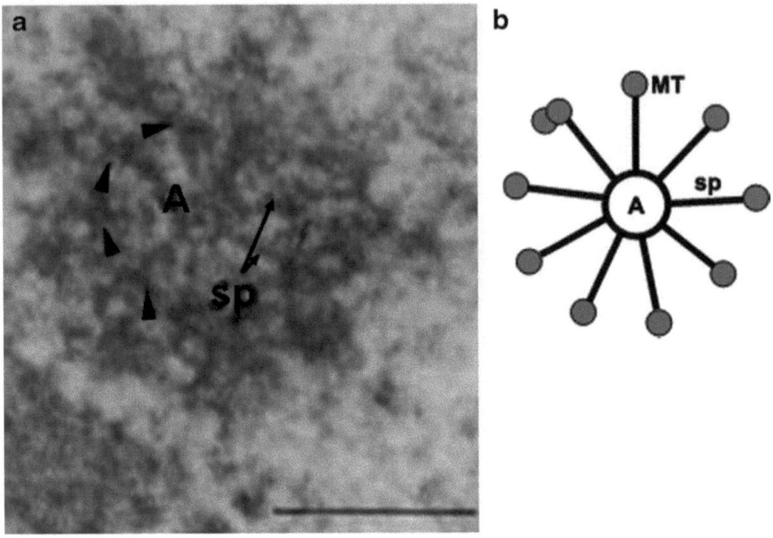

Fig. 1.10 Procentriole in S phase. (**a**) EM cross-section of the *cartwheel*, ×160,000 (With permission from Vorobjev and Chentsov 1982). (**b**) Schematic drawing. *A*, axis, *sp*, spokes of a cartwheel. *Arrowheads*: single and double MT attached to the ends of the spokes

pursue the ninefold symmetry of the triplet microtubules. They protrude from the microtubule sets like the blades of a turbine, and are oriented toward the distal end (Fig. 1.16). As is obvious in electron micrographs, the lumen of the centriole appears empty at its proximal end on one-third of the total length whereas an amorphous substance fills the centriolar lumen at the distal end (Fig. 1.8A).

During centriole duplication, procentrioles are formed near the proximal ends of the mother as well as the daughter centriole. At the earliest stage they look like a short cylinder with uneven stubs. Electron microscope studies revealed that the procentriole wall is made up of amorphous material in which nine single or double MT are immersed. These microtubules are not symmetrically arranged at this stage. Inside the procentriole, at its proximal end, is an axis with spokes, the so-called cartwheel. The microtubules of the procentriole are attached to the end of the nine spokes (Fig. 1.10). After the centriole reaches normal length, it loses the axis with spokes (Vorobjev and Chentsov 1980, 1982; Paintrand et al. 1992; Chretien et al. 1997; Kenney et al. 1997; Ibrahim et al. 2009). A careful review of the literature revealed, that the cartwheel is a feature of procentrioles but is no longer present in adult centrioles (Alvey 1986). Recently, cryo-electron tomography of purified centrosomes of human cells proved the presence of the cartwheel in procentrioles but its absence in the G1 phase of the cell cycle (Guichard et al. 2010). Thus, the cartwheel is a non-permanent structure restricted to procentrioles.

Centriole duplication during the cell cycle takes place not only in cells comprising a centrosome with two adjacent centrioles but also in cells harbouring a primary cilium as observed in epithelial cells (Vorobjev and Chentsov 1982). In this case, the mature

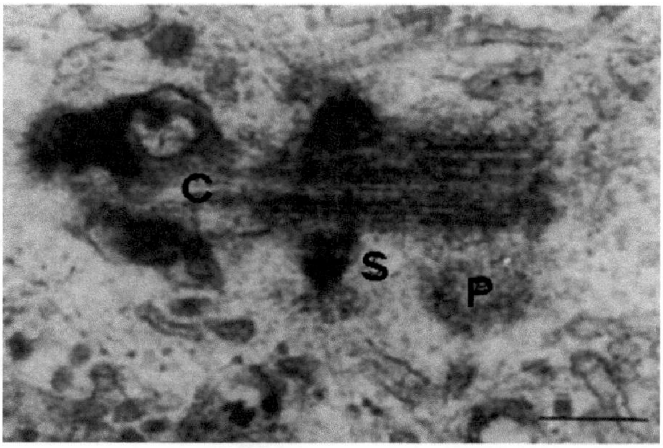

Fig. 1.11 Replicating centriole in *S* phase. *P* procentriole, *s* subdistal appendages, *C* cilium. *Bar*, 0.2 μm. ×80,000 (With permission from Vorobjev and Chentsov 1982)

centriole, now transformed into a basal body and its associated primary cilium grows a procentriole from its proximal region (Fig. 1.11). Thus freeing centrioles by disassembly of the primary cilium seems not to be necessary for centriole duplication.

1.3.2 Molecular Mechanisms of Centriole Duplication

The centrosome comprises a huge number of proteins. Along with intrinsic centrosomal proteins that build up centrioles and the PCM, several proteins associate only temporarily with the centrosome to regulate cell cycle progression. During the past decade, great progress has been made unravelling the molecular composition of the centrosome and elucidating the function of centrosomal proteins (Andersen et al. 2003; Jakobsen et al. 2011). This work has provided insight into the molecular mechanisms underlying the assembly of centrioles. Centriole duplication at an evolutionary scale has been outlined in several excellent comprehensive reviews (Loncarek and Khodjakov 2009; Azimzadeh and Marshall 2010; Sluder and Khodjakov 2010; Nigg and Stearns 2011). The following summary will therefore focus on centriole formation in vertebrate cells and the relevant proteins (Figs. 1.12 and 1.13).

The major proteins constituting centrioles/basal bodies are αβ-tubulin dimers, tektins, and Sp77 and Sp83 (Steffen et al. 1994; Hinchcliffe and Linck 1998; Stephens and Lernieux 1998). Post-translational modification of tubulin strongly influences microtubule stability. Centrioles and basal bodies both contain highly modified tubulin, especially detyrosinated, acetylated, and polyglutamylated tubulin as well as Δ2-tubulin (Piperno et al. 1987; Bobinnec et al. 1998; for review see: Ikegami and Setou 2010; Janke and Bulinski 2012). Detyrosination is the removal of the C-terminal tyrosine of α-tubulin. Detyrosinated tubulin can then be further subjected to removal of the C-terminal glutamate producing Δ2-tubulin. Polyglutamylation is the progressive addition of Glu residues onto the γ-carboxyl group of Glu residues in the C-terminal region of polymerized tubulin. Polyglutamylation of α/β-tubulin seems

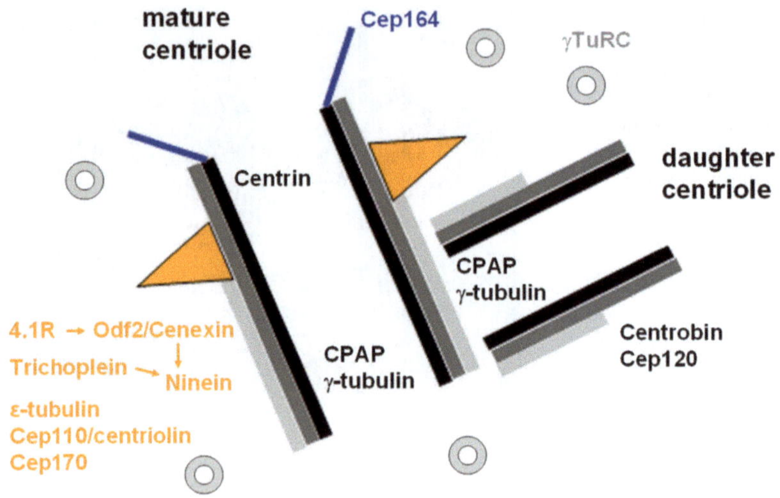

Fig. 1.12 Localization of centrosomal proteins

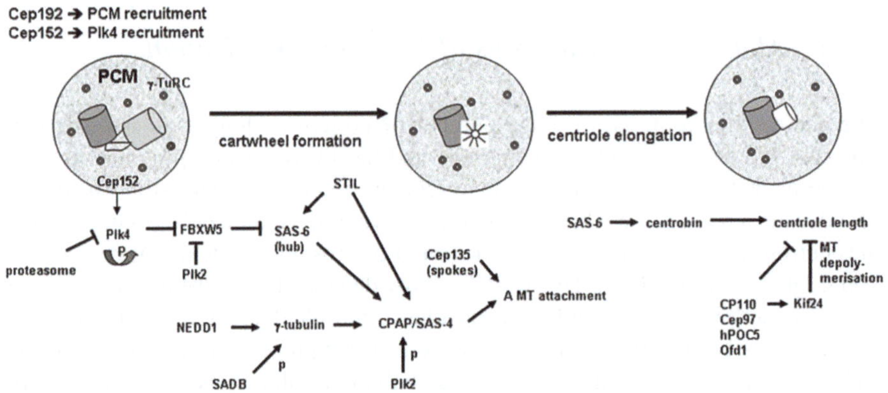

Fig. 1.13 Overview of the protein interaction network that controls centriole duplication. For clarity, cartwheel formation and centriole duplication is shown only for one centriole of a pair. See text for explanation

to be an early event in centriole/basal body assembly (Million et al. 1999). Tubulin detyrosination as well as polyglutamylation seem to stabilize MT. γ-tubulin is a major protein of the pericentriolar matrix but additionally is tightly bound to the microtubules of the centriolar cylinder and localizes to the proximal center of centrioles (Fuller et al. 1995).

The main feature of intrinsic centrosomal proteins is their coiled-coil structure that promotes protein-protein interactions to generate the electron dense pericentriolar matrix (PCM). The PCM not only harbours γ-TuRCs to nucleate MTs but contributes to centriole duplication (Dammermann et al. 2004; Loncarek et al. 2008). PCM recruitment around the centrioles might be mediated by Cep192 (centrosomal protein of 192 kDa) (Gomez-Ferreria et al. 2007; Zhu et al. 2008). Centrosomal attachment of the γ-TuRC is additionally mediated by CDK5RAP2 (Fong et al. 2008). However, the key regulator of centriole duplication is the Polo-like kinase 4 (Plk4). Inhibition of Plk4 prevents centriole duplication whereas overexpression can trigger centriole amplification (Habedanck et al. 2005). Plk4 activity depends on autophosphorylation and is regulated by the proteasome (Cunha-Ferreira et al. 2009; Rogers et al. 2009; Guderian et al. 2010; Holland et al. 2010; Silibourne et al. 2010). Recruitment of Plk4 to the centrosome is mediated by the PCM protein Cep152. Depletion of Cep152 not only affects Plk4 recruitment but results in failure of CPAP (centrosomal P4.1-associated protein; also named SAS-4 or CENPJ, centromere protein J) and SAS-6 localization to the centriole (Cizmecioglu et al. 2010; Dzhindzhev et al. 2010; Hatch et al. 2010). As described above, the assembly of the procentriole starts with the formation of the cartwheel composed of a central hub to which nine spokes are attached (Fig. 1.10). For the initial steps of centriole assembly, SAS-6 is essential and might be required to build the central hub (reviewed in: Gönczy 2012). Consistently, HsSAS-6 (human SAS-6) is the earliest protein localized to the emerging procentriole in human cells (Strnad et al. 2007). However, in rat tracheal multiciliated cells SAS-6 could also be found at the proximal end of basal bodies and at the proximal region of ciliary axonemes (Kleylein-Sohn et al. 2007; Strnad et al. 2007; Vladar and Stearns 2007; Strnad and Gönczy 2008). In human cells, SAS-6 (HsSAS-6) recruits the tubulin-binding protein centrobin to allow for centriole elongation (Gudi et al. 2011). Human SAS-6 (HsSAS-6) is a substrate of the SCF-FBXW5 ubiquitin ligase complex that targets HsSAS-6 for degradation. Phosphorylation of FBXW5 by Plk2 and Plk4 suppresses its activity to allow centriole duplication (Puklowski et al. 2011; Cizmecioglu et al. 2012). The central hub seems to be critical in establishing the ninefold symmetry whereas the radial spokes seem to be important to specify the centriole diameter (see: Strnad and Gönczy 2008). A putative protein of the radial spokes is Cep135 which is essential for centriole assembly (Kleylein-Sohn et al. 2007; Vladar and Stearns 2007). As was evident from EM data, the A microtubule attaches first to the spokes of the cartwheel followed by assembly of B and C microtubules. Moreover, it was suggested that the A microtubule is nucleated by γ-TuRC allowing growth from the proximal to the distal end of the centriole, whereas B and C microtubules are never capped and are therefore assembled at variable positions along the existing microtubule to grow in both directions (Guichard et al. 2010). This would explain why γ-tubulin is essential for centriole duplication. Recruitment of γ-tubulin to the centrosome is NEDD1 dependent and depletion of NEDD1 like γ-tubulin depletion inhibits centriole duplication (Haren et al. 2006). Moreover, γ-tubulin mediated centriole duplication is activated by Ser/Thr kinase activity of SADB to phosphorylate Ser 131 of γ-tubulin (Alvarado-Kristensson et al. 2009). γ-tubulin interacts with CPAP and both proteins are located within the proximal lumen of the centrioles. CPAP might be responsible for the attachment of the A MT to the spokes of the cartwheel (Hung et al. 2000; Kohlmaier et al. 2009). The activity of CPAP in centriole elongation depends on its phosphorylation by Polo-like kinase 2 (Plk2) (Chang et al. 2010). Recruitment of CPAP to the base of the nascent procentriole depends on HsSAS-6 and recruitment of both proteins is mediated by STIL (SCL/TAL1 interrupting locus), a protein mutated in primary microcephaly (Kohlmaier et al. 2009; Tang et al. 2011; Arquint et al. 2012; Vulprecht et al. 2012). CP110, Cep97, hPOC5, and Ofd1 (Oral-facial-digital type I protein) are important for centriole length control and the formation of primary cilia, most likely by inhibiting microtubule growth (Chen et al. 2002; Ferrante et al. 2006; Spektor et al. 2007; Azimzadeh et al. 2009; Schmidt et al. 2009; Tang et al. 2009; Singla et al. 2010). CP110 interacts with the centriolar kinesin Kif24 that possesses microtubule depolymerization activity (Kobayashi et al. 2011). Finally, maturation of procentrioles is controlled by activated Polo-like kinase 1 (Plk1) phosphorylated at T210 (Loncarek and Khodjakov 2010).

To the distal lumen of centrioles/basal bodies localizes another conserved protein, centrin, which is a calcium binding phosphoprotein of the EF-hand superfamily. Centrin seems to be involved in the early events of pro-centriole formation (Paoletti et al. 1996; Laoukili et al. 2000). Two proteins specifically enriched at the daughter centriole, centrobin and Cep120, have also been implicated in centriole duplication since centriole duplication is inhibited by depletion of either Cep120 or centrobin (Zou et al. 2005; Mahjoub et al. 2010; Gudi et al. 2011). Additionally, depletion of the coiled-coil protein SPICE (spindle and centriole-associated protein), or Cep63 and its interacting protein Cep152 affected centriole duplication (Archinti et al. 2010; Hatch et al. 2010).

MT are nucleated by γ-TuRCs and anchored at the subdistal appendages of the mother centriole. Anchoring is mediated by ninein, a protein of the subdistal appendages which itself is recruited by Odf2/Cenexin (Lange and Gull 1995; Mogensen et al. 2000; Nakagawa et al. 2001). Both ninein and Odf2 localization to subdistal appendages depends on the multifunctional structural protein 4.1R as observed by 4.1R knock down. Additionally, MT anchoring and organization in interphase is reduced by 4.1R depletion, thus supporting the MT anchoring function of subdistal appendages as observed by electron microscopy (Krauss et al. 2008). Recruitment of ninein by Odf2 seems to be mediated by Trichoplein, a keratin filament-binding protein (Ibi et al. 2011). Further proteins of the subdistal appendages are ε-tubulin (Chang et al. 2003), centriolin/Cep110 (Ou et al. 2002; Gromley et al. 2003), and the Polo-like kinase 1 (Plk1) interacting protein Cep170 (Guarguaglini et al. 2005) (Fig. 1.12). Cep170 and the splice variant hCenexin1 of Odf2 recruit Plk1 at the centrosome, whose activity is then important for proper recruitment of the major PCM proteins pericentrin and γ-tubulin and the formation of bipolar spindles in mitosis (Soung et al. 2009; Zhang et al. 2009).

Cep164 (centrosomal protein of 164 kDa) specifically localizes to the distal appendages, persists throughout mitosis, and is indispensable for primary cilia formation (Graser et al. 2007). Likewise, Ofd1 depletion impairs formation of distal appendages and ciliogenesis (Singla et al. 2010).

Despite duplication of the centriole pair during S phase the two resulting centriole doublets function as a single microtubule-organizing center until the onset of mitosis. Physical linkage of both oldest centrioles seems to be achieved by interaction of

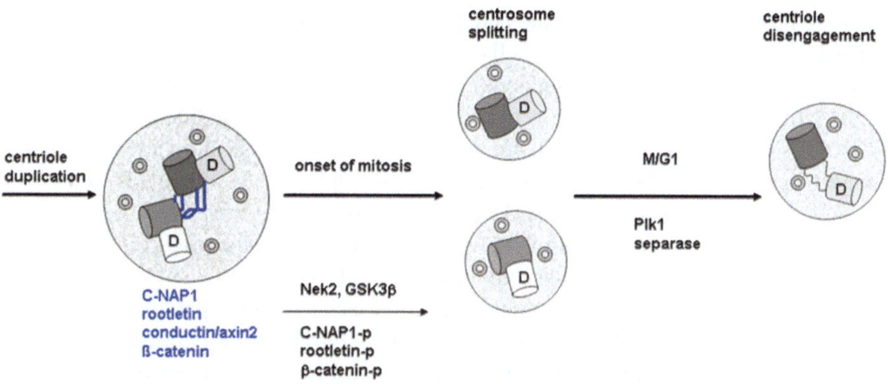

Fig. 1.14 Overview of the protein network that controls centrosome splitting at the onset of mitosis to generate the bipolar mitotic spindle, and centriole disengagement that is a prerequisite for centriole duplication in interphase. *D* daughter centriole

distantly related proteins C-Nap1 and rootletin (Fig. 1.14). At the onset of mitosis, NIMA-related kinase 2 (Nek2) phosphorylates C-Nap1 and rootletin leading to their displacement and in turn the separation of the duplicated centrosomes for bipolar spindle formation (Bahe et al. 2005). Additionally, C-Nap1 recruits conductin/axin2 to the centrosome to control Nek2- and GSK3β-mediated phosphorylation of β-catenin. Inhibition of β-catenin phosphorylation or degradation causes centrosome splitting (Bahmanyar et al. 2008; Hadjihannas et al. 2010). Finally, centriole duplication demands previous loosening of the tightly engaged centrioles by the end of mitosis to G1 phase (Tsou and Stearns 2006a). Centriole disengagement is mediated by Polo-like kinase 1 (Plk1) at G2 or early M phase and by the protease separase at anaphase eventually cleaving cohesion ring complexes that keep the centrioles together (Tsou and Stearns 2006b; Tsou et al. 2009; Nakamura et al. 2009; Schöckel et al. 2011).

1.4 Primary Cilia

1.4.1 Structure of Primary Cilia

Generally, primary cilia are very small entities with a length of about 1–4 μm only. For a long time, it was therefore difficult to prove the presence of a single cilium on each cell. These difficulties have been overcome by application of antibody probes that specifically highlight the cilium. In 1987 Wehland and Weber found that primary cilia contain detyrosinated α-tubulin (Wehland and Weber 1987). This led to the generation of antibodies directed against modified tubulin especially detyrosinated and acetylated α-tubulin, which is a marker of stable MTs (Westermann and Weber 2003), that are now common tools for the detection of primary cilia (Wheatley et al. 1996). Primary cilia are found on most cells regardless of the analysed material (cultured cells, organ cultures or tissue sections). Thus, all tissues contain cells bearing a primary cilium but not necessarily every cell is ciliated (discussed in: Seeley and Nachury 2010). Examples of cell types that are not ciliated in situ include adipocytes and hepatocytes and blood cells, i.e. monocytes and macrophages, lymphocytes, red blood cells, and neutrophils, (Scherft and Daems 1967; Wheatley 1969, 1982, 1995; Alieva and Vorobjev 2004; see the complete list of cells and tissues bearing primary cilia at http://www.bowserlab.org/primarycilia/cilialist.html).

Most primary cilia are of 9+0 structure, i.e. they lack the central pair of microtubules, and they lack dynein on their A microtubules (Fig. 1.15a). Primary cilia, additionally, project from the distal end of the basal body, the former mother centriole. Often, the basal body and its associated daughter centriole are deeply invaginated in the cell giving rise to a specialized membraneous structure, the ciliary pocket (Fig. 1.15b) (Molla-Herman et al. 2010; reviewed in: Rohatgi and Snell 2010; Ghossoub et al. 2011). When adopting an apical position, primary cilia project from the cell surface into the environment and they do so with a definite polarity. The ciliary polarity is seen in human diploid fibroblasts and 3T3 cell lines in which cilia are found underneath the cells projecting towards the substratum. In contrast, cilia of canine kidney epithelial cells, PtK1 cells, or human renal proximal tube cells are extending into the medium (Wheatley et al. 1996; Jensen et al 1979). Primary cilia are generally immotile. The idea, therefore, that primary cilia are sensory structures has been proposed already very early (Munger 1958; Barnes 1961; Sorokin 1962; Albrecht-Buehler 1977; Albrecht-Buehler and Bushnell 1979).

Cultured 3T3 cells often harbour primary cilia artificially enclosed by the cytoplasm that is in contrast to most tissue cells in which the primary cilium emerges to the outside of the cell (Albrecht-Buehler and Bushnell 1980). However, they feature the usual structural organization of primary cilia. Additionally, the primary cilium was usually found near a nuclear indentation with a nearly constant distance between the basal body and the nuclear membrane of 1.2 μm. Although not yet confirmed by electron microscopy, a linkage between nucleus and basal body has therefore been suggested (Albrecht-Buehler and Bushnell 1980). Although both centrioles of a pair are associated even during cilium formation, the daughter centriole apparently changes its position, suggesting a somehow loose association (Sorokin 1962; Vorobjev and Chentsov 1982). Variability in angle and distance between both centrioles of a pair was also observed in human heart tissue (Myklebust et al. 1977).

When cilium formation starts the mature centriole has been transformed into the basal body. Therefore, appendages that have formerly ascribed to the mother centriole are also found on the basal body. Subdistal appendages extend from some point along the side of the mother centriole/basal body into the centrosomal matrix (see e.g. Figs. 1.11, 1.15b, and 1.19). These appendages have been named basal feet when referring to the basal body. The subdistal appendages of the mother centriole occur in varying numbers, but often two or three are seen in the same cross-sectional plane. Each subdistal appendage makes contact at its base with two adjacent microtubule triplets of the centriole. Likewise, the basal body of primary cilia usually possesses more than one basal foot alongside its centriolar wall (Hagiwara et al. 2002; Odor and Blandau 1985).

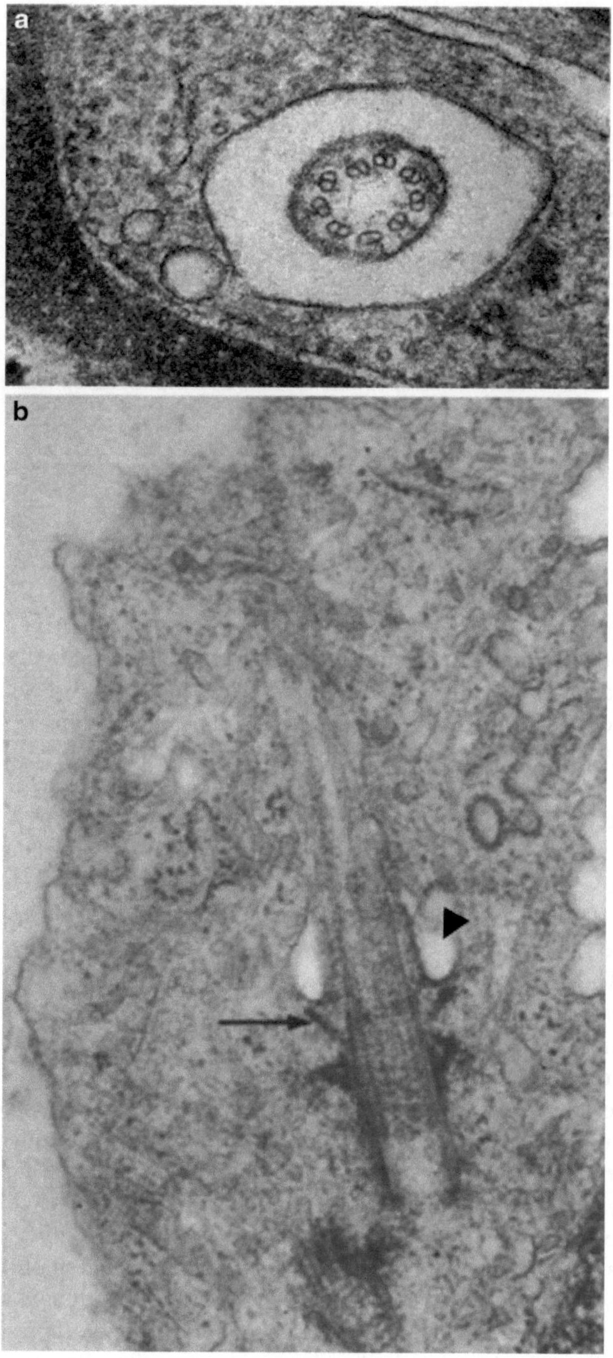

Fig. 1.15 (**a**) Transverse section through the basal region of a primary cilium shaft of 3T6 fibroblasts showing typical 9+0 organization. Due to dislocations of one peripheral doublet deviations from the typical 9+0 arrangement are frequently seen in more distal parts of the primary cilium. ×77,000 (With kind permission from Wheatley 1972). (**b**) Primary cilium in the human heart. Subdistal appendages and distal appendages (*arrow*) making contact to the cell membrane emerge from the basal body. A ciliary pocket (*arrowhead*) is formed due to the deep invagination of the cilium in the cell. ×45,000 (With permission from Myklebust et al. 1977)

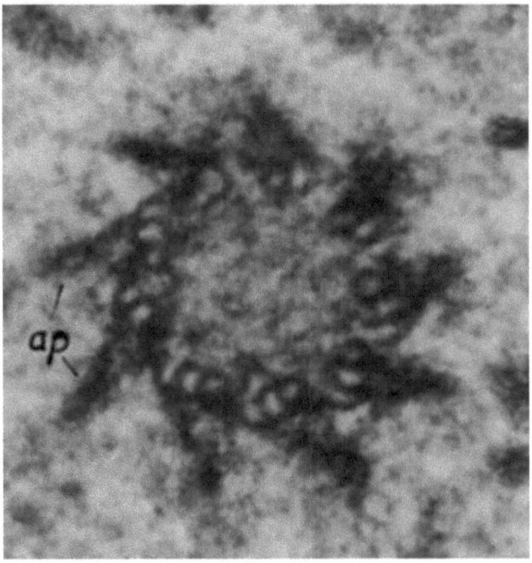

Fig. 1.16 Distal part of the mother centriole producing cilia in the G0 phase. Symmetrical arrangement of distal appendages (*ap*). ×140,000 (With permission from Vorobjev and Chentsov 1982)

In contrast, only one basal foot was usually observed at the basal body of motile cilia (Odor and Blandau 1985). Basal feet and subdistal appendages are of conical shape and show a distinctive cross-banding (Sorokin 1962; Rattner and Phillips 1973). Moreover, basal feet and subdistal appendages are the focal point of microtubles. Thus, basal feet and subdistal appendages are more or less synonymous items that are used in the particular context of the basal body and the mother centriole, respectively.

At the distal end of both mother centriole and basal body appendages are found that have been named context-dependent distal appendages or transition fibers, respectively. The distal appendages/transition fibers locate strictly symmetrically, adopting the ninefold structure of the underlying microtubule organization (Fig. 1.16). The transition fibers of the basal body are actually sheet-like projections and have therefore also been named alar sheets. These appendages not only contact the cell membrane but anchor the basal body to the plasma membrane (Anderson 1972). Moreover, the transition fibers/alar sheets might function as a "ciliary pore" to restrict vesicle and macromolecule exchange between the cytoplasm and the ciliary lumen (Deane et al. 2001). Ultrastructurally, the cilium displays several subregions (reviewed in: Fisch and Dupuis-Williams 2011). These are the transition zone linking the cilium to the basal body, followed by the doublet zone and finally by the singlet zone including the tip structure. Whereas the doublet zone harbours MT doublets, the singlet zone is characterized by the termination of the incomplete B tubule towards the tip of the cilium leaving the singlet A tubule and, in motile cilia, the central pair. At the ciliary tip growth and resorption of the axoneme takes place. The ciliary tip therefore is the place where intraflagellar transport (IFT) is regulated and remodelled (see Sect. 1.4.3). Accordingly, the ciliary tip features a complex ultrastructural organization. Most remarkable, the concentration of electron dense material at the tip of primary cilia has been observed more than 30 years ago (Albrecht-Buehler and Bushnell 1980).

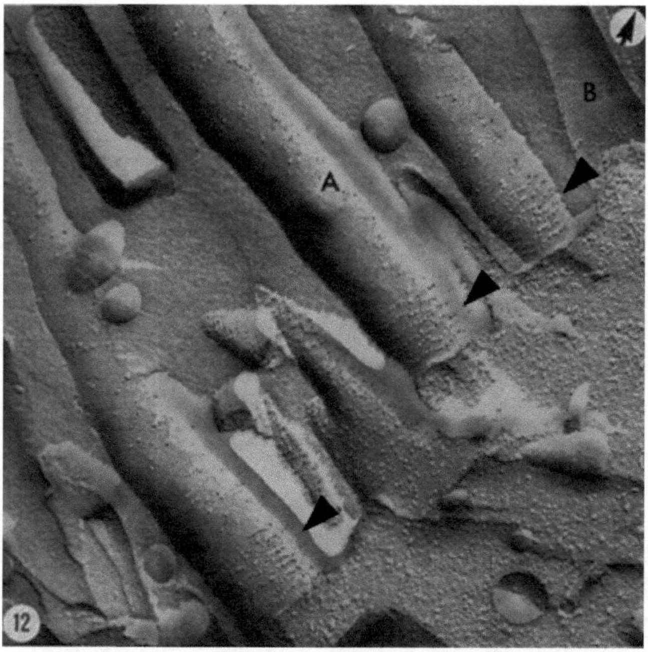

Fig. 1.17 Freeze-etching reveals the ciliary necklace (*arrowheads*) in cilia from rat tracheal epithelium. Six particle strands are found in the necklace region. ×80,000 (With permission from Gilula and Satir 1972)

The transition zone defines the region where the basal body passes into the axoneme of the cilium. In this region, transition fibers anchor the basal body to the membrane. The region of the ciliary membrane adjacent to the junction with the cell membrane is the ciliary necklace. In electron micrographs a distinct pattern with an electron dense appearance at the base of the ciliary shaft and a periodic striation extending up the ciliary shaft characterizes the ciliary necklace (Fig. 1.17). This ordered array of electron dense structures reflects intramembrane particles that are part of a protein complex that links axonemal microtubules to the membrane. Tethering of the MT doublets of the axoneme to the ciliary membrane at the transition zone is also visible by the presence of Y-shaped bridges in TEM cross-sections (e.g. Fig. 1.1b arrowhead; Wheatley 1972). The ciliopathy protein Cep290 is localized to this region and seems to be involved in the tethering of MT to the membrane (Craige et al. 2010). Freeze-fracture electron microscopy revealed the presence of the ciliary necklace in motile as well as in primary cilia but not in sperm flagella (Gilula and Satir 1972; Fig. 1.17). It has been postulated that the necklace functions as a selective barrier for ciliary entrance (Hu et al. 2010). The ciliary membrane forms a compartment that is separated from the cytoplasm and the cell membrane and is specialized to harbour receptors for e.g. Sonic Hedgehog, platelet-derived growth factor (PDGF), serotonin, or somatostatin (reviewed in: Satir and Christensen 2007; Anderson et al. 2008). Therefore trafficking to primary cilia, selective

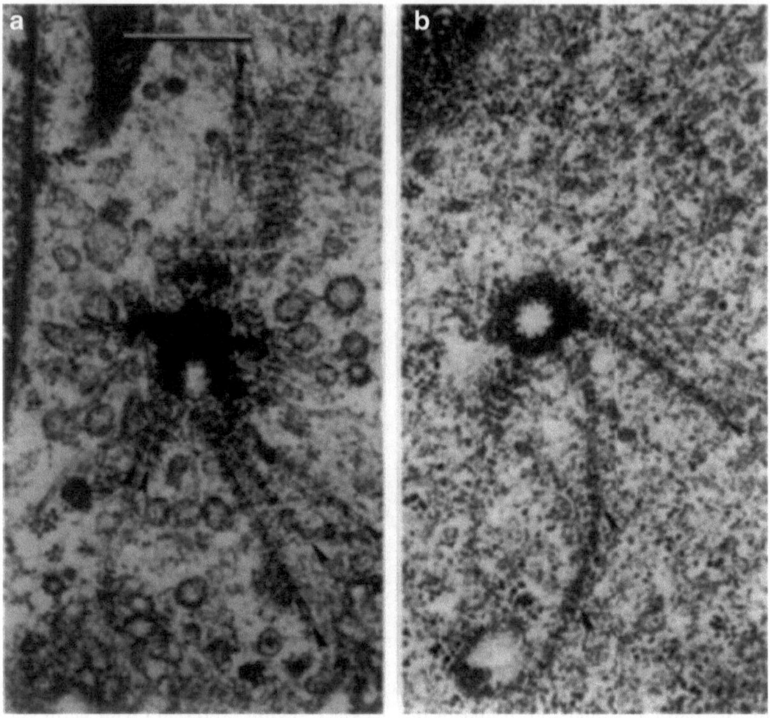

Fig. 1.18 A striated "rootlets" connected to basal centrioles in 3T3 cells. *Bar*, 0.5 μm (With permission from Albrecht-Buehler and Bushnell 1980)

access and retention of ciliary proteins are critical for ciliary function (reviewed in: Garcia-Gonzalo and Reiter 2012). Recently, a protein complex at the necklace including Septin 2 (Sept2) has been identified, which is essential for cilia formation and the location of signaling receptors in the ciliary membrane. Disruption of this "ciliopathy" complex impairs the diffusion barrier of the necklace (Chih et al. 2012). Furthermore, as observed in the model organism *C. elegans,* MKS (Meckel-Gruber syndrome) and NPHP (nephronophthisis) protein complexes that cause the ciliopathies Meckel-Gruber syndrome (MKS) and nephronophthisis (NPHP) when mutated are essential for the early stage of ciliogenesis, i.e. the docking of the basal body to the ciliary membrane, and the establishment of the ciliary gate (Williams et al. 2011). The idea, that the transition zone with the distal appendages/transition fibers bound to the cell membrane, and the necklace region function as a "ciliary pore", as suggested by Deane et al. in 2001, has been substantiated further by the discovery that IFT kinesin-2 KIF17 interacts with the nuclear import protein importin-ß2 and is inhibited by RanGTP (Dishinger et al. 2010). Furthermore, in analogy to nuclear import mechanisms, nucleoporins have been located at the ciliary base (Kee et al. 2012).

At the proximal end of the basal body in NIH3T3 cells striated fibers arise that are reminiscent of the striated rootlets observed on basal bodies of motile and primary cilia (Fig. 1.18). Rootlets might anchor the basal body/cilium complex in the cell

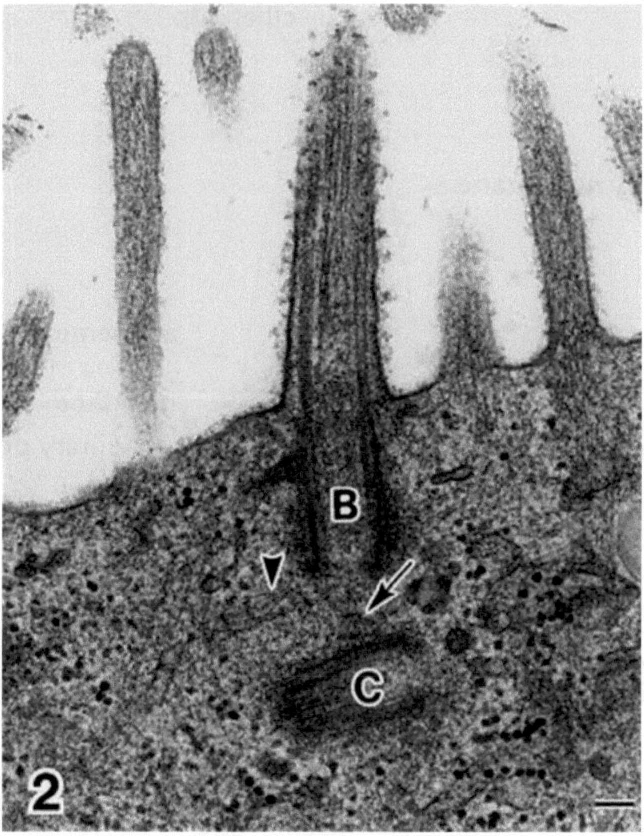

Fig. 1.19 A striated connector (*arrow*) interlinks the basal body (*B*) of a solitary cilium in a secretory cell of the human oviduct epithelium with its associated centriole (*C*). A striated rootlet (*arrow head*) is also associated with the proximal end of the basal body. *Scale bar* = 1.0 μm (From: Hagiwara et al. 2002)

(Odor and Blandau 1985; Harrison 1989; Hagiwara et al. 2008). One major protein of the rootlets of ciliated cells is Rootletin, a coiled-coil protein of 220-kDa (Yang et al. 2002). Moreover, the basal body and its associated centriole are interlinked by fibrous structures similar to those observed between the pair of centrioles in interphase cells (Paintrand et al. 1992; Hagiwara et al. 2002). As noticed in solitary cilia of the secretory cells of the human oviduct epithelium these fibers show a cross-banding pattern and are hence termed the striated connector (Hagiwara et al. 2002; reviewed in Hagiwara et al. 2008) (Fig. 1.19). The striated connector shares a protein component with the striated rootlets that emerge from the base of the basal body (Hagiwara et al. 2000, 2002). A scheme of a cilium is shown in Fig. 1.20.

Even though the vertebrate primary cilium is described as 9+0, serial sectioning revealed that this organization lasts only for the proximal part of the primary cilium shaft. Further distally (as the axoneme tapers down), dislocation of

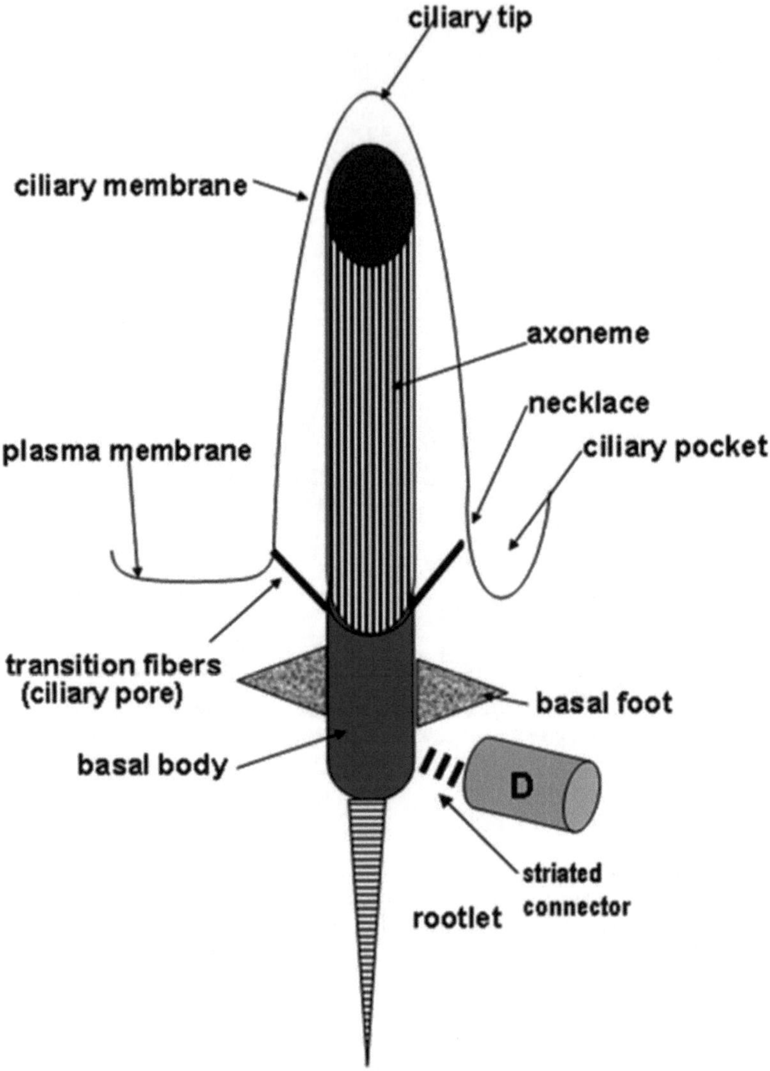

Fig. 1.20 Scheme of the cilium. A clear ciliary pocket is not present in every cilium and is therefore illustrated on one side only. *D* daughter centriole. See text for explanation

microtubule doublets leads to disturbance of 9+0 structure and eventually to a reduction of MT doublets at the most distal region (Fig. 1.21). Displacement of MT doublets occurs more or less randomly (Gluenz et al. 2010). Variations of 9+0 organization have been observed in primary cilia of different tissues and are not confined to the kidney primary cilia (e.g. sensory cilia in olfactory receptors of man; Moran et al. 1982; and primary cilia of the central nervous system, Dahl 1963; Webber and Lee 1975).

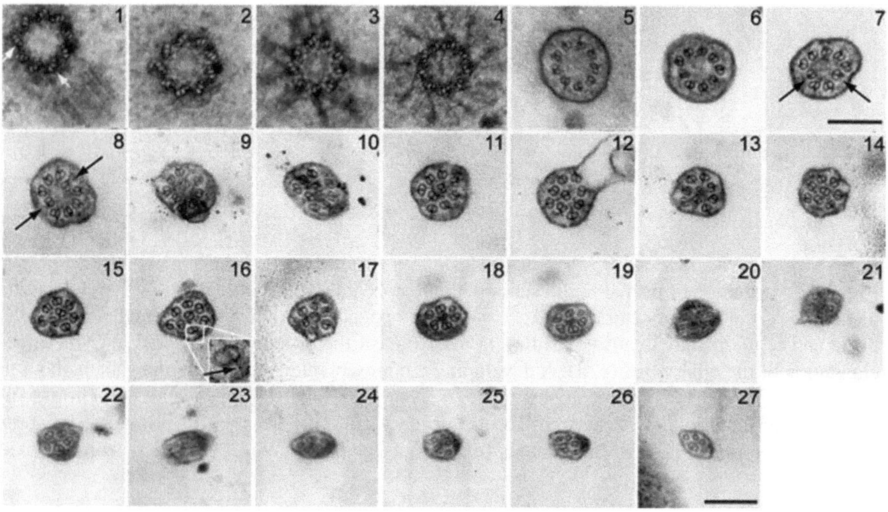

Fig. 1.21 TEM serial sectioning through a primary cilium of kidney IMCD3 cells starting from the proximal part of the basal body (*1*; longitudinal sectioning of its associated daughter centriole is visible) to the tip of the primary cilium (*27*). Subdistal appendages (in *3*) and distal appendages (in *4*) are visible. A strict 9+0 organization of the primary cilium is found only at the proximal end but became disturbed by displacement of microtubule doublets. *Scale bars* = 200 nm (With permission from Gluenz et al. 2010)

1.4.2 Ultrastructural Observations on Primary Cilia Generation

Formation of primary cilia always starts from the mature centriole. The daughter centriole, albeit associated with the mature centriole, never sprouts a primary cilium. Moreover, in cultured mammalian cells primary cilia grow asynchronously in sister cells. The cell that received the oldest mother centriole generated more than two cell cycles before usually grows a cilium first. Concomitantly, in the two mother centrioles of a pair of sister cells asymmetric localization of Odf2/Cenexin is observed. The oldest centriole always harbours a higher amount of Odf2, suggesting that a sufficient amount is necessary for initiation of ciliogenesis (Anderson and Stearns 2009).

Electron microscope studies on the formation of primary cilia in fibroblasts and smooth muscle cells of mammalian and chicken origin revealed great similarities. Based on these observations Sorokin provided a model for cilium formation which is subdivided into three phases (Sorokin 1962) (Figs. 1.22 and 1.23). Cilium formation starts with the appearance of a solitary vesicle (the so called primary ciliary vesicle) at the distal end of the mature centriole (phase I). Growth of the distal end of the mature centriole against the membrane of the primary vesicle results in the formation of a ciliary bud, and invagination and flattening of the primary vesicle. As a result, two layers of membrane from the primary vesicle surround the bud,

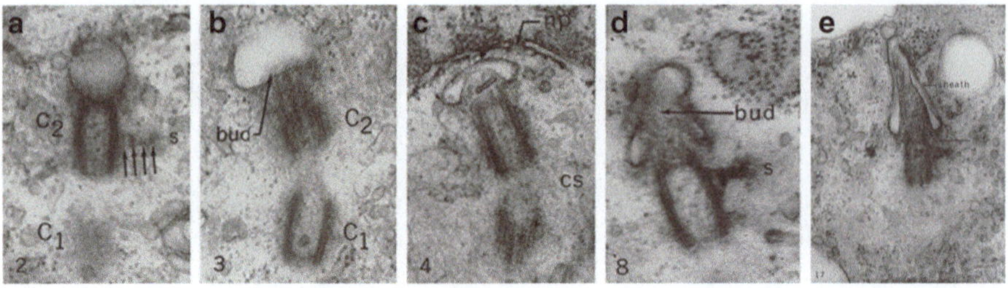

Fig. 1.22 Formation of primary cilia starts with the appearance of a primary ciliary vesicle at the distal end of the mature centriole (*C2*; *S*, subdistal appendage; **a**). The ciliary bud forms at the distal end of the mature centriole leading to distortion of the primary ciliary vesicle (**b**). Further elongation of the ciliary bud (**c, d**). A developing cilium with elongated shaft and sheath about to emerge from the cell (**e**). The mature centriole (*C2*) harbours subdistal appendages (*S*). The associated daughter centriole (*C1*) does not form a primary cilium. Formation of primary cilia starts inside the cytoplasm next to the nucleus (**c**; np, nuclear pore) (With permission from Sorokin 1962)

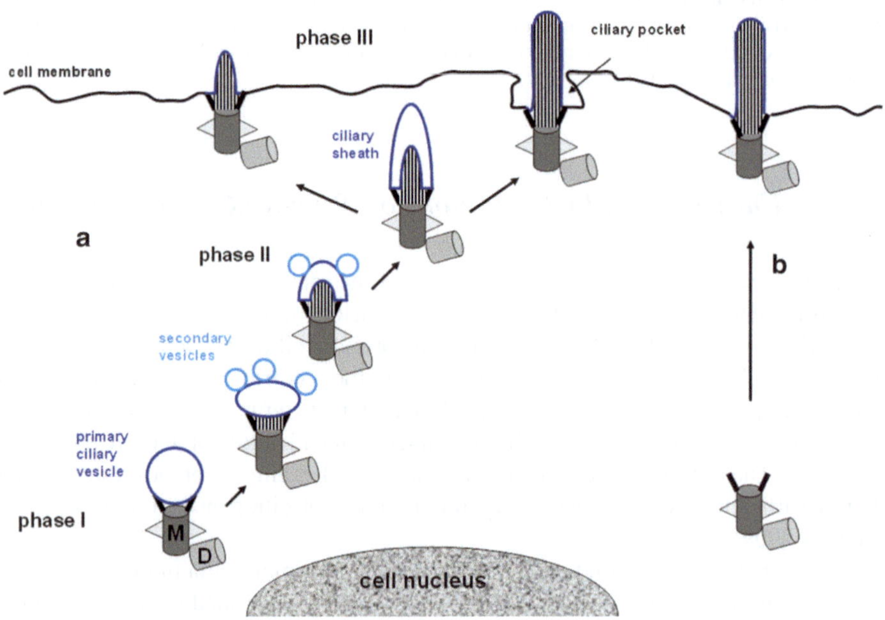

Fig. 1.23 Overview of primary cilium formation. See text for explanation. *M* mother centriole with distal and subdistal appendages, *D* daughter centriole. (**a**) Cilium formation starts deep in the cytoplasma and the basal body/cilium complex migrates towards the cell membrane. Fusion of the ciliary membrane with the cell membrane exposes the cilium to the extracellular space. (**b**) The mother centriole first migrates to and anchors at the cell membrane followed by ciliary growth

thus forming the ciliary sheath. By elongation of the ciliary bud and concurrent lengthening of the sheath the cilium grows towards the cell surface (phase II). The sheath develops by repeated formation of secondary ciliary vesicles distal to the primary vesicle and subsequent fusion. When the sheath reaches the cell surface, its outer membrane fuses with the cell membrane whereas the inner membrane still covers the cilium. Membrane fusion thus results in exposition of the cilium to the extracellular environment (phase III). However, the primary cilium is often deeply invaginated and encircled by two membrane layers that form the ciliary pocket (Molla-Herman et al. 2010; reviewed in: Rohatgi and Snell 2010; Ghossoub et al. 2011). Since formation of a primary cilium can start at the mother centriole located in the cytoplasm next to the nucleus, migration of the mature centriole to the cell surface is not necessary for sprouting a cilium. However, alternatively, the mother centriole first migrates and anchors at the cell membrane followed by sprouting a cilium into the extracellular space.

Cells express cilia throughout interphase but never in mitosis. Formation of cilia already starts in G1 phase, and they are resorbed very late in the cell cycle when cells enter prophase (Rieder et al. 1979; Wheatley et al. 1996; Alieva and Vorobjev 2004). Even though an inverse relationship between the frequency of ciliated cells and the proliferative activity of a cell culture has been observed as well as a coincidence of deciliation with the initiation of DNA synthesis in Balb/c and Swiss3T3 cells, cilia may also be present in S phase (Fonte et al. 1971; Wheatley 1971; Tucker et al. 1979; Vorobjev and Chentsov 1982; Alieva and Vorobjev 2004). Primary cilia were found in epithelial cells in S phase even though only at low frequency (see Fig. 1.11). Furthermore, these primary cilia were entirely intracytoplasmatic and never reached the cell surface as primary cilia formed in G0 arrested cells (Fig. 1.24) (Vorobjev and Chentsov 1982). In cultured cells ciliogenesis seems to be largely dependent on cell-cell and cell-substratum contacts. Therefore, cells grown as adherent cell layers are able to generate a primary cilium, whereas growth of cells in liquid suspension is not compatible with cilium expression (Wheatley et al. 1996). However, cells arrested in G0 phase have a higher incidence of primary cilia than cycling cells (Mori et al. 1979).

Resorption of cilia not only might free the basal body to function as centrosome in cell cycle progression but otherwise might be causative for re-entering the cell cycle. Ciliary disassembly by recruitment of phosphorylated Tctex-1 to the transition zone before S-phase entry is essential for cell cycle progression. It was therefore suggested that cilia prevent cell cycle progression by acting as a brake (Li et al. 2011). In support of this model it has also been shown that the induction of abnormally long cilia coincides with a delay in G1 to S transition. Lack of the mother centriole protein Nde1 (nuclear distribution gene E homologue 1), which otherwise suppresses cilia formation, increased ciliary length and resulted in a delay in cell cycle re-entry (Kim et al. 2011). Since Nde1 by binding to the dynein light chain LC8, and Tctex-1 both interact with cytoplasmic dynein these findings suggest that cytoplasmic dynein might affect proliferation possibly via ciliary length control.

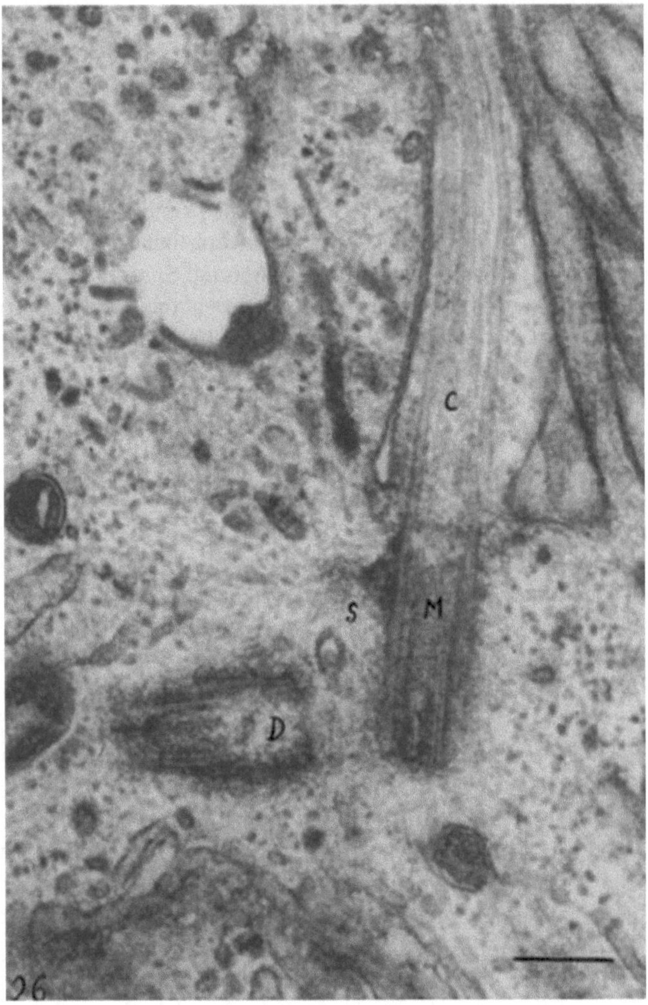

Fig. 1.24 Centrioles and primary cilium in G0 phase. *M* mother centriole, *D* daughter centriole, *S* subdistal appendages, *C* cilium (With permission from Vorobjev and Chentsov 1982)

1.4.3 Molecular Mechanisms of Ciliogenesis

Formation of cilia depends on several proteins that have formerly been found to be present on centrioles/basal bodies (an overview of ciliary assembly and disassembly is given in Fig. 1.25). Depletion of the distal appendage protein Cep164 prevents cilium assembly but the mechanism of its action is not clear (Graser et al. 2007). Cep97, CP110, and CPAP have opposed functions in centriole duplication and ciliogenesis. CP110 is recruited by Cep97 to the distal end of the growing centriole to

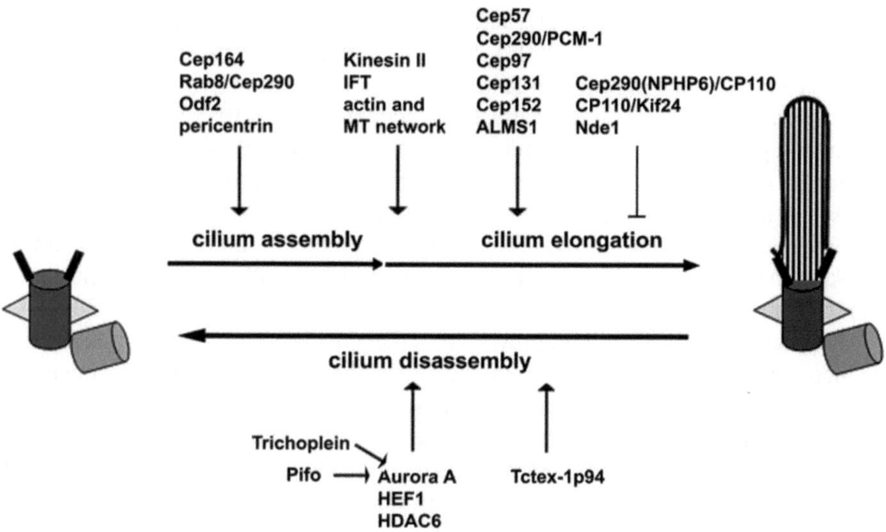

Fig. 1.25 Overview of proteins involved in promoting cilium formation and in its disassembly. See text for explanation

inhibit MT growth thus limiting centriolar length. Both proteins therefore act coordinately to suppress the assembly of the ciliary axoneme (Spektor et al. 2007). Depletion of CP110 or Cep97 caused overly long centrioles but did not promote the formation of a primary cilium ensheathed by a membrane (Spektor et al. 2007; Kohlmaier et al. 2009; Schmidt et al. 2009; Tang et al. 2009). CP110 and Cep97 associate with Kif24, a member of the kinesin-13 subfamily. Depletion of Kif24 supported the formation of primary cilia but did not affect the length of centrioles, as opposed to CP110 or Cep97 depletion. Kif24 may affect ciliogenesis by controlling microtubule dynamics (Kobayashi et al. 2011). Suppression of ciliogenesis by CP110 strictly depends on its interaction with Cep290 (also named NPHP6, nephronophthisis protein 6) although ablation of Cep290 prevents ciliogenesis. Cep290/NPHP6 is one of currently 11 members of proteins that all localize to cilia or centrosomes and when mutated cause nephronophthisis, a recessive cystic kidney disease. Cep290 and CP110 interact with Rab8a, a small GTPase required for cilia assembly by promoting the docking and fusion of membranous vesicles. The available data suggest that Cep290 together with Rab8a promote ciliogenesis whereas binding to CP110 abrogates its function (Tsang et al. 2008). Additionally, Cep290 interacts with the centriolar satellite protein PCM-1 to promote ciliogenesis (Kim et al. 2008). Studies in *Chlamydomonas reinhardtii* revealed that Cep290 tethers MT to the flagellar membrane at the transition zone. It was therefore suggested that Cep290 is involved in the function of the ciliary pore to regulate entry of proteins into the ciliary compartment (Craige et al. 2010). Two other Rab GTPases in addition to Rab8a, Rab-17 and Rab-23, and their cognate GTPase-activating proteins

(GAPs), XM_037557, TBC1D7, and EVI5like, are involved in primary cilium formation (Yoshimura et al. 2007). Moreover, Rab8a specifically interacts with Odf2/cenexin, which in turn is essential for cilium formation (Ishikawa et al. 2005; Yoshimura et al. 2007). Additionally, Rab8 cooperates with BBS proteins (encoded by a group of genes that are mutated in the ciliopathy Bardet-Biedl-syndrome, reviewed in: Beales 2005) in ciliary membrane formation (Nachury et al. 2007). CPAP on the other hand promotes MT assembly, leading to overly long centrioles when overexpressed (Kohlmaier et al. 2009; Schmidt et al. 2009; Tang et al. 2009). CPAP (SAS-4) recruits HYLS-1 to the outer centriole wall to enable anchoring of centrioles at the plasma membrane and ciliogenesis (Dammermann et al. 2009).

In polarized cells the cilium grows out at the apical site. Apicobasal polarization is therefore a prerequisite for correct docking of the basal body to the plasma membrane and ciliogenesis (Pan et al. 2007). Disorganization of the apical actin and MT networks as observed by knock down of planar cell polarity components might be causative for abrogated ciliogenesis (Oishi et al. 2006; Park et al. 2006, 2008). Protein complexes involved in basal body docking seem to encompass proteins affected in Meckel-Gruber syndrome (MKS1, meckelin, Cep290, and RPGRIP1L) and proteins affected in nephronophthisis (NPHP4 and Nephrocystin/NPHP1) (Mollet et al. 2005; Roepman et al. 2005; Dawe et al. 2007b). A link between the apical actin cytoskeleton and the basal body is provided by meckelin (also named MKS3) that interacts with the actin-binding protein filamin A, and loss of filamin A causes defects in basal body positioning and ciliogenesis (Adams et al. 2012). As a modulator of ciliary length control, the light chain subunit of cytoplasmic dynein Tctex-1 (or DYNLT1) was identified. Depletion of Tctex-1 as well as suppression of dynein heavy chain-2 (DHC2) resulted in longer cilia (Palmer et al. 2011). However, Tctex-1 additionally executes dynein-independent functions in actin remodelling (Chuang et al. 2005). When phosphorylated at Thr-94 Tctex-1 is recruited to the ciliary transition zone before S-phase entry. Phospho(T94)Tctex-1 in turn affects actin cytoskeleton rearrangement and ciliary resorption (Li et al. 2011). Another protein that associates with cytoplasmic dynein, nuclear distribution gene E homologue 1 (Nde1), localizes to the mother centriole and suppresses cilia formation (Kim et al. 2011). The inhibitory role of the actin network in ciliogenesis and cilium length control was substantiated by a large functional genomic screen to identify modulators of ciliogenesis (Kim et al. 2010).

Elongation of the growing microtubule axoneme and ciliary maintenance is achieved by bidirectional intraflagellar transport (IFT) in all types of cilia (Rosenbaum and Witman 2002; Pedersen and Rosenbaum 2008). Multiprotein complexes, the IFT particles, transport ciliary components from the cell body to the tip of the cilium (anterograde transport) and turn-over products back to the cell body (retrograde) (Qin et al. 2004). Anterograde transport is powered by kinesin-2 motor complexes (reviewed in: Scholey 2008). In human and mouse the two motor subunits of the heterotrimeric kinesin-II complex are Kif3A and Kif3B (Scholey 1996). Kinesin-II is essential for assembly of primary cilia. Deletion of either Kif3A or Kif3B causes embryonic lethality in mice and conditional deletion of Kif3A in the kidney results in impaired assembly of primary cilia (Nonaka et al. 1998; Marszalek et al. 1999;

Takeda et al. 1999; Lin et al. 2003). Retrograde transport towards the basal body is mediated by cytoplasmic dynein 2, and its impaired function resulted in stumpy cilia (May et al. 2005). Kinesin and dynein mediate bidirectional transport in association with IFT particles. IFT particles are subdivided into IFT-B, comprising 14 known proteins (IFT20, IFT22, IFT25, IFT27, IFT46, IFT52, IFT54, IFT57, IFT70, IFT74/IFT72, IFT80, IFT81, IFT88 (also known as Tg737 or polaris) and IFT172), and IFT-A, a complex of six additional IFTs (IFT144/WDR19, IFT140, IFT139/Ttc21b, IFT122, IFT121/WDR35, and IFT43). Both complexes play distinct parts in ciliary transport. Whereas IFT complex B contributes to anterograde transport, IFT-A is required for retrograde transport. Therefore, IFT-B is essential for ciliary assembly and maintenance whereas the roles of IFT-A complexes in ciliogenesis appear to be more complex. All IFT components are highly conserved among ciliated eukaryotes and are essential for the assembly of almost all eukaryotic cilia and flagella (Avidor-Reiss et al. 2004; Li et al. 2004; reviewed in: Pedersen et al. 2008; Ishikawa and Marshall 2011). Disruption of either kinesin-2 or individual IFTs in vertebrates eliminates the primary cilium, resulting in diverse developmental and cell signaling defects, known as ciliopathies (reviewed in: Nigg and Raff 2009). IFT therefore contributes to cilia formation and stability not only in primary cilia but also in motile cilia. IFT proteins form a complex with pericentrin at the base of primary and motile cilia and this location is dependent on both components at the same time because depletion of anyone of these components mislocalizes the other and inhibits primary cilia assembly (Jurczyk et al. 2004).

Ciliary length is under control of Cep57, Cep131, Cep152, and ALMS1 (a gene mutated in Alström syndrome, a rare pleiotropic condition) since depletion of any of these proteins resulted in stunted cilia (Graser et al. 2007). Additionally, pharmacological agents and environmental changes can modify ciliary length. For example, lithium treatment as well as a reduction in intracellular Ca^{2+} concentration cause primary cilia to extend in length (reviewed in: Miyoshi et al. 2011; Besschetnova et al. 2010). Ciliary MTs are highly enriched in post-translational modification of α/β-tubulin, namely acetylation (especially K40 of α-tubulin), polyglutamylation and polyglycylation (Piperno and Fuller 1985; Bré et al. 1996; Ikegami et al. 2010). Additionally, detyrosinated tubulin and Δ2-tubulin are presumably found on the B-tubule of the axoneme. Studies of post-translational modifications of tubulin in several model organisms revealed that glycylation is involved in axonemal stability whereas glutamylation most likely affects the beating behaviour of motile cilia. Glutamylation and glycylation of tubulin may regulate each other. However, glutamylation defects of the ciliary axoneme caused by mutations in the centrosomal protein Cep41 are involved in Joubert syndrome. Cep41 regulates ciliary entry of the polyglutamylase enzyme TTLL6 (Lee et al. 2012). Although axonemal tubulin is highly acetylated, its effect on MT assembly and stability is not clear (reviewed in: Janke and Bulinski 2012).

Early work already provided a link between cell cycle exit and the formation of primary cilia. Thus serum-starvation is widely used to promote ciliogenesis in cultured cells (Tucker et al. 1979). The reduced amount of growth factors might trigger cell cycle exit thus increasing the proportion of cells in G0 that form a primary cilium. Alternatively, in serum-starved medium the amount of factors that inhibit ciliogenesis might be reduced, resulting in promotion of ciliogenesis. Inhibition of ciliogenesis seems to be mediated by an intact PI(3)K signaling pathway and may be counteracted by the von-Hippel-Lindau tumor suppressor (Thoma et al. 2007; Lolkema et al. 2008). The von-Hippel-Lindau protein (VHL) is a component of an E3 ubiquitin ligase complex. It targets hypoxia-inducible factor (HIF) for destruction in the presence of oxygen. VHL seems to promote ciliogenesis at least in some cells since the VHL syndrome is characterized by the lack of primary cilia in renal cysts (Esteban et al. 2006). PI(3)K signaling inactivates GSK3β via phosphorylation, and loss of pVHL together with inhibition of GSK3β abrogates ciliogenesis. GSK3β and pVHL thus might both function in promoting ciliogenesis (Thoma et al. 2007). Their influence on the formation of cilia might be explained by the fact that both pVHL and GSK3β affect microtubule orientation and stability (discussed in: Santos and Reiter 2008).

Disassembly of the primary cilium is mediated by Aurora A kinase and its activator HEF1 (enhancer of filamentation 1). Aurora A phosphorylates and activates histone deacetylase 6 (HDAC6) that in turn deacetylates axonemal tubulin leading to regression of the primary cilium (Matsuyama et al. 2002; Pugacheva et al. 2007). However, since deletion of the *Hdac6* gene in mice causes only mild effects but no gross abnormalities as expected when cilia are affected, the influence of HDAC6 on cilia formation is debated (Zhang et al. 2008). Aurora A is also activated by the basal body and ciliary necklace protein Pitchfork (Pifo) as well as by Trichoplein resulting in ciliary disassembly (Kinzel et al. 2010; Inoko et al. 2012). Ciliary disassembly is additionally controlled by ubiquitination, and by the lipid 5-phosphatase Inpp5e (Huang et al. 2009; Jacoby et al. 2009).

1.5 Motile Cilia

1.5.1 Ultrastructure of Motile Cilia

Motile cilia are generally found on specialized epithelia lining the airways, paranasal sinuses, the oviduct, and the ventricular system of the brain. They are usually present in large numbers and feature a coordinated beating. Motile cilia are generally of a 9+2 structure with nine peripheral MT doublets and two centrally located singlets (Figs. 1.26 and 1.27). MT doublets consist of a complete A tubule (assembled from 13 protofilaments) and an incomplete B tubule (assembled from 10 protofilaments). Peripheral doublet microtubules are interlinked by nexin bridges that are involved in the bending motions (in Hagiwara et al. 2008).

Two singlet microtubules are found in the center of the axoneme encircled by the central sheath. Radial spokes connect the doublet MT with the central apparatus. Outer and inner dynein arm complexes are present at the complete A tubule. Dynein arms are large multiprotein complexes consisting of polypeptides of different sizes: heavy (HC; 400–500 kDa), intermediate (IC; 45–110 kDa) and light chains

Fig. 1.26 Cross-sections of the shafts of ependymal cilia of the rat brain showing typical 9+2 structure. ×67,000 (With permission from Brightman and Palay 1963)

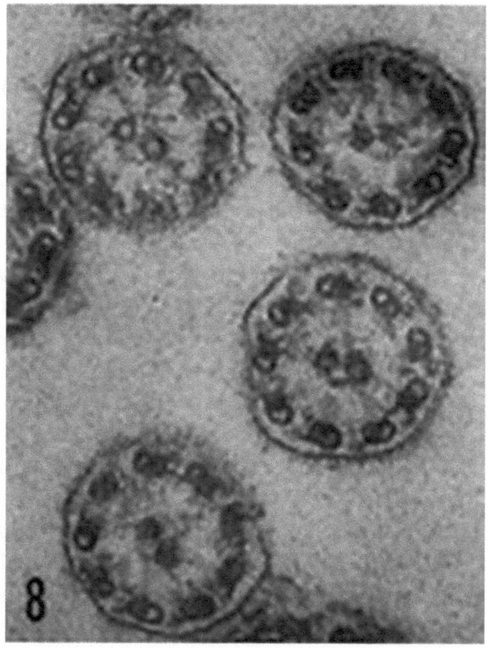

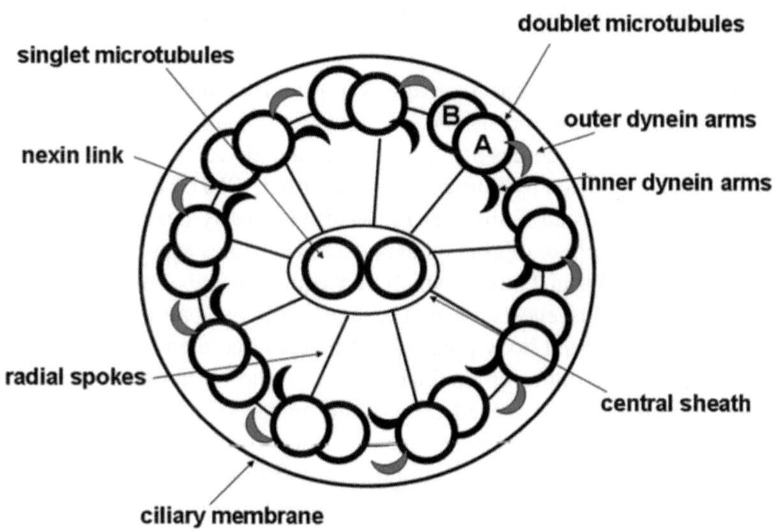

Fig. 1.27 Structure of a 9+2 axoneme

(LC; 8–55 kDa). The ATPase activity resides in the HC and provides the energy for ciliary motion. Dynein arms are attached to the MT by specialized docking complexes (Takada et al. 2002). The central apparatus and the radial spokes contain kinases and phosphatases and execute a regulatory function on dynein arms (Porter and Sale 2000; Smith 2002; reviewed in: Ibanez-Tallon et al. 2003). Moreover, the mouse serin-threonine kinase Stk36 (also known as *fused*, Fu), which is involved in Hedgehog signal transduction, is essential for construction of the central pair apparatus (Wilson et al. 2009).

However, motile cilia with a 9+0 structural organization of primary cilia are found on the node cells, as well as on prechordal and notochordal plate cells. These cilia protrude as single entities on the ventral surface into the yolk sac cavity. Nodal monocilia are motile and are important for establishment of left-right asymmetry by generation of leftward flow of the extraembryonic fluid. That they are involved in left-right axis determination has been suggested very early and could be substantiated by the generation of mice with a deletion in the kinesin motor protein KIF3B missing nodal cilia but not basal bodies (Sulik et al. 1994; Nonaka et al. 1998). Motion of the nodal cilia is quite distinct from the whip-like back-and-forth motion of typical 9+2 cilia, as they perform a vortical motion by moving of the distal end of the cilium in a circle around the axis of rotation. This special motion might explain why nodal cilia are of 9+0 structure. However, the existence of two kinds of monocilia on nodal cells serving different functions has also been postulated (McGrath et al. 2003; reviewed in: Lee and Anderson 2008).

1.5.2 Basal Body Formation During Multiciliogenesis

One difference between the basal apparatus of primary cilia and motile 9+2 cilia is the presence of the daughter centriole. Motile 9+2 cilia are associated with a single centriole or basal body whereas primary cilia are mostly associated with two centrioles, as originally reported by Barnes in 1961 (Barnes 1961). The single centriole of motile 9+2 cilia suggested a different mode of basal body generation. In epithelia that harbour a high number of cilia on every cell, centrioles must first divide repeatedly to form many basal bodies, which subsequently elaborate the motile cilia. Before the onset of ciliogenesis, the cell harbours only one pair of centrioles but has to generate 200–300 basal body precursor structures that then transform into basal bodies. Thus, in multiciliated epithelial cells, e.g. of the vertebrate respiratory tracts and oviduct, ciliogenesis is preceded by the formation of multiple centrioles from precursor structures (Dirksen 1991).

Generation of centrioles generally follows two different pathways that have both been described in the Rhesus monkey oviduct as well as in tracheal epithelia of the rat (Sorokin 1968; Anderson and Brenner 1971). In the centriolar pathway, which is a minor pathway, multiple procentrioles are generated at right angles to the centriole along its axis. These procentrioles rapidly develop into mature basal bodies by elongation and expansion and eventually detach from the centriole (Fig. 1.28).

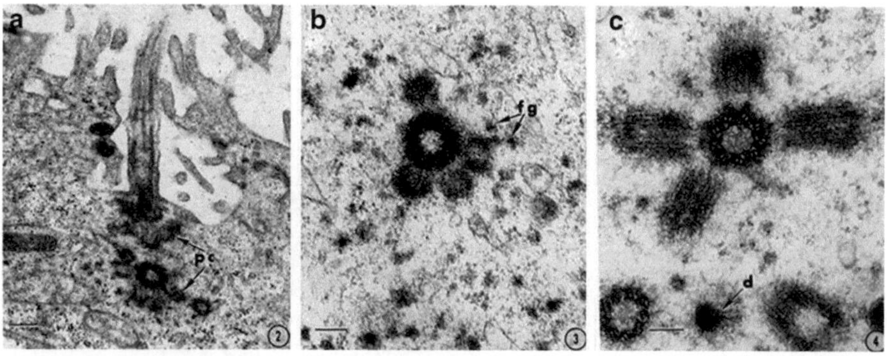

Fig. 1.28 Centriolar pathway of centriole/basal body formation in the rhesus monkey oviduct. (**a**) Formation of procentrioles (*pc*) from the basal body and its associated centriole. ×28,000. (**b**) Centriole generating four procentrioles. ×61,000. (**c**) Procentrioles have been grown in length. ×64,000 (From Anderson and Brenner 1971)

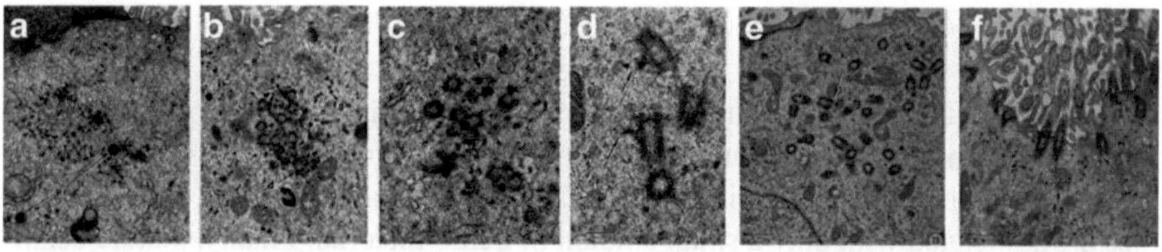

Fig. 1.29 Acentriolar pathway of centriole/basal body formation in the rhesus monkey oviduct. (**a**) Aggregation of fibrous granules with occasional procentrioles (*pc*). ×28,000. (**b**) Procentrioles within aggregate of fibrous granules. ×28,000. (**c**) Formation of four procentrioles. ×28,000. (**d**) Maturation to basal bodies by the acquisition of basal feet (*arrows*) on their walls. ×35,000. (**e**) Random arrangement of basal bodies in the apical region. ×17,000. (**f**) Basal bodies forming cilia. ×17,000 (From Anderson and Brenner 1971)

The major pathway in multiplication of basal bodies is the acentriolar pathway (Fig. 1.29). In this pathway precursor structures are formed independently of the presence of mature centrioles. In the acentriolar pathway fibrous granules are temporarily formed that later aggregate and generate procentrioles. First, amorphous clouds of filamentous material appear that enclose electron-dense particles of varying size. These precursor structures are called fibrogranular masses or deuterosomes, amongst others. All multiciliated cells investigated so far form deuterosomes of fundamental morphological similarity. Deuterosomes have been identified during ciliogenesis in multiciliated epithelial cells of different tissues and in diverse vertebrate species (reviewed in: Dirksen 1991). However, immature centrioles are never found associated. These amorphous fibrogranular masses or deuterosomes then aggregate to form large electron dense masses of ~75 to over 400 nm in size to which immature centrioles or procentrioles finally associate. As with centriole elongation and maturation these

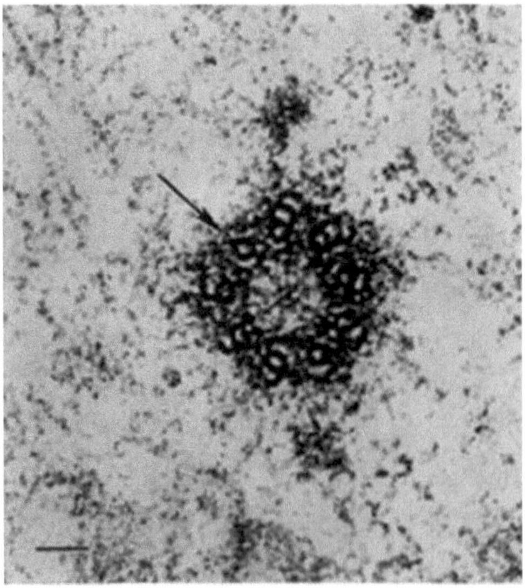

Fig. 1.30 Cartwheel structure in a procentriole of the acentriolar pathway. The A tubule is formed first followed by progressive formation of B and C tubules. ×112,000 (From Anderson and Brenner 1971)

dense masses then drop off. Initially as many as nine immature procentrioles are associated with these condensation forms, but the number of centrioles subsequently decreases concomitant with their elongation and maturation. At the end, mature centrioles are associated with only a small amount of dense material. Finally, mature centrioles migrate towards the cell apex freeing the dense material. The formation of centrioles from deuterosomes suggests that deuterosomes are centriolar precursors consisting mainly of deposited centriolar material. Because basal body formation and ciliogenesis are asynchronous processes, a cell may contain basal bodies and cilia at all stages from fully developed basal bodies at the apical surface of the cell with outgrowing cilia, to dense aggregates with or without immature centrioles deeper in the cell.

Procentriole formation traverses through a cartwheel stage as already described for centriole duplication in cycling cells (Fig. 1.30, compare Fig. 1.10). The cartwheel is no longer present in the basal body undergoing cilium formation. Centrioles/basal bodies show the same conserved structure of nine triplet MT as known from centrioles/basal bodies of cycling cells. The formation of the ninefold symmetry traverses through intermediate structures containing singlets and doublets and therefore resembles the formation of daughter centrioles. When the triplet configuration is established centrioles start to grow in width and length. Eventually, transformation of the centriole to the basal body is accomplished by the addition of accessory structures. Usually one, sometimes two basal feet form in the mid-region of the basal body at right angles to the wall. Transition fibers extend from the C tubules at the apical region. However, they assume their final arrangement until the basal body-cilium relationship is established (Anderson and Brenner 1971).

Ciliogenesis starts when basal bodies have migrated to the cell surface. Cilia then grow out from the distal regions of basal bodies. Ciliary budding thus happens at the

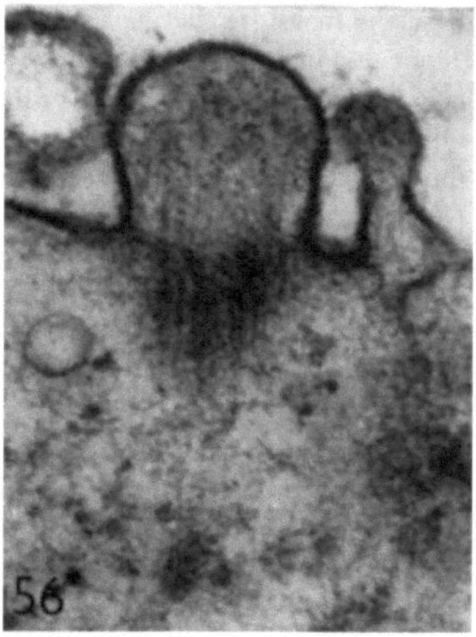

Fig. 1.31 A ciliary bud in the bronchus of a 21-day foetal rat lung. ×58,000 (With permission from Sorokin 1968)

apical cell membrane, not inside the cytoplasm as observed in fibroblasts undergoing ciliogenesis (Sorokin 1962). Sprouting of cilia from the apical membrane into the epithelium-lined cavity has also been observed in epithelial cells in which basal bodies normally reside in the apical cytoplasm (Fig. 1.31; Sorokin 1968).

Motile cilia of the oviduct epithelium of man and rabbit are different from single cilia on secretory cells of the oviduct epithelium. Motile cilia are longer than single cilia. Additionally, they have only one basal foot emerging from the basal body and apparently lack rootlets (Odor and Blandau 1985; Hagiwara et al. 2002). The single basal foot projects laterally from the basal body and points into the direction of ciliary beating stroke (Boisvieux-Ulrich and Sandoz 1991). Since motile cilia function to propel cells and fluids over the epithelial surface their orchestrating beating is essential for efficient flow direction. In the brain ependyma it has been shown that basal bodies first dock apically with random orientation and then reorient in a common direction. Reorientation depends on hydrodynamic forces and the planar cell polarity (PCP) protein Vangl2 (Guirao et al. 2010; reviewed in: Marshall and Kintner 2008; Kishimoto and Sawamoto 2012). Apical docking of basal bodies to the plasma membrane in multiciliated cells requires planar cell polarity components Dishevelled (Dvl) and Inturned (Park et al. 2008).

Key regulators of centriole duplication in cycling cells seem to be involved in centriole assembly pathway in multiciliated cells as well. Basal bodies of multiciliated cells contain SAS-6 and depletion of SAS-6 prevents centriole assembly (Vladar and Stearns 2007). Simultaneous overexpression of key proteins for centriole duplication, Plk4/SAS6/SAS4 in CHO cells results in the formation of fibrogranular aggregates as already observed by Dirksen (1991). Eventually, multiple centrioles are assembled around a parent centriole reminiscent of the formation of multiple centrioles in normal ciliated trachea/oviduct cells by the centriolar pathway (Sorokin 1968; Dirksen 1991).

It seems therefore that centriole duplication in cycling cells as well as centriole generation in multiciliated cells by the centriolar pathway rely on a common mechanism (Kuriyama 2009). Recently, microRNAs have been identified as key regulators of multiciliogenesis in vertebrates. The microRNA miR449 strongly accumulates in multiciliated cells and promotes centriole multiplication by repression of the Delta/Notch pathway (Marcet et al. 2011). Additionally, repression of microRNA processors prevents the formation and elongation of primary cilia (Moser et al. 2011).

1.6 Summary

Generally speaking, cilia are hair-like projections extending from the cell surface into the cellular environment. All cilia are anchored in the cell by their basal bodies. Motile cilia are mostly of 9+2 structure and are found in large numbers on the surface of specialized epithelia. Their orchestrated beating is essential for movement of cells and fluids. Each motile cilium is associated with one basal body only without an associated centriole. Generation of basal bodies in multiciliated cells occurs mostly by the acentriolar pathway. In contrast, most cells of the vertebrate body harbour one single cilium called the primary cilium. Primary cilia are mostly of 9+0 structure and are immotile. However, primary cilia are important mechanosensors and/or chemosensors, and are essential for development and tissue homeostasis in adulthood. Primary cilia are generated from the distal end of the former mother centriole that has been transformed into the basal body. The basal body of the primary cilium is associated with its daughter centriole. Duplication of centrioles occurs synchronously with the cell cycle and uses the existing centrioles as templates. Furthermore, motile cilia of 9+0 structure and immotile cilia of 9+2 structure are also existent. Formation of cilia strictly depends on the presence of the basal bodies but their exclusive presence is not sufficient. Cilia are sophisticated structures and may require more than 1,000 different proteins for their generation and function including those of the basal body. Thus, lots of proteins in addition to those highly enriched at the basal body affect ciliogenesis and have been implicated in cilia-related diseases when mutated.

Acknowledgement I gratefully acknowledge proof reading by Niko Prpic-Schäper.

References

AbouAlaiwi WA, Takahashi M, Mell BR, Jones TJ, Ratnam S, Kolb RJ, Nauli SM (2009) Ciliary polycystin-2 is a mechanosensitive calcium channel involved in nitric oxide signaling cascades. Circ Res 10:860–869

Adams M, Simms RJ, Abdelhamed Z, Dawe HR, Szymanska K, Logan CV, Wheway G, Pitt E, Gull K, Knowles MA, Blair E, Cross SH, Sayer JA, Johnson CA (2012) A meckelin-filamin A interaction mediates ciliogenesis. Hum Mol Genet 21:1272–1286

Afzelius BA (1976) A human syndrome caused by immotile cilia. Science 193:317–319
Albrecht-Buehler G (1977) Phago-kinetic tracks of 3T3 cells: parallels between the orientation of track segments and of cellular structures which contain actin or tubulin. Cell 12:333–339
Albrecht-Buehler G, Bushnell A (1979) The orientation of centrioles in migrating 3T3 cells. Exp Cell Res 120:111–118
Albrecht-Buehler G, Bushnell A (1980) The ultrastructure of primary cilia in quiescent 3T3 cells. Exp Cell Res 126:427–437
Alieva IB, Vorobjev IA (2004) Vertebrate primary cilia: a sensory part of centrosomal complex in tissue, but a "sleeping beauty" in cultured cells? Cell Biol Int 28:139–150
Alvarado-Kristensson M, Rodriguez MJ, Silio V, Valpuesta JM, Carrera AC (2009) SADB phosphorylation of γ-tubulin regulates centrosome duplication. Nat Cell Biol 11:1081–1092
Alvey PL (1985) An investigation of the centriole cycle using 3T3 and CHO cells. J Cell Sci 78:147–162
Alvey PL (1986) Do adult centrioles contain cartwheels and lie at right angles to each other? Cell Biol Int Rep 10:589–598
Anderson RG (1972) The three-dimensional structure of the basal body from the rhesus monkey oviduct. J Cell Biol 54:246–265
Anderson RGW, Brenner RM (1971) The formation of basal bodies (centrioles) in the rhesus monkey oviduct. J Cell Biol 50:10–34
Anderson CT, Stearns T (2009) Centriole age underlies asynchronous primary cilium growth in mammalian cells. Curr Biol 19:1498–1502
Andersen JS, Wilkinson CJ, Mayor T, Mortensen P, Nigg EA, Mann M (2003) Proteomic characterization of the human centrosome by protein correlation profiling. Nature 426:570–574
Anderson CT, Castillo AB, Brugmann SA, Helms JA, Jacobs CR, Stearns T (2008) Primary cilia: cellular sensors for the skeleton. Anat Rec 291:1074–1078
Archinti M, Lacasa C, Teixido-Travesa N, Luders J (2010) SPICE-a previously uncharacterized protein required for centriole duplication and mitotic chromosome congression. J Cell Sci 123:3039–3046
Arellano JI, Guadiana SM, Breunig JJ, Rakic P, Sarkisian MR (2012) Development and distribution of neuronal cilia in mouse neocortex. J Comp Neurol 520:848–873
Arquint C, Sonnen KF, Stierhof Y-D, Nigg EA (2012) Cell-cycle-regulated expression of STIL controls centriole number in human cells. J Cell Sci 125:1342–1352
Avidor-Reiss T, Maer AM, Koundakjian E, Polyanovsky A, Keil T, Subramaniam S, Zucker CS (2004) Decoding cilia function: defining specialized genes required for compartmentalized cilia biogenesis. Cell 117:527–539
Azimzadeh J, Marshall WF (2010) Building the centriole. Curr Biol 20:R816–R825
Azimzadeh J, Hergert P, Delouvée A, Euteneuer U, Formstecher E, Khodjakov A, Bornens M (2009) hPOC5 is a centrin-binding protein required for assembly of full-length centrioles. J Cell Biol 185:101–114
Badano JL, Mitsuma N, Beales PL, Katsanis N (2006) The ciliopathies: an emerging class of human genetic disorders. Annu Rev Genomics Hum Genet 7:125–148
Bahe S, Stierhof YD, Wilkinson CJ, Leiss F, Nigg EA (2005) Rootletin forms centriole-associated filaments and functions in centrosome cohesion. J Cell Biol 171:27–33
Bahmanyar S, Kaplan DD, DeLuca JG, Giddings TH Jr, O'Toole ET, Winey M, Salmon ED, Casey PJ, Nelson WJ, Barth AIM (2008) β-catenin is a Nek2 substrate involved in centrosome separation. Genes Dev 22:91–105
Barnes BG (1961) Ciliated secretory cells in the pars distalis of the mouse hypophysis. J Ultrastruct Res 5:453–467
Beales PL (2005) Lifting the lid on Pandora's box: the Bardet-Biedl syndrome. Curr Opin Genet Dev 15:315–323
Besschetnova TY, Kolpakova-Hart E, Guan Y, Zhou J, Olsen BR, Shah JV (2010) Identification of signaling pathways regulating primary cilium length and flow-mediated adaptation. Curr Biol 20:182–187
Bishop GA, Berbari NF, Lewis JS, Mykytyn K (2007) Type III adenylyl cyclase localizes to primary cilia throughout the adult mouse brain. J Comp Neurol 505:562–571

Blinzinger K (1962) Elektronenmikroskopische Untersuchungen am Ependym der Hirnventrikel des Goldhamsters (Mesocricetus aureatus). Acta Neuropathol 1:527–532

Bobinnec Y, Moudjou M, Fouquet JP, Desbruyères E, Eddé B, Bornens M (1998) Glutamylation of centriole and cytoplasmic tubulin in proliferating non-neuronal cells. Cell Motil Cytoskeleton 39:223–232

Boisvieux-Ulrich E, Sandoz D (1991) Determination of ciliary polarity precedes differentiation in the epithelial cells of quail oviduct. Biol Cell 72:3–14

Bornens M (2002) Centrosome composition and microtubule anchoring mechanisms. Curr Opin Cell Biol 14:25–34

Bré MH, Redeker V, Quibell M, Darmanaden-Delorme J, Bressac C, Cosson J, Huitorel P, Schmitter JM, Rossier J, Johnson T, Adoutte A, Levilliers N (1996) Axonemal tubulin polyglycylation probed with two monoclonal antibodies: widespread evolutionary distribution, appearance during spermatozoan maturation and possible function in motility. J Cell Sci 109:727–738

Brightman MW, Palay SL (1963) The fine structure of ependyma in the brain of the rat. J Cell Biol 19:415–439

Bruni JE, Montemurro DG, Clattenburg RE, Singh RP (1972) A scanning electron microscopic study of the ependymal surface of the third ventricle of the rabbit, rat, mouse, and human brain. Anat Rec 174:407–420

Carvalho-Santos Z, Azimzadeh J, Pereira-Leal JB, Bettencourt-Dias M (2011) Tracing the origins of centrioles, cilia, and flagella. J Cell Biol 194:165–175

Chamberlain JG (1973) Analysis of developing ependymal and choroidal surfaces in rat brains using scanning electron microscopy. Dev Biol 31:22–30

Chang P, Giddings TM, Winey M, Stearns T (2003) ε-tubulin is required for centriole duplication and microtubule organization. Nat Cell Biol 5:71–76

Chang J, Cizmecioglu O, Hoffmann I, Rhee K (2010) PLK2 phosphorylation is critical for CPAP function in procentriole formation during the centrosome cycle. EMBO J 29:2395–2406

Chapin HC, Caplan MJ (2010) The cell biology of polycystic kidney disease. J Cell Biol 191:701–710

Chen Z, Indjeian VB, McManus M, Wang L (2002) Dynlacht BD (2002) CP110, a cell cycle-dependent CDK substrate, regulates centrosome duplication in human cells. Dev Cell 3:339–350

Chih B, Liu P, Chinn Y, Chalouni C, Komuves LG, Hass PE, Sandoval W, Peterson AS (2012) A ciliopathy complex at the transition zone protects the cilia as a privileged membrane domain. Nat Cell Biol 14:61–72

Chretien D, Buendia B, Fuller SD, Karsenti E (1997) Reconstruction of the centrosome cycle from cryoelectron micrographs. J Struct Biol 120:117–133

Christensen ST, Pedersen LB, Schneider L, Satir P (2007) Sensory cilia and integration of signal transduction in human health and disease. Traffic 8:97–109

Chuang JZ, Yeh TY, Bollati F, Conde C, Canavosio F, Caceres A, Sung CH (2005) The dynein light chain Tctex-1 has a dynein-independent role in actin remodelling during neurite outgrowth. Dev Cell 9:75–86

Cizmecioglu O, Arnold M, Bahtz R, Settele F, Ehret L, Haselmann-Weiß U, Antony C, Hoffmann I (2010) Cep152 acts as a scaffold for recruitment of Plk4 and CPAP to the centrosome. J Cell Biol 191:731–739

Cizmecioglu O, Krause A, Bahtz R, Ehret L, Malek N, Hoffmann I (2012) Plk2 regulates centriole duplication through phosphorylation-mediated degradation of Fbxw7 (human Cdc4). J Cell Sci 125:981–992

Cohen E, Binet S, Meininger V (1988) Ciliogenesis and centriole formation in the mouse embryonic nervous system. An ultrastructural analysis. Biol Cell 62:165–169

Collin SP, Collin HB (2004) Primary cilia in vertebrate corneal endothelial cells. Cell Biol Int 28:125–130

Corbit KC, Aanstad P, Singla V, Norman AR, Stainier DY, Reiter JF (2005) Vertebrate smoothened functions at the primary cilium. Nature 437:1018–1021

Craige B, Tsao CC, Diener DR, Hou Y, Lechtreck KF, Rosenbaum JL, Witman GB (2010) CEP290 tethers flagellar transition zone microtubules to the membrane and regulates flagellar protein content. J Cell Biol 190:927–940
Croyle MJ, Lehman JM, O'Connor AK, Wong SY, Malarkey EB, Iribarne D, Dowdle WE, Schoeb TR, Verney ZM, Athar M, Michaud EJ, Reiter JF, Yoder BK (2011) Role of epidermal primary cilia in the homeostasis of skin and hair follicles. Dev Stem Cells 138:1675–1685
Cunha-Ferreira I, Rodrigues-Martins A, Bento I, Riparbelli M, Zhang W, Laue E, Callaini G, Glover DM, Bettencourt-Dias M (2009) The SCF/Slimb ubiquitin ligase limits centrosome amplification through degradation of SAK/PLK4. Curr Biol 19:43–49
Dahl HA (1963) Fine structure of cilia in rat cerebral cortex. Z Zellforsch 60:369–386
Dammermann A, Müller-Reichert T, Pelletier L, Habermann B, Desai A, Oegema K (2004) Centriole assembly requires both centriolar and pericentriolar material proteins. Dev Cell 7:815–829
Dammermann A, Pemble H, Mitchell BJ, McLeod I, Yates JR III, Kintner C, Desai AB, Oegema K (2009) The hydrolethalus syndrome protein HYLS-1 links core centriole structure to cilia formation. Genes Dev 23:2046–2059
Dawe HR, Farr H, Gull K (2007a) Centriole/basal body morphogenesis and migration during ciliogenesis in animal cells. J Cell Sci 120:7–15
Dawe HR, Smith UM, Cullinane AR, Gerelli D, Cox P, Badano JL, Blair-Reid S, Sriram N, Katsanis N, Attie-Bitach T, Afford SC, Copp AJ, Kelly DA, Gull K, Johnson CA (2007b) The Meckel-Gruber syndrome proteins MKS1 and meckelin interact and are required for primary cilium formation. Hum Mol Genet 16:173–186
De Robertis E (1956) Morphogenesis of the retinal rods. An electron microscope study. J Biophys Biochem Cytol 2:209–225
Deane JA, Cole DG, Seeley ES, Diener DR, Rosenbaum JL (2001) Localization of intraflagellar transport protein IFT52 identifies basal body transitional fibers as the docking site for IFT particles. Curr Biol 11:1586–1590
Del Bigio MR (2010) Ependymal cells: biology and pathology. Acta Neuropathol 119:55–73
Dirksen ER (1991) Centriole and basal body formation during ciliogenesis revisited. Biol Cell 72:31–38
Dishinger JF, Kee HL, Jenkins PM, Fan S, Hurd TW, Hammond JW, Truong YN, Margolis B, Martens JR, Verhey KJ (2010) Ciliary entry of the kinesin-2 motor KIF17 is regulated by importin-ß2 and RanGTP. Nat Cell Biol 12:703–710
Doxsey S (2001) Re-evaluating centrosome function. Nat Rev Mol Cell Biol 2:688–698
Dzhindzhev NS, Yu QD, Weiskopf K, Tzolovsky G, Cunha-Ferreira I, Riparbelli M, Rodrigues-Martins A, Bettencourt-Dias M, Callaini G, Glover DM (2010) Asterless is a scaffold for the onset of centriole assembly. Nature 467:714–718
Esteban MA, Harten SK, Tran MG, Maxwell PH (2006) Formation of primary cilia in the renal epithelium is regulated by the von Hippel-Lindau tumor suppressor protein. J Am Soc Nephrol 17:1801–1806
Ezratty E, Stokes N, Chai S, Shah A, Williams S, Fuchs E (2011) A role for the primary cilium in Notch signaling and epidermal differentiation during skin development. Cell 145:1129–1141
Fawcett DW, Porter KR (1954) A study of the fine structure of ciliated epithelia. J Morphol 94:221–281
Ferrante MI, Zullo A, Barra A, Bimonte S, Messaddeq N, Studer M, Dollé P, Franco B (2006) Oral-facial-digital type I protein is required for primary cilia formation and left-right axis specification. Nat Genet 38:112–117
Fisch C, Dupuis-Williams P (2011) Ultrastructure of cilia and flagella – back to the future! Biol Cell 103:249–270
Fliegauf M, Benzing T, Omran H (2007) When cilia go bad: cilia defects and ciliopathies. Nat Rev Mol Cell Biol 8:880–893
Fong KW, Choi YK, Rattner JB, Qi RZ (2008) CDK5RAP2 is a pericentriolar protein that functions in centrosomal attachment of the gamma-tubulin ring complex. Mol Biol Cell 19:115–125

Fonte VG, Searls RL, Hilfer SR (1971) The relationship of cilia with cell division and differentiation. J Cell Biol 49:226–229
Fuchs JL, Schwark HD (2004) Neuronal primary cilia: a review. Cell Biol Int 28:111–118
Fuller S, Gowen B, Reinsch S, Sawyer A, Buendia B, Wepf R, Karsenti E (1995) The core of the mammalian centriole contains gamma-tubulin. Curr Biol 5:1384–1393
Gallagher BC (1980) Primary cilia of the corneal endothelium. Am J Anat 159:475–484
Ganner A, Lienkamp S, Schäfer T, Romaker D, Wegierski T, Park TJ, Spreitzer S, Simons M, Gloy J, Kim E, Wallingford JB, Walz G (2009) Regulation of ciliary polarity by the APC/C. Proc Natl Acad Sci USA 106:17799–17804
Garcia-Gonzalo FR, Reiter JF (2012) Scoring a backstage pass: mechanisms of ciliogenesis and ciliary access. J Cell Biol 197:697–709
Gerdes M, Davis EE, Katsanis N (2009) The vertebrate primary cilium in development, homeostasis, and disease. Cell 137:32–45
Ghossoub R, Molla-Herman A, Bastin P, Benmerah A (2011) The ciliary pocket: a once-forgotten membrane domain at the base of cilia. Biol Cell 103:131–144
Gilula NB, Satir P (1972) The ciliary necklace. A ciliary membrane specialization. J Cell Biol 53:494–509
Gluenz E, Höög JL, Smith AE, Dawe HR, Shaw MK, Gull K (2010) Beyond 9+0: non-canonical axoneme structures characterize sensory cilia from protists to humans. FASEB J 24:3117–3121
Goetz SC, Anderson KV (2010) The primary cilium: a signaling center during vertebrate development. Nat Rev Genet 11:331–344
Gomez-Ferreria MA, Rath U, Buster DW, Chanda SK, Caldwell JS, Rines DR, Sharp DJ (2007) Human Cep192 is required for mitotic centrosome and spindle assembly. Curr Biol 17:1960–1966
Gönczy P (2012) Towards a molecular architecture of centriole assembly. Nat Rev Mol Cell Biol 13:425–435
Graser S, Stierhof YD, Lavoie SB, Gassner OS, Lamla S, Clech ML, Nigg EA (2007) Cep164, a novel centriole appendage protein required for primary cilium formation. J Cell Biol 179:321–330
Gromley A, Jurczyk A, Sillibourne J, Halilovic E, Mogensen M, Groisman I, Blomberg M, Doxsey S (2003) A novel human protein of the maternal centriole is required for the final stages of cytokinesis and entry into S phase. J Cell Biol 161:535–545
Guarguaglini G, Duncan PI, Stierhof YD, Holmström T, Duensing S, Nigg EA (2005) The forkhead-associated domain protein Cep170 interacts with Polo-like kinase 1 and serves as a marker for mature centrioles. Mol Biol Cell 16:1095–1107
Guderian G, Westendorf J, Uldschmid A, Nigg EA (2010) Plk4 trans-autophosphorylation regulates centriole number by controlling ßTrCP-mediated degradation. J Cell Sci 123:2163–2169
Gudi R, Zou C, Li J, Gao Q (2011) Centrobin-tubulin interaction is required for centriole elongation and stability. J Cell Biol 193:711–725
Gueth-Hallonet C, Antony C, Aghion J, Santa-Maria A, Lajoie-Mazenc I, Wright M, Maro B (1993) γ-Tubulin is present in acentriolar MTOCs during early mouse development. J Cell Sci 105:157–166
Guichard P, Chretien D, Marco S, Tassin AM (2010) Procentriole assembly revealed by cryo-electron tomography. EMBO J 29:1565–1572
Guirao B, Meunier A, Mortaud S, Aguilar A, Corsi JM, Strehl L, Hirota Y, Desoeuvre A, Boutin C, Han YG, Mirzadeh Z, Cremer H, Montcouquiol M, Sawamoto K, Spassky N (2010) Coupling between hydrodynamic forces and planar cell polarity orients mammalian motile cilia. Nat Cell Biol 12:341–350
Habedanck R, Stierhof YD, Wilkinson CJ, Nigg EA (2005) The Polo kinase Plk4 functions in centriole duplication. Nat Cell Biol 7:1140–1146
Hadjihannas MV, Brückner M, Behrens J (2010) Conductin/axin2 and Wnt signalling regulates centrosome cohesion. EMBO Rep 11:317–324
Hagiwara H, Kano A, Aoki T, Ohwada N (2000) Immunocytochemistry of the striated rootlets associated with solitary cilia in human oviductal secretory cells. Histochem Cell Biol 114:205–212

Hagiwara H, Harada S, Maeda S, Aoki T, Ohwada N, Takata K (2002) Ultrastructure and immunohistochemical study of the basal apparatus of solitary cilia in the human oviduct epithelium. J Anat 200:89–96

Hagiwara H, Ohwada N, Aoki T, Suzuki T, Takata K (2008) The primary cilia of secretory cells in the human oviduct mucosa. Med Mol Morphol 41:193–198

Handel M, Schulz S, Stanarius A, Schreff M, Erdtmann-Vourliotis M, Schmidt H, Wolf G, Hollt V (1999) Selective targeting of somatostatin receptor 3 to neuronal cilia. Neuroscience 89:909–926

Haren L, Remy MH, Bazin I, Callebaut I, Wright M, Merdes A (2006) NEDD1-dependent recruitment of the gamma-tubulin ring complex to the centrosome is necessary for centriole duplication and spindle assembly. J Cell Biol 172:505–515

Harrison F (1989) Primary cilia associated with striated rootlets in granulated and folliculo-stellate cells of the avian adenohypophysis. Anat Embryol 180:543–547

Hatch EM, Kulukian A, Holland AJ, Cleveland DW, Stearns T (2010) Cep152 interacts with Plk4 and is required for centriole duplication. J Cell Biol 191:721–729

Haycraft CJ, Banizs B, Aydin-Son Y, Zhang Q, Michaud EJ, Yoder BK (2005) Gli2 and gli3 localize to cilia and require the intraflagellar transport protein polaris for processing and function. PLoS Genet 1:e53

Hierck BP, Van der Heiden K, Alkemade FE, Van de Pas S, Van Thienen JV, Groenendijk BCW, Bax WH, Van der Laarse A, DeRuiter MC, Horrevoets AJG, Poelmann RE (2008) Primary cilia sensitize endothelial cells for fluid shear stress. Dev Dyn 237:725–735

Hinchcliffe E, Linck R (1998) Two proteins isolated from sea urchin sperm flagella: structural components common to the stable microtubules of axonemes and centrioles. J Cell Sci 111:585–595

Holland AJ, Lan W, Niessen S, Hoover H, Cleveland DW (2010) Polo-like kinase 4 kinase activity limits centrosome overduplication by autoregulating its own stability. J Cell Biol 188:191–198

Hoyer-Fender S (2010) Centriole maturation and transformation to basal body. Semin Cell & Dev Biol 21:142–147

Hoyer-Fender S (2011) Centrosomes in fertilization, early embryonic development, stem cell division, and cancer. Atlas Genet Cytogenet Oncol Haematol. http://atlasgeneticsoncology.org/Deep/CentrosomeStemCellID20105.html

Hu Q, Milenkovic L, Jin H, Scott MP, Nachury MV, Spiliotis ET, Nelson WJ (2010) A septin diffusion barrier at the base of the primary cilium maintains ciliary membrane protein distribution. Science 329:436–439

Huang P, Schier AF (2009) Dampened Hedgehog signaling but normal Wnt signaling in zebrafish without cilia. Development 136:3089–3098

Huang K, Diener DR, Rosenbaum JL (2009) The ubiquitin conjugation system is involved in the disassembly of cilia and flagella. J Cell Biol 186:601–613

Huangfu D, Anderson KV (2005) Cilia and Hedgehog responsiveness in the mouse. Proc Natl Acad Sci USA 102:11325–11330

Hung LY, Tang CJ, Tang TK (2000) Protein 4.1 R-135 interacts with a novel centrosomal protein (CPAP) which is associated with the gamma-tubulin complex. Mol Cell Biol 20:7813–7825

Ibanez-Tallon I, Heintz N, Omran H (2003) To beat or not to beat: roles of cilia in development and disease. Hum Mol Genet 12:R27–R35

Ibi M, Zou P, Inoko A, Shiromizu T, Matsuyama M, Hayashi Y, Enomoto M, Mori D, Hirotsune S, Kiyono T, Tsukita S, Goto H, Inagaki M (2011) Trichoplein controls microtubule anchoring at the centrosome by binding to Odf2 and ninein. J Cell Sci 124:857–864

Ibrahim R, Messaoudi C, Chichon FJ, Celati C, Marco S (2009) Electron tomography study of isolated human centrioles. Microsc Res Tech 72:42–48

Ikegami K, Setou M (2010) Unique post-translational modifications in specialized microtubule architecture. Cell Struct Funct 35:15–22

Ikegami K, Sato S, Nakamura K, Ostrowski LE, Setou M (2010) Tubulin polyglutamylation is essential for airway ciliary function through the regulation of beating asymmetry. Proc Natl Acad Sci USA 107:10490–10495

Inoko A, Matsuyama M, Goto H, Ohmuro-Matsuyama Y, Hayashi Y, Enomoto M, Ibi M, Urano T, Yonemura S, Kiyono T, Izawa I, Inagaki M (2012) Trichoplein and Aurora A block aberrant primary cilia assembly in proliferating cells. J Cell Biol. doi:10.1083/jcb.201106101

Ishikawa H, Marshall WF (2011) Ciliogenesis: building the cell's antenna. Nat Rev Mol Cell Biol 12:222–234

Ishikawa H, Kubo A, Tsukita S, Tsukita S (2005) Odf2-deficient mother centrioles lack distal/subdistal appendages and the ability to generate primary cilia. Nat Cell Biol 7:517–524

Jacoby M, Cox JJ, Gayral S, Hampshire DJ, Ayub M, Blockmans M, Pernot E, Kisseleva MV, Compère P, Schiffmann SN, Gergely F, Riley JH, Pérez-Morga D, Woods CG, Schurmanns S (2009) INPP5e mutations cause primary cilium signaling defects, ciliary instability and ciliopathies in human and mouse. Nat Genet 41:1027–1031

Jakobsen L, Vanselow K, Skogs M, Toyoda Y, Lundberg E, Poser I, Falkenby LG, Bennetzen M, Westendorf J, Nigg EA, Uhlen M, Hyman AA, Andersen JS (2011) Novel asymmetrically localizing components of human centrosomes identified by complementary proteomics methods. EMBO J 30:1520–1535

Janke C, Bulinski JC (2012) Post-translational regulation of the microtubule cytoskeleton: mechanisms and functions. Nat Rev Mol Cell Biol 12:773–786

Jenkins PM, McEwen DP, Martens JR (2009) Olfactory cilia: linking sensory cilia function and human disease. Chem Senses 34:451–464

Jensen CG, Jensen LCW, Rieder CL (1979) The occurrence and structure of primary cilia in a subline of Potorous tridactylus. Exp Cell Res 123:444–449

Jurczyk A, Gromley A, Redick S, San Agustin J, Witman G, Pazour GJ, Peters DJM, Doxsey S (2004) Pericentrin forms a complex with intraflagellar transport proteins and polycystin-2 and is required for primary cilia assembly. J Cell Biol 166:637–643

Kee HL, Dishinger TL, Blasius CJ, Liu B, Margolis B, Verhey KJ (2012) A size-exclusion permeability barrier and nucleoporins characterize a ciliary pore complex that regulates transport into cilia. Nat Cell Biol 14:431–437

Kellogg DR, Moritz M, Alberts BM (1994) The centrosome and cellular organization. Annu Rev Biochem 63:639–674

Kenney J, Karsenti E, Gowen B, Fuller SD (1997) Three-dimensional reconstruction of the mammalian centriole from cryoelectron micrographs: the use of common lines for orientation and alignment. J Struct Biol 120:320–328

Kim J, Krishnaswami SR, Gleeson JG (2008) CEP290 interacts with the centriolar satellite component PCM-1 and is required for Rab8 localization to the primary cilium. Hum Mol Genet 17:3796–3805

Kim J, Lee JE, Heynen-Genel S, Suyama E, Ono K, Lee KY, Ideker T, Aza-Blanc P, Gleeson JG (2010) Functional genomic screen for modulators of ciliogenesis and cilium length. Nature 464:1048–1051

Kim S, Zaghloul NA, Bubenshchikova E, Oh EC, Rankin S, Katsanis N, Obara T, Tsiokas L (2011) Nde1-mediated inhibition of ciliogenesis affects cell cycle re-entry. Nat Cell Biol 13:351–360

Kinzel D, Boldt K, Davis EE, Burtscher I, Trumbach D, Diplas B, Attie-Bitach T, Wurst W, Katsanis N, Ueffing M, Lickert H (2010) Pitchfork regulates primary cilia disassembly and left-right asymmetry. Dev Cell 19:66–77

Kishimoto N, Sawamoto K (2012) Planar polarity of ependymal cilia. Differentiation 83:S86–S90

Kleylein-Sohn J, Westendorf J, Le Clech M, Habedanck R, Stierhof YD, Nigg EA (2007) Plk-4 induced centriole biogenesis in human cells. Dev Cell 13:190–202

Klinkerfuss GH (1964) An electron microscopic study of the ependyma and subependymal glia of the lateral ventricle of the cat. Am J Anat 115:71–100

Kobayashi T, Tsang WY, Li J, Lane W, Dynlacht BD (2011) Centriolar kinesin Kif24 interacts with CP110 to remodel microtubules and regulate ciliogenesis. Cell 145:914–925

Kohlmaier G, Loncarek J, Meng X, McEwen BF, Mogensen MM, Spektor A, Dynlacht BD, Khodjakov A, Gönczy P (2009) Overly long centrioles and defective cell division upon excess of the SAS-4-related protein CPAP. Curr Biol 19:1012–1018

Krauss SW, Spence JR, Bahmanyar S, Barth AIM, Go MM, Czerwinski D, Meyer AJ (2008) Downregulation of protein 4.1R, a mature centriole protein, disrupts centrosomes, alters cell cycle progression, and perturbs mitotic spindles and anaphase. Mol Cell Biol 28:2283–2294

Kuriyama R (2009) Centriole assembly in CHO cells expressing Plk4/SAS6/SAS4 is similar to centriogenesis in ciliated epithelial cells. Cell Motil Cytoskeleton 66. doi:10.1002/cm.20368

Lancaster MA, Schroth J, Gleeson JG (2011) Subcellular spatial regulation of canonical Wnt signalling at the primary cilium. Nat Cell Biol 13:700–707

Lange BM, Gull K (1995) A molecular marker for centriole maturation in the mammalian cell cycle. J Cell Biol 130:919–927

Laoukili J, Perret E, Middendorp S, Houcine O, Guennou C, Marano F, Bornens M, Tournier F (2000) Differential expression and cellular distribution of centrin isoforms during human ciliated cell differentiation in vitro. J Cell Sci 113:1355–1364

Lee JD, Anderson KV (2008) Morphogenesis of the node and notochord: the cellular basis for the establishment and maintenance of left-right asymmetry in the mouse. Dev Dyn 237:3464–3476

Lee JE, Silhavy JL, Zaki MS, Schroth J, Bielas SL, Marsh SE, Olvera J, Brancati F, Iannicelli M, Ikegami K, Schlossman AM, Merriman B, Attié-Bitach T, Logan CV, Glass IA, Cluckey A, Louie CM, Lee JH, Raynes HR, Rapin I, Castroviejo IP, Setou M, Barbot C, Boltshauser E, Nelson SF, Hildebrandt F, Johnson CA, Doherty DA, Valente EM, Gleeson JG (2012) CEP41 is mutated in Joubert syndrome and is required for tubulin glutamylation at the cilium. Nat Genet 44:193–199

Lehman J, Laag E, Michaud EJ, Yoder BK (2009) An essential role for dermal primary cilia in hair follicle morphogenesis. J Invest Dermatol 129:438–448

Li JB, Gerdes JM, Haycraft CJ, Fan Y, Teslovich TM, May-Simera H, Li H, Blacque OE, Li L, Leitch CC, Lewis RA, Green JS, Parfrey PS, Leroux MR, Davidson WS, Beales PL, Guay-Woodford LM, Yoder BK, Stormo GD, Katsanis N, Dutcher SK (2004) Comparative genomics identifies a flagellar and basal body proteome that includes the BBS5 human disease gene. Cell 117:541–552

Li A, Saito M, Chuang JZ, Tseng YY, Dedesma C, Tomizawa K, Kaitsuka T, Sung CH (2011) Ciliary transition zone activation of phosphorylated Tctex-1 controls ciliary resorption, S-phase entry and fate of neural progenitors. Nat Cell Biol 13:402–411

Lienkamp S, Ganner A, Walz G (2012) Inversin, Wnt signaling and primary cilia. Differentiation 83:S49–S55

Lin F, Hiesberger T, Cordes K, Sinclair AM, Goldstein LS, Somlo S, Igarashi P (2003) Kidney-specific inactivation of the KIF3A subunit of kinesin-II inhibits renal ciliogenesis and produces polycystic kidney disease. Proc Natl Acad Sci USA 100:5286–5291

Liu A, Wang B, Niswander LA (2005) Mouse intraflagellar transport proteins regulate both the activator and repressor functions of Gli transcription factors. Development 132:3103–3111

Lolkema MP, Mans DA, Ulfman LH, Volpi S, Voest EE, Giles RH (2008) Allele-specific regulation of primary cilia function by the von Hippel-Lindau tumor suppressor. Eur J Hum Genet 16:73–78

Loncarek J, Khodjakov A (2009) Ab ovo or de novo? Mechanisms of centriole duplication. Mol Cells 27:135–142

Loncarek J, Khodjakov A (2010) Centriole reduplication during prolonged interphase requires procentriole maturation governed by Plk1. Curr Biol 20:1277–1282

Loncarek J, Hergert P, Magidson V, Khodjakov A (2008) Control of daughter centriole formation by the pericentriolar material. Nat Cell Biol 10:322–328

Lu CJ, Du H, Wu J, Jansen DA, Jordan KL, Xu N, Sieck GC, Qian Q (2008) Non-random distribution and sensory functions of primary cilia in vascular smooth muscle cells. Kidney Blood Press Res 31:171–184

Magloire H, Couble ML, Romeas A, Bleicher F (2004) Odontoblast primary cilia: facts and hypotheses. Cell Biol Int 28:93–99

Mahjoub MR, Xie Z, Stearns T (2010) Cep120 is asymmetrically localized to the daughter centriole and is essential for centriole assembly. J Cell Biol 191:331–346

Malone AM, Anderson CT, Tummala P, Kwon RY, Johnston TR, Stearns T, Jacobs CR (2007) Primary cilia mediate mechanosensing in bone cells by a calcium-independent mechanism. Proc Natl Acad Sci USA 96:13325–13330

Marcet B, Chevalier B, Luxardi G, Coraux C, Zaragosi LE, Cibois M, Robbe-Sermesant K, Jolly T, Cardinaud B, Moreilhon C, Giovannini-Chami L, Nawrocki-Raby B, Birembaut P, Waldmann R, Kodjabachian L, Barbry P (2011) Control of vertebrate multiciliogenesis by miR-449 through direct repression of the Delta/Notch pathway. Nat Cell Biol 13:693–699

Marshall WF, Kintner C (2008) Cilia orientation and the fluid mechanics of development. Curr Opin Cell Biol 20:48–52

Marszalek JR, Ruiz-Lozano P, Roberts E, Chien KR, Goldstein LS (1999) Situs inversus and embryonic ciliary morphogenesis defects in mouse mutants lacking the KIF3A subunit of kinesin-II. Proc Natl Acad Sci USA 96:5043–5048

Masyuk AI, Masyuk TV, Splinter PL, Huang BQ, Stroope AJ, LaRusso NF (2006) Cholangiocyte cilia detect changes in luminal fluid flow and transmit them into intracellular Ca^{2+} and cAMP signaling. Gastroenterology 131:911–920

Matsuyama A, Shimazu T, Sumida Y, Saito A, Yoshimatsu Y, Seigneurin-Berny D, Osada H, Komatsu Y, Nishino N, Khochbin S, Horinouchi S, Yoshida M (2002) In vivo destabilization of dynamic microtubules by HDAC6-mediated deacetylation. EMBO J 21:6820–6831

May SR, Ashique AM, Karlen M, Wang B, Shen Y, Zarbalis K, Reiter J, Ericson J, Peterson AS (2005) Loss of the retrograde motor for IFT disrupts localization of Smo to cilia and prevents the expression of both activator and repressor functions of Gli. Dev Biol 287:378–389

McGlashan SR, Jensen CG, Poole CA (2006) Localization of extracellular matrix receptors on the chondrocyte primary cilium. J Histochem Cytochem 54:1005–1014

McGrath J, Somlo S, Makova S, Tian X, Brueckner M (2003) Two populations of node monocilia initiate left-right asymmetry in the mouse. Cell 114:61–73

Menco BPM, Cunningham AM, Qasba P, Levy N, Reed RR (1997) Putative odour receptors localize in cilia of olfactory receptor cells in rat and mouse: a freeze-substitution ultrastructure study. J Neurocytol 26:297–312

Million K, Larcher J, Laoukili J, Bourguignon D, Marano F, Tounier F (1999) Polyglutamylation and polyglycylation of alpha- and beta-tubulins during in vitro ciliated cell differentiation of human respiratory epithelial cells. J Cell Sci 112:4357–4366

Miyamoto T, Porazinski S, Wang H, Borovina A, Ciruna B, Shimizu A, Kajii T, Kikuchi A, Furutani-Seiki M, Matsuura S (2011) Insufficiency of BUBR1, a mitotic spindle checkpoint regulator, causes impaired ciligenesis in vertebrates. Hum Mol Genet 20:2058–2070

Miyoshi K, Kasahara K, Miyazaki I, Asanuma M (2011) Factors that influence primary cilium length. Acta Med Okayama 65:279–285

Mogensen MM, Malik A, Piel M, Bouckson-Castaing V, Bornens M (2000) Microtubule minus-end anchorage at centrosomal and non-centrosomal sites: the role of ninein. J Cell Sci 113:3013–3023

Molla-Herman A, Ghossoub R, Blisnick T, Meunier A, Serres C, Silbermann F, Emmerson C, Romeo K, Bourdoncle P, Schmitt A, Saunier S, Spassky N, Bastin P, Benmerah A (2010) The ciliary pocket: an endocytic membrane domain at the base of the primary and motile cilia. J Cell Sci 123:1785–1795

Mollet G, Silbermann F, Delous M, Salomon R, Antignac C, Saunier S (2005) Characterization of the nephrocystin/nephrocystin-4 complex and subcellular localization of nephrocystin-4 to primary cilia and centrosomes. Hum Mol Genet 14:645–656

Moran DT, Rowley JC III, Jafek BW, Lovell MA (1982) The fine structure of the olfactory mucosa in man. J Neurocytol 11:721–746

Mori Y, Akedo H, Tanigaki Y, Tanaka K, Okada M (1979) Ciliogenesis in tissue-cultured cells by the increased density of cell population. Exp Cell Res 120:435–439

Moritz M, Braunfeld MB, Sedat JW, Alberts B, Agard DA (1995) Microtubule nucleation by gamma-tubulin-containing rings in the centrosome. Nature 378:638–640

Moser JJ, Fritzler MJ, Rattner JB (2011) Repression of GW/P body components and the RNAi microprocessor impacts primary ciliogenesis in human astrocytes. BMC Cell Biol 12:37

Munger BL (1958) A light and electron microscopic study of cellular differentiation in the pancreatic islets of the mouse. Am J Anat 103:275–311
Myklebust R, Engedal H, Sætersdal TS, Ulstein M (1977) Primary 9+0 cilia in the embryonic and the adult human heart. Anat Embryol 151:127–139
Nachury MV, Loktev AV, Zhang Q, Wetslake CJ, Peranen J, Merdes A, Slusarski DC, Scheller RH, Bazan JF, Sheffield VC, Jackson PK (2007) A core complex of BBS proteins cooperates with the GTPase Rab8 to promote ciliary membrane biogenesis. Cell 129:1201–1213
Nakagawa Y, Yamane Y, Okanoue T, Tsukita S (2001) Outer dense fiber 2 is a widespread centrosome scaffold component preferentially associated with mother centrioles: its identification from isolated centrosomes. Mol Biol Cell 12:1687–1697
Nakamura A, Arai H, Fujita N (2009) Centrosomal Ali1 and cohesion function in separase-regulated centriole disengagement. J Cell Biol 187:607–614
Nakayama Y, Kohno K (1974) Number and polarity of the ependymal cilia in the central canal of some vertebrates. J Neurocytol 3:449–458
Nauli SM, Alenghat FJ, Luo Y, Williams E, Vassilev P, Li X, Elia AEH, Lu W, Brown EM, Quinn SJ, Ingber DE, Zhou J (2003) Polycystin 1 and 2 mediate mechanosensation in the primary cilium of kidney cells. Nat Genet 33:129–137
Nauli SM, Kawanabe Y, Kaminski JJ, Pearce WJ, Ingber DE, Zhou J (2008) Endothelial cilia are fluid shear sensors that regulate calcium signaling and nitric oxide production through polycystin-1. Circulation 117:1161–1171
Nauli SM, Jin X, Hierck BP (2011) The mechanosensory role of primary cilia in vascular hypertension. Int J Vasc Med. doi:10.1155/2011/376281
Nigg EA, Raff JW (2009) Centrioles, centrosomes, and cilia in health and disease. Cell 139:663–678
Nigg EA, Stearns T (2011) The centrosome cycle: centriole biogenesis, duplication and inherent asymmetries. Nat Cell Biol 13:1154–1160
Nonaka S, Tanaka Y, Okada Y, Takeda S, Harada A, Kanai Y, Kido M, Hirokawa N (1998) Randomization of left-right asymmetry due to loss of nodal cilia generating leftward flow of extraembryonic fluid in mice lacking KIF3B motor protein. Cell 95:829–837
Ocbina PJ, Tuson M, Anderson KV (2009) Primary cilia are not required for normal canonical Wnt signaling in the mouse embryo. PLoS One 4:e6839
Odor DL, Blandau RJ (1985) Observations on the solitary cilium of rabbit oviductal epithelium: its motility and ultrastructure. Am J Anat 174:437–453
Oishi I, Kawakami Y, Raya A, Callol-Massot C, Belmonte JCI (2006) Regulation of primary cilia formation and left-right patterning in zebrafish by a noncanonical Wnt signaling mediator, duboraya. Nat Genet 38:1316–1322
Ou YY, Mack GJ, Zhang M, Rattner JB (2002) CEP110 and ninein are located in a specific domain of the centrosome associated with centrosome maturation. J Cell Sci 115:1825–1835
Paintrand M, Moudjou M, Delacroix H, Bornens M (1992) Centrosome organization and centriole architecture: their sensitivity to divalent cations. J Struct Biol 108:107–128
Palmer KJ, MacCarthy-Morrogh L, Smyllie N, Stephens DJ (2011) A role for Tctex-1 (DYNLT1) in controlling primary cilium length. Eur J Cell Biol 90:865–871
Pan J, You Y, Huang T, Brody SL (2007) RhoA-mediated apical actin enrichment is required for ciliogenesis and promoted by Foxj1. J Cell Sci 120:1868–1876
Paoletti A, Moudjou M, Paintrand M, Salisbury J, Bornens M (1996) Most of centrin in animal cells is not centrosome-associated and centrosomal centrin is confined to the distal lumen of centrioles. J Cell Sci 109:3089–3102
Park TJ, Haigo SL, Wallingford JB (2006) Ciliogenesis defects in embryos lacking inturned or fuzzy function are associated with failure of planar cell polarity and Hedgehog signalling. Nat Genet 38:303–311
Park TJ, Mitchell BJ, Abitua PB, Kintner C, Wallingford JB (2008) Dishevelled controls apical docking and planar polarization of basal bodies in ciliated epithelial cells. Nat Genet 40:871–879
Pazour GJ, Rosenbaum JL (2002) Intraflagellar transport and cilia-dependent diseases. Trends Cell Biol 12:551–555

Pazour GJ, Dickert BL, Vucica Y, Seeley ES, Rosenbaum JL, Witman GB, Cole DG (2000) Chlamydomonas IFT88 and its mouse homologue, polycystic kidney disease gene tg737, are required for assembly of cilia and flagella. J Cell Biol 151:709–718

Pedersen LB, Rosenbaum JL (2008) Intraflagellar transport (IFT): role in ciliary assembly, resorption and signalling. Curr Top Dev Biol 85:23–61

Pedersen LB, Veland IR, Schroder JM, Christensen ST (2008) Assembly of primary cilia. Dev Dyn 237:1993–2006

Pickett-Heaps J (1971) The autonomy of the centriole: fact or fallacy? Cytobios 3:205

Piel M, Meyer P, Khodjakov A, Rieder CL, Bornens M (2000) The respective contributions of the mother and daughter centrioles to centrosome activity and behaviour in vertebrate cells. J Cell Biol 149:317–329

Piperno G, Fuller MT (1985) Monoclonal antibodies specific for an acetylated form of α-tubulin recognize the antigen in cilia and flagella from a variety of organisms. J Cell Biol 101:2085–2094

Piperno G, LeDizet M, Chang XJ (1987) Microtubules containing acetylated α-tubulin in mammalian cells in culture. J Cell Biol 104:289–302

Poole CA, Flint MH, Beaumont BW (1985) Analysis of the morphology and function of primary cilia in connective tissues: a cellular cybernetic probe? Cell Motil 5:175–193

Poole CA, Jensen CG, Snyder JA, Gray CG, Hermanutz VL, Wheatley DN (1997) Confocal analysis of primary cilia structure and colocalization with the Golgi apparatus in chondrocytes and aortic smooth muscle cells. Cell Biol Int 21:483–494

Poole CA, Zhang ZJ, Ross JM (2001) The differential distribution of acetylated and detyrosinated alpha-tubulin in the microtubular cytoskeleton and primary cilia of hyaline cartilage chondrocytes. J Anat 199:393–405

Porter ME, Sale WS (2000) The 9+2 axoneme anchors multiple inner arm dyneins and a network of kinases and phosphatases that control motility. J Cell Biol 151:37–42

Praetorius HA, Spring KR (2001) Bending the MDCK cell primary cilium increases intracellular calcium. J Membr Biol 184:71–79

Praetorius HA, Spring KR (2002) Removal of the MDCK cell primary cilium abolishes flow sensing. J Membr Biol 191:69–76

Pugacheva EN, Jablonski SA, Hartman TR, Henske EP, Golemis EA (2007) HEF1-dependent Aurora A activation induces disassembly of the primary cilium. Cell 129:1351–1363

Puklowski A, Homsi Y, Keller D, May M, Chauhan S, Kossatz U, Grünwald V, Kubicka S, Pich A, Manns MP, Hoffmann I, Gönczy P, Malek NP (2011) The SCF-FBXW5 E3-ubiquitin ligase is regulated by PLK4 and targets HsSAS-6 to control centrosome duplication. Nat Cell Biol 13:1004–1009

Purkinje J (1836) Über Flimmerbewegungen im Gehirn. Müller's Arch Anat U Physiol 3:289

Qin H, Diener DR, Geime S, Cole DG, Rosenbaum JL (2004) Intraflagellar transport (IFT) cargo: IFT transports flagellar precursors to the tip and turn-over products to the cell body. J Cell Biol 164:255–266

Rattner JB, Phillips SG (1973) Independence of centriole formation and DNA synthesis. J Cell Biol 57:359–372

Raychowdhury MK, Ramos AJ, Zhang P, McLaughin M, Dai XQ, Chen XZ, Montalbetti N, Del Rocio CM, Ausiello DA, Cantiello HF (2009) Vasopressin receptor-mediated functional signaling pathway in primary cilia of renal epithelial cells. Am J Physiol Renal Physiol 296:F87–F97

Rieder CL, Jensen CG, Jensen LCW (1979) The resorption of primary cilia during mitosis in a vertebrate (PtK1) cell line. J Ultrastruct Res 68:173–185

Robbins E, Jentzsch G, Micali A (1968) The centriole cycle in synchronized HeLa cells. J Cell Biol 36:329–339

Roepman R, Letteboer SJ, Arts HH, van Beersum SE, Lu X, Krieger E, Ferreira PA, Cremers FP (2005) Interaction of nephrocystin-4 and RPGRIP1 is disrupted by nephronophthisis or Leber congenital amaurosis-associated mutations. Proc Natl Acad Sci USA 102:18520–18525

Rogers GC, Rusan NM, Roberts DM, Peifer M, Rogers SL (2009) The SCF^{Slimb} ubiquitin ligase regulates Plk4/Sak levels to block centriole reduplication. J Cell Biol 184:225–239

Rohatgi R, Snell WJ (2010) The ciliary membrane. Curr Opin Cell Biol 22:541–546

Rohatgi R, Milenkovic L, Scott MP (2007) Patched 1 regulates hedgehog signaling at the primary cilium. Science 317:372–376
Rosenbaum JL, Witman GB (2002) Intraflagellar transport. Nat Rev Mol Cell Biol 3:813–825
Santos N, Reiter JF (2008) Building it up and taking it down: the regulation of vertebrate ciliogenesis. Dev Dyn 237:1972–1981
Satir P, Christensen T (2007) Overview of structure and function of mammalian cilia. Annu Rev Physiol 69:377–400
Scherft JP, Daems WT (1967) Single cilia in chondrocytes. J Ultrastruct Res 19:546–555
Schmidt TI, Kleylein-Sohn J, Westendorf J, Le Clech M, Lavoie SB, Stierhof YD, Nigg EA (2009) Control of centriole length by CPAP and CP110. Curr Biol 19:1005–1011
Schneider L, Clement CA, Teilmann SC, Pazour GJ, Hoffmann EK, Satir P, Christensen ST (2005) PDGFRalphaalpha signaling is regulated through the primary cilium in fibroblasts. Curr Biol 15:1861–1866
Schneider L, Cammer M, Lehman J, Nielsen SK, Guerra CF, Veland IR, Stock C, Hoffmann EK, Yoder BK, Schwab A, Satir P, Christensen ST (2010) Directional cell migration and chemotaxis in wound healing response to PDGF-AA are coordinated by the primary cilium in fibroblasts. Cell Physiol Biochem 25:279–292
Schöckel L, Möckel M, Mayer B, Boos D, Stemmann O (2011) Cleavage of cohesion rings coordinates the separation of centrioles and chromatids. Nat Cell Biol 13:966–972
Scholey JM (1996) Kinesin-II, a membrane traffic motor in axons, axonemes, and spindles. J Cell Biol 133:1–4
Scholey JM (2008) Intraflagellar transport motors in cilia: moving along the cell's antenna. J Cell Biol 180:23–29
Seeley ES, Nachury MV (2010) The perennial organelle: assembly and disassembly of the primary cilium. J Cell Sci 123:511–518
Shah AS, Ben-Shahar Y, Moninger TO, Kline JN, Welsh MJ (2009) Motile cilia of human airway epithelia are chemosensory. Science 325:1131–1134
Silibourne JE, Tack F, Vloemans N, Boeckx A, Thambirajah S, Bonnet P, Ramaekers FCS, Bornens M, Grand-Perret T (2010) Autophosphorylation of Polo-like kinase 4 and its role in centriole duplication. Mol Biol Cell 21:547–561
Simons M, Gloy J, Ganner A, Bullerkotte A, Bashkurov M, Krönig C, Schermer B, Benzing T, Cabello OA, Jenny A, Mlodzik M, Polok B, Driever W, Obara T, Walz G (2005) Inversin, the gene product mutated in nephronophthisis type II, functions as a molecular switch between Wnt signaling pathways. Nat Genet 37:537–543
Singla V, Romaguera-Ros M, Garcia-Verdugo JM, Reiter JF (2010) Ofd1, a human disease gene, regulates the length and distal structure of centrioles. Dev Cell 18:410–424
Siroky BJ, Ferguson WB, Fuson AL, Xie Y, Fintha A, Komlosi P, Yoder BK, Schwiebert EM, Guay-Woodford LM, Bell PD (2006) Loss of primary cilia results in deregulated and unabated apical calcium entry in ARPKD collecting duct cells. Am J Physiol Renal Physiol 290:F1320–F1328
Sluder G, Khodjakov A (2010) Centriole duplication: analogue control in a digital age. Cell Biol Int 34:1239–1245
Smith EF (2002) Regulation of flagellar dynein by the axonemal central apparatus. Cell Motil Cytoskeleton 52:33–42
Sorokin S (1962) Centrioles and the formation of rudimentary cilia by fibroblasts and smooth muscle cells. J Cell Biol 15:363–377
Sorokin S (1968) Reconstructions of centriole formation and ciliogenesis in mammalian lungs. J Cell Sci 3:207–230
Sotelo JR, Trujillo-Cenoz O (1958) Electron microscope study on the development of ciliary components of the neural epithelium of the chicken embryo. Z Zellforsch 49:1–12
Soung N, Park J, Yu L, Lee KH, Lee J, Bang JK, Veenstra TD, Rhee K, Lee KS (2009) Plk1-dependent and –independent roles of an ODF2 splice variant, hCenexin1, at the centrosome of somatic cells. Dev Cell 16:539–550
Spektor A, Tsang WY, Khoo D, Dynlacht BD (2007) Cep97 and CP110 suppress a cilia assembly program. Cell 130:678–690

Steffen W, Fajer EA, Linck RW (1994) Centrosomal components immunologically related to tektins from ciliary and flagellar microtubules. J Cell Sci 107:2095–2105
Stephens R, Lernieux N (1998) Tektins as structural determinants in basal bodies. Cell Motil Cytoskeleton 40:379–392
Strnad P, Gönczy P (2008) Mechanisms of procentriole formation. Trends Cell Biol 18:389–396
Strnad P, Leidel S, Vinogradova T, Euteneuer U, Khodjakov A, Gönczy P (2007) Regulated HsSAS-6 levels ensure formation of a single procentriole per centriole during the centrosome duplication cycle. Dev Cell 13:203–213
Sulik K, Dehart DB, Inagaki T, Carson JL, Vrablic T, Gesteland K, Schoenwolf GC (1994) Morphogenesis of the murine node and notochordal plate. Dev Dyn 201:260–278
Svedbergh B, Bill A (1972) Scanning electron microscopic studies of the corneal endothelium in man and monkeys. Acta Ophthalmol 50:321–336
Szöllösi D, Calarco P, Donahue RP (1972) Absence of centrioles in the first and second meiotic spindles of mouse oocytes. J Cell Sci 11:521–541
Takada S, Wilkerson CG, Wakabayashi K, Kamiya R, Witman GB (2002) The outer dynein arm-docking complex: composition and characterization of a subunit (Oda1) necessary for outer arm assembly. Mol Biol Cell 13:1015–1029
Takeda S, Yonekawa Y, Tanaka Y, Okada Y, Nonaka S, Hirokawa N (1999) Left-right asymmetry and kinesin superfamily protein KIF3A: new insights in determination of laterality and mesoderm induction by kif3A-/- mice analysis. J Cell Biol 145:825–836
Tang CJ, Fu RH, Wu KS, Hsu WB, Tang TK (2009) CPAP is a cell-cycle regulated protein that controls centriole length. Nat Cell Biol 11:825–831
Tang CJ, Lin S, Hsu W, Lin Y, Wu C, Lin Y, Chang C, Wu K, Tang TK (2011) The human microcephaly protein STIL interacts with CPAP and is required for procentriole formation. EMBO J 30:4790–4804
Tasouri E, Tucker KL (2011) Primary cilia and organogenesis: is Hedgehog the only sculptor? Cell Tissue Res 345:21–40
Tennyson VM, Pappas GD (1962) An electron microscope study of ependymal cells of the fetal, early postnatal and adult rabbit. Z Zellforsch 56:595–618
Thoma CR, Frew IJ, Hoerner CR, Montani M, Moch H, Krek W (2007) pVHL and GSK3beta are components of a primary cilium-maintenance signalling network. Nat Cell Biol 9:588–595
Tsang WY, Bossard C, Khanna H, Peränen J, Swaroop A, Malhotra V, Dynlacht BD (2008) CP110 suppresses primary cilia formation through its interaction with CEP290, a protein deficient in human ciliary disease. Dev Cell 15:187–197
Tsou MF, Stearns T (2006a) Controlling centrosome number: licenses and blocks. Curr Opin Cell Biol 18:74–78
Tsou MF, Stearns T (2006b) Mechanism limiting centrosome duplication to once per cell cycle. Nature 442:947–951
Tsou MB, Wang WJ, Yule KA, Uryu K, Stearns T, Jallepalli PV (2009) Polo kinase and separase regulate the mitotic licensing of centriole duplication in human cells. Dev Cell 17:344–354
Tucker RW, Pardee AB, Fujiwara K (1979) Centriole ciliation is related to quiescence and DNA synthesis in 3T3 cells. Cell 17:527–535
Veland IR, Awan A, Pedersen L, Yoder BK, Christensen ST (2009) Primary cilia and signaling pathways in mammalian development, health and disease. Nephron Physiol 111:39–53
Vladar EK, Stearns T (2007) Molecular characterization of centriole assembly in ciliated epithelial cells. J Cell Biol 178:31–42
Vorobjev IA, Chentsov YS (1980) The ultrastructure of centriole in mammalian tissue culture cells. Cell Biol Int Rep 4:1037–1044
Vorobjev IA, Chentsov YS (1982) Centrioles in the cell cycle. I. Epithelial cells. J Cell Biol 98:938–949
Vulprecht J, David A, Tibelius A, Castiel A, Konotop G, Liu F, Bestvater F, Raab MS, Zentgraf H, Izraeli S, Krämer A (2012) STIL is required for centriole duplication in human cells. J Cell Sci 125:1353–1362

Wann AKT, Zuo N, Haycraft CJ, Jensen CG, Poole CA, McGlashan SR, Knight MM (2012) Primary cilia mediate mechanotransduction through control of ATP-induced Ca^{2+} signaling in compressed chondrocytes. FASEB J 26:1–9

Webber WA, Lee J (1975) Fine structure of mammalian renal cilia. Anat Rec 182:339–344

Wehland J, Weber K (1987) Turnover of the carboxyterminal tyrosine of a-tubulin and means of reaching elevated levels of detyrosination in living cells. J Cell Sci 88:185–203

Westermann S, Weber K (2003) Post-translational modifications regulate microtubule function. Nat Rev Mol Cell Biol 4:938–947

Wheatley DN (1969) Cilia in cell-cultured fibroblasts. 1. On their occurrence and relative frequencies in primary and established cell lines. J Anat 105:351–362

Wheatley DN (1971) Cilia in cell-cultured fibroblasts. III. Relationship between mitotic activity and cilium frequency in mouse 3T6 fibroblasts. J Anat 110:367–382

Wheatley DN (1972) Cilia in cell-cultured fibroblasts. IV. Variation within the mouse 3T6 fibroblastic cell line. J Anat 113:83–93

Wheatley DN (1982) Centrioles: the central enigma of cell biology. Premium Press, New York

Wheatley DN (1995) Primary cilia in normal and pathological tissues. Pathobiology 63:222–238

Wheatley DN, Wang AM, Strugnell GE (1996) Expression of primary cilia in mammalian cells. Cell Biol Int 20:73–81

Whitfield JF (2008) The solitary (primary) cilium – a mechanosensory toggle switch in bone and cartilage cells. Cell Signal 20:1019–1024

Willaredt MA, Hasenpusch-Theil K, Gardner HAR, Kitanovic I, Hirschfeld-Warneken VC, Gojak CP, Gorgas K, Bradford CL, Spatz J, Wölfl S, Theil T, Tucker KL (2008) A crucial role for primary cilia in cortical morphogenesis. J Neurosci 28:12887–12900

Williams CL, Li C, Kida K, Inglis PN, Mohan S, Semenec L, Bialas NJ, Stupay RM, Chen N, Blacque OE, Yoder BK, Leroux MR (2011) MKS and NPHP modules cooperate to establish basal body/transition zone membrane associations and ciliary gate function during ciliogenesis. J Cell Biol 192:1023–1041

Wilson CW, Nguyen CT, Chen MH, Yang JH, Gacayan R, Huang J, Chen JN, Chuang PT (2009) Fused has evolved divergent roles in vertebrate Hedgehog signalling and motile ciliogenesis. Nature 459:98–102

Yang J, Liu X, Yue G, Adamian M, Bulgakov O, Li T (2002) Rootletin, a novel coiled-coil protein, is a structural component of the ciliary rootlet. J Cell Biol 159:431–440

Yoshimura S, Egerer J, Fuchs E, Haas AK, Barr FA (2007) Functional dissection of Rab GTPases involved in primary cilium formation. J Cell Biol 178:363–369

Zhang Y, Kwon S, Yamaguchi T, Cubizolles F, Rousseaux S, Kneissel M, Cao C, Li N, Cheng H, Chua K, Lombard D, Mizeracki A, Matthias G, Alt FW, Khochbin S, Matthias P (2008) Mice lacking histone deacetylase 6 have hyperacetylated tubulin but are viable and develop normally. Mol Cell Biol 28:1688–1701

Zhang X, Chen Q, Feng J, Hou J, Yang F, Liu J, Jiang Q, Zhang C (2009) Sequential phosphorylation of Nedd1 by Cdk1 and Plk1 is required for targeting of the γTuRC to the centrosome. J Cell Sci 122:2240–2251

Zhu F, Lawo S, Bird A, Pinchev D, Ralph A, Richter C, Mueller-Reichert T, Kittler R, Hyman AA, Pelletier L (2008) The mammalian SPD-2 ortholog Cep192 regulates centrosome biogenesis. Curr Biol 18:136–141

Zou C, Li J, Bai Y, Gunning WT, Wazer DE, Band V, Gao Q (2005) Centrobin: a novel daughter centriole-associated protein that is required for centriole duplication. J Cell Biol 171:437–445

Chapter 2
Primary Cilia, Sonic Hedgehog Signaling, and Spinal Cord Development

Laura E. Mariani and Tamara Caspary

Contents

Abstract Embryonic spinal cord development requires Sonic hedgehog (Shh) signaling to define ventral motor neuron and interneuron progenitor domains during neural patterning. Shh signaling is inextricably linked to primary cilia, and mutations that disrupt cilia structure and/or function lead to abnormal Shh signaling. The embryonic spinal cord is highly sensitive to perturbations in Shh activity and displays abnormal patterning phenotypes when Shh signaling is up- or downregulated.

L.E. Mariani
Graduate Program in Neuroscience, Emory University, 30322 Atlanta, GA, USA

T. Caspary (✉)
Department of Human Genetics, Emory University, 30322 Atlanta, GA, USA
e-mail: tcaspar@emory.edu

K.L. Tucker and T. Caspary (eds.), *Cilia and Nervous System Development and Function*,
DOI 10.1007/978-94-007-5808-7_2,

Mutations in a variety of different cilia genes lead to neural tube patterning phenotypes that provide useful information about the role of different proteins in transducing Shh signals. Here we discuss Shh-dependent spinal cord development and describe what is currently known about the molecular mechanisms regulating Shh signaling in the neural tube.

Keywords Primary cilia • Sonic hedgehog • Patched • Smoothened • Gli • Sufu • Kif7 • PKA • Neural tube patterning • Mouse genetics • Forward genetic screens • Intraflagellar transport • Protein trafficking

2.1 Introduction

Primary cilia are linked to a variety of biochemical pathways, with the Sonic hedgehog (Shh) signaling pathway especially notable for its integral connection to cilia structure and function (reviewed in: Berbari et al. 2009; Eggenschwiler and Anderson 2007; Wong and Reiter 2008). Without primary cilia, Shh signaling cannot occur (Huangfu et al. 2003; Huangfu and Anderson 2005), and genetic mutations that disrupt ciliogenesis or other important cilia functions typically cause defects in Shh signaling, as well. Many such mutations are incompatible with life, because Shh is so important for embryonic development, but in animal models they have revealed much about the complex interaction between cilia and developmental signaling.

Shh plays many roles in the developing nervous system (reviewed in: Sánchez-Camacho and Bovolenta 2009; Hatten and Roussel 2011), where it was first shown to possess an essential function in patterning the embryonic spinal cord, also known as the neural tube (Echelard et al. 1993). During development, there is a gradient of Shh activity in the neural tube that confers different cell fates upon neural progenitors at distinct positions along the dorsoventral axis. The resulting progenitor domains are highly consistent among embryos and also highly sensitive to changes in Shh activity (reviewed in: Jessell 2000; Briscoe and Ericson 2001; Lupo et al. 2006; Dessaud et al. 2008), making the pattern of the ventral neural tube an informative readout of any perturbations to the Shh signaling pathway. Indeed, such perturbations are what provided the initial link between cilia and Shh signaling (Huangfu et al. 2003).

In recent years, descriptive in vivo studies on the relationship between cilia and Shh activity in the neural tube were complemented by cellular and molecular experiments that give a more mechanistic explanation of the role of cilia in Shh signaling. We now know that Shh signals are transduced via a series of proteins that dynamically enter and exit the primary cilium in a ligand-dependent manner (Corbit et al. 2005; Haycraft et al. 2005; Rohatgi et al. 2007), ultimately resulting in either the activation or repression of Shh target genes. If the primary cilium is absent or abnormal, Shh signaling is adversely affected (Huangfu et al. 2003; Huangfu and Anderson 2005; Liu et al. 2005). Furthermore, an entire suite of specialized molecules regulates the trafficking and transport of Shh pathway components and other ciliary proteins into and within cilia (Haycraft et al. 2005; Kovacs et al. 2008; Liu et al. 2005; May et al. 2005; Seo et al. 2011). Disruption of any aspect of this molecular system can lead to Shh signaling defects and abnormal neural development.

In the first part of this chapter, we explore how Shh signaling patterns the neural tube under normal conditions, as well as how this system allows for particularly elegant investigation of Shh signaling in vivo. We focus our discussion primarily on mammalian models, since much of the work on cilia and developmental signaling has been conducted using genetic techniques in the mouse. We also describe in more detail how Shh signaling occurs at the level of individual cells and how primary cilia are essential for the proper regulation of the proteins that constitute the Shh pathway. Finally, we survey the effects of mutations that affect ciliogenesis, ciliary protein transport, and other important processes in primary cilia function to summarize what is currently known and what remains to be explored in terms of the complex interactions between cilia and the Shh pathway.

2.2 Shh Signaling in the Developing Spinal Cord

2.2.1 Patterning the Neural Tube

In vertebrates, the nervous system arises through a process known as neurulation, during which the undifferentiated neuroepithelium of the ectoderm thickens into the neural plate and folds in upon itself to form the neural tube (reviewed in: Colas and Schoenwolf 2001; Copp et al. 2003). The most rostral portion of the neural tube develops into the brain; the more caudal portion gives rise to the spinal cord. As the cells of the neural tube begin to differentiate, the tissue is patterned by opposing gradients of biochemical signaling activity across the dorsoventral axis. Dorsal patterning is primarily governed by BMP and Wnt cues secreted from the roof plate (reviewed in: Lee and Jessell 1999; Caspary and Anderson 2003; Chizhikov and Millen 2005; Liu and Niswander 2005), whereas ventral patterning is regulated by Shh secreted from the notochord and floor plate. Shh is one member of the Hedgehog family of proteins. In mammals, this family also includes Indian hedgehog, which is essential for skeletal development (reviewed in: Mackie et al. 2011; Whitfield 2008) and Desert hedgehog, which is essential for development and maintenance of the male germ line (Bitgood et al. 1996). Although the vertebrate Hedgehog ligands play distinct roles in development, they use similar mechanisms: Indian hedgehog has been shown to act via primary cilia and signals through many of the same downstream molecules as Shh (Whitfield 2008).

Shh signaling is necessary for the formation of six distinct domains in the ventral neural tube (floor plate, p3, pMN, p2, p1, p0) (Fig. 2.1). The floor plate is the most ventral part of the neural tube and serves as a developmental signaling center, providing a secondary source of Shh ligand during neural patterning, as well as cues that mediate axon guidance later in development, along with Slit and Netrin.

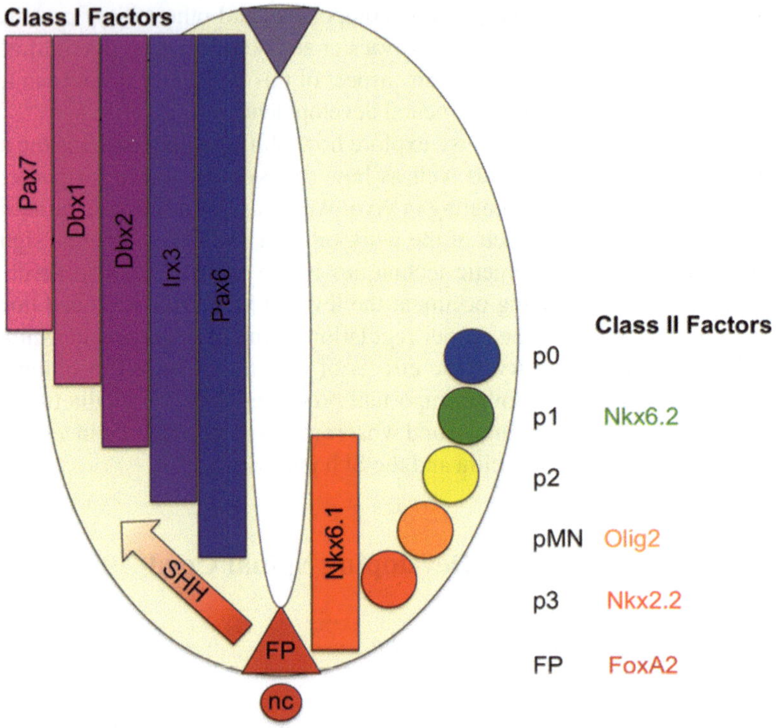

Fig. 2.1 Specification of ventral neural progenitor domains by Shh. Shh produced in the notochord (*nc*) and floor plate (*FP*) leads to the repression of Class I transcription factors and the expression of Class II transcription factors in well-defined dorsoventral domains across the neural tube (*rectangles*). Combinatorial expression of these factors defines the six distinct progenitor domains of the ventral neural tube: FP, p3, pMN, p2, p1, and p0. Nkx6.1 is expressed across multiple ventral progenitor domains; other Class II factors (Nkx6.2, Olig2, Nkx2.2, FoxA2) are restricted to individual domains and can thus be used as molecular markers for specific progenitor cell types (*circles*)

The progenitor cells in the pMN domain give rise to motor neurons (MNs) and subsequently oligodendrocytes. Cells in the other ventral progenitor domains differentiate into different classes of spinal interneurons.

During neural development, Shh is first produced by the notochord, a rod-shaped structure derived from mesoderm that runs the length of the ventral face of the neural tube. Diffusion of Shh ligand from the notochord to the ventral midline of the neural tube induces the formation of the floor plate, which also begins to produce Shh (Martí et al. 1995; Roelink et al. 1995). In both the notochord and the floor plate, Shh protein must be activated through catalytic cleavage, N-terminal palmitoylation, and C-terminal cholesterol modification (Bumcrot et al. 1995; Chen et al. 2004; Pepinsky et al. 1998; Porter et al. 1996), before it is released into the extracellular space through a process that depends on the transmembrane protein Dispatched1 (Caspary et al. 2002; Kawakami et al. 2002; Ma et al. 2002; Tian et al. 2005).

In the simplest terms, Shh can be conceived of as a classic morphogen, meaning the ligand confers spatial information onto differentiating cells via a concentration gradient: cells in the most ventral part of the neural tube are close to the source of Shh, and are therefore exposed to higher concentrations of ligand than more dorsal cells. Proper patterning of the ventral neural tube depends on the variable level of Shh signaling activity along the dorsoventral axis, corresponding to the graded concentration of Shh ligand (Ericson et al. 1997). Ventral cell types (floor plate, p3) require a high level of Shh signaling to induce their differentiation, mediolateral cell types (pMN, p2, p1, p0) require less Shh activity, and dorsal cell fates are repressed by Shh activity.

2.2.2 *Shh Signaling Pathway Components*

Differentiating neural progenitors respond to Shh through a multi-step molecular signaling cascade that results in the expression of distinct target genes in different classes of Shh-responsive cells. Ultimately, Shh signaling regulates target gene expression via the Gli family of transcription factors, which can act either as transcriptional activators or repressors (Bai et al. 2004; Ding et al. 1998; Lei et al. 2004; Litingtung and Chiang 2000; Matise et al. 1998). Other critical regulators of Gli transcriptional activity include the Shh receptor, Patched1 (Ptch1) (Denef et al. 2000; Marigo et al. 1996; Taipale et al. 2002); the major effector of Shh activity, Smoothened (Smo) (Alcedo et al. 1996; Stone et al. 1996; van den Heuvel and Ingham 1996); the Gli inhibitor, Suppressor of Fused (Sufu) (Ding et al. 1999; Kogerman et al. 1999; Svärd et al. 2006); an atypical kinesin, Kif7 (Cheung et al. 2009; Endoh-Yamagami et al. 2009; Liem et al. 2009); and protein kinase A (PKA) (Epstein et al. 1996; Hammerschmidt et al. 1996; Pan et al. 2009; Tuson et al. 2011) (Fig. 2.2). As the Shh signal is transduced, all of the aforementioned proteins localize to primary cilia in a highly regulated manner.

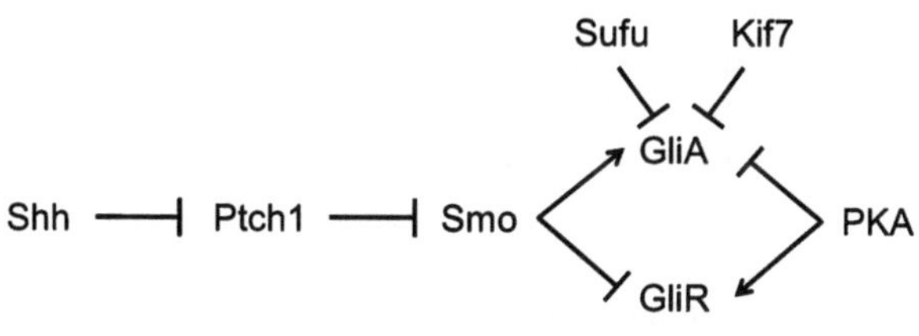

Fig. 2.2 Genetic interactions between core Shh signaling pathway components. Genetic experiments ablating individual components of the Shh pathway have shown whether these proteins primarily activate or inhibit downstream Shh signaling. Shh ligand serves to inhibit the activity of its ligand Ptch1, which in turn inhibits the major Shh effector Smo. Activated Smo leads to the production of GliA, which in turn transcribes Shh target genes. In the absence of Shh, Smo is inhibited and GliR is formed. Other major inhibitors of GliA include Sufu and Kif7. PKA is required for formation of both GliA and GliR, but its most prominent function in Shh signaling is to inhibit Gli activity in the absence of ligand

In the absence of Shh, Ptch1 inhibits Smo, thereby acting as a brake on Shh signaling (Denef et al. 2000; Taipale et al. 2002). When Shh binds to Ptch1, the inhibition of Smo is relieved, triggering activation of Gli proteins and transcription of Shh target genes. The opposing functions of Ptch1 and Smo in Shh signaling become clear when these proteins are ablated in mice. *Ptch1*$^{-/-}$ embryos exhibit a ventralized neural tube in which cells throughout the tissue express markers of the floor plate, indicating a maximal level of Shh signaling in the absence of the inhibitory receptor (Goodrich et al. 1997). *Smo*$^{-/-}$ embryos, on the other hand, are insensitive to Shh ligand and fail to develop any Shh-dependent cell types in the neural tube (Wijgerde et al. 2002). Similar experiments in mouse genetics have used neural tube patterning phenotypes to examine the role of many other proteins in the Shh pathway.

Under normal conditions, Shh-dependent neural patterning arises from differing levels of Gli activator and repressor activity along the dorsoventral axis of the neural tube (Lei et al. 2004; Motoyama et al. 2003; Sasaki et al. 1999). In vertebrates, the Gli family of transcription factors consists of three members: Gli1 functions solely as a transcriptional activator, while Gli2 and Gli3 contain both activator and repressor domains (Bai et al. 2002; Ding et al. 1998; Park et al. 2000; Persson et al. 2002). Gli2 serves as the primary activator and cleaved Gli3 the primary repressor of Shh target genes in the mammalian neural tube (Sasaki et al. 1999). The regulation of Gli2 and Gli3 is critical for the proper transduction of Shh signaling. In the absence of Shh activity, Gli3 is proteolytically cleaved into its repressor form (GliR) through a PKA-dependent mechanism (Wang et al. 2000), and Gli2 is targeted for degradation. Sufu binds to Glis in the absence of Shh signaling, serving both to stabilize the Gli proteins and inhibit their activation (Chen et al. 2009; Tukachinsky et al. 2010; Wang et al. 2010). When Shh is present, the ensuing Shh signaling cascade blocks the cleavage of Gli3 into GliR and stabilizes full-length Gli2, which can then be converted into its activator form (GliA) through an as-yet-unknown mechanism. The atypical kinesin Kif7 was recently revealed to be necessary for the formation of both GliA and GliR via a cilia-dependent process (Cheung et al. 2009; Endoh-Yamagami et al. 2009; Liem et al. 2009). In addition, PKA regulates not just the processing of Gli3 into GliR, but also the activation of Gli2 (Tuson et al. 2011). Ultimately, since Shh pathway activity mirrors the gradient of Shh ligand across the dorsoventral axis, the end result of this tightly regulated signaling cascade is that GliA levels are high in the ventral neural tube, whereas GliR predominates in the dorsal neural tube. These opposing gradients of activator and repressor establish the progenitor domains in the ventral neural tube.

The functions of individual Gli proteins in neural tube patterning have also been studied through genetic ablation experiments. *Gli1*$^{-/-}$ mice are phenotypically normal, indicating that Gli1 is dispensable for Shh signaling in mammals. In fact, *Gli1* is itself a Shh target gene, and it has been shown that all Gli1 expression requires Shh signaling (Bai et al. 2002). In the absence of Gli1, it is likely that the other Gli proteins (primarily Gli2) compensate for its function. *Gli2*$^{-/-}$ mouse embryos, in contrast, display reduced Shh activity: they fail to form a floor plate and show a reduction in the number of v3 progenitors, indicating that the highest levels of Shh activity cannot be achieved without Gli2. This indicates Gli2's role as the primary activator of Shh target gene transcription in the neural tube (Ding et al. 1998; Matise et al. 1998; Park et al. 2000). Finally, *extra toes* (Xt^J) mouse mutants, which lack Gli3 function due to a genomic deletion in the *Gli3* locus, exhibit mediolateral progenitors in more dorsal regions of the neural tube. These embryos show an expansion of the Shh-responsive domain in the absence of the primary repressor (Persson et al. 2002). Epistasis experiments indicate that Gli3 does have some role as an activator; however, $Gli2^{-/-};Gli3^{Xt/Xt}$ double mutant embryos completely lack v3 cells, suggesting the v3 progenitors that are present in $Gli2^{-/-}$ embryos are induced by GliA derived from Gli3 (Motoyama 2003). The neural tube's sensitivity to such slight changes in the balance of GliA and GliR makes this an excellent system for testing hypotheses about Shh signaling through genetic manipulations.

2.2.3 Maintaining Neural Patterning

Progenitor domains in the developing spinal cord are defined by the expression of two classes of Shh-responsive transcription factors (Briscoe 2000). Class I factors, such as Pax7, Pax6, Dbx1, Dbx2, and Irx3, are constitutively expressed by neural progenitors, but are repressed by Shh signaling. Class II factors, such as FoxA2, Nkx6.1, Nkx6.2, Nkx2.2, and Olig2, require Shh signaling for their expression (Briscoe and Ericson 2001). The proteins in each class differ in their sensitivity to Shh activity, such that unique combinations of these factors are expressed in each progenitor domain depending on the ratio of GliA and GliR to which cells are exposed. For example, Nkx6.1, Pax6, and Olig2 are all expressed in the pMN domain, but in the more dorsal p2 domain, Olig2 expression does not occur, while Irx3 expression is permitted. Cross-repressive interactions between pairs of class I and class II proteins establish the boundaries between progenitor domains (for example, Irx3 and Olig2 repress each other's expression). As development proceeds, these unique transcription factor expression profiles regulate the expression of cell type-specific genes that promote differentiation into distinct classes of spinal cord neurons. In this way, class I and class II transcription factors form a gene regulatory network that solidifies the pattern within the neural tube.

Recent work provides further evidence for a model in which graded Shh activity initiates patterning, but other mechanisms refine and maintain the ventral progenitor domains. Studies using a GFP reporter of Gli activity show that the level of signaling in individual progenitor domains changes during the course of development, while their transcriptional profile remains constant. This indicates that proper patterning depends on more than the absolute level of Shh signaling in a cell. Indeed, a model describing the cross-repressive gene regulatory network between the Shh-responsive transcription factors Nkx2.2, Olig2, and Pax6 can explain how these factors remain confined to particular dorsoventral domains, even in the face of variable levels of Shh ligand and GliA/GliR activity during development. At any given time, neural progenitor cells are influenced both by their current level of Shh signaling and their current transcriptional profile. For instance, due to its status as the strongest repressor in the gene regulatory network, once Nkx2.2 is expressed, it is able to prevent the expression of other Shh-responsive genes (e.g., Olig2), despite fluctuations in levels of Shh signaling. Conversely, cells that might have initially expressed Olig2 at lower levels of Shh activity are able to alter their transcriptional profiles to express Nkx2.2 in response to increased Shh signaling (Balaskas et al. 2012).

Although a classic morphogen model would suggest that cell fates in the ventral neural tube are defined solely by the amount of Shh in the extracellular environment (which in turn determines the levels of GliA and GliR), recent studies reveal that Shh-dependent neural patterning is more complex. This is not so surprising, however, when one considers the inherent complexity of the developing spinal cord. Throughout development, the tissue of the neural tube grows and expands, meaning progenitor cells are constantly shifting their position relative to the source of Shh in the notochord and floor plate. Furthermore, the Shh gradient does not remain constant: the overall amount of ligand increases over time (Chamberlain et al. 2008), such that Shh levels that would be sufficient to induce ventral progenitors earlier in development can be found at more dorsal positions later in development, without inducing a change in dorsal cell fate. Thus, other mechanisms beyond the interpretation of the Shh ligand gradient seem necessary to ensure that proper patterning is established and maintained in the ventral neural tube.

One such mechanism is temporal adaptation. Shh-responsive cells create negative feedback by upregulating Ptch1, the major inhibitor of Shh activity, so cells exposed to a constant concentration of Shh reduce their GliA response over time. Higher concentrations of Shh induce higher initial GliA activity, meaning the duration of GliA-dependent signaling is longer in cells exposed to the highest levels of Shh. Importantly, the duration of Shh signaling is critical for proper specification of progenitor domains. In vivo, the p3 domain has been shown to transiently express Olig2, a marker of the more dorsal pMN domain, before expressing Nkx2.2. This result is also seen in vitro, where neural tube explants exposed to Shh express Olig2 and Nkx2.2 sequentially, and the normal duration-dependent response to Shh is found to require Ptch1 upregulation (Dessaud et al. 2007).

In summary, the mammalian neural tube makes a remarkable system for studying the regulation of Shh signaling in vivo. Since neural progenitor cells along the dorsoventral axis express molecular markers in a highly stereotyped fashion, and we have a solid understanding of the connections between levels of Shh ligand, signal transduction through the primary components of the pathway, and the ultimate patterns of gene expression that define neural progenitor domains, the assessment of neural patterning in mutant embryos gives us many informative clues as to how a given gene regulates Shh signaling, especially when combined with epistasis experiments to show whether the gene of interest acts upstream or downstream of known Shh signaling components. In recent years, genetic experiments like these have shown that primary cilia are essential for the proper function of the Shh pathway and that the perturbation of cilia leads to a variety of Shh signaling defects that can be assessed via their effects on neural tube patterning.

2.3 Molecular Mechanisms of Shh Signaling in Cilia

2.3.1 Requirement of Cilia for Sonic Hedgehog Signaling

Primary cilia are seen on virtually every type of vertebrate cell and are required for Shh signaling. The connection between primary cilia and Shh signaling is unique to vertebrates (reviewed in: Huangfu and Anderson 2006; Ingham et al. 2011) and was first revealed through in vivo analysis of mouse mutants with defects in cilia genes. More recently, however, new experiments have told us much about the molecular biology of cilia-dependent Shh signaling within individual cells. All major components of the Shh signaling pathway are known to localize in or near primary cilia, and Shh pathway proteins are trafficked in and out of cilia depending on the activation state of the pathway (Fig. 2.3). In general, disrupting ciliogenesis or ciliary protein transport also leads to disruption of Shh signaling.

Cell biology experiments point to an extremely complex mechanism through which a large number of effectors regulate Shh signaling in cilia. These molecules can be disrupted independently of one another, leading to abnormal Shh signaling activity through similar overlapping but ultimately distinct mechanisms. In such cases, the neural tube makes an appealing system for analyzing the roles of individual molecules in the regulation of the Shh pathway, since we can observe shifts in the patterning of ventral neural progenitor domains when a gene is manipulated in mouse embryos and use the patterning phenotype to make inferences about the effects of the mutation on Shh signaling.

Shh signaling activity depends on the proper localization of the pathway's component proteins to primary cilia: Ptch1, Smo, Gli2, Gli3, Sufu, and Kif7 are all known to move into and out of the cilium during Shh signaling (Corbit et al. 2005; Haycraft et al. 2005; Liem et al. 2009; Rohatgi et al. 2007; Tukachinsky et al. 2010). Indeed, without primary cilia, no Shh signaling can occur. Cilia are required for the formation of both GliA and GliR (Haycraft et al. 2005; Huangfu and Anderson 2005; Liu et al. 2005; May et al. 2005); cells without cilia receive no transcriptional signals from the Shh pathway. Despite this, mediolateral progenitors requiring only low levels of Shh signaling are still observed in some mutants with no cilia (Huangfu and Anderson 2005). The progenitors form due to the loss of GliR-dependent repression of mediolateral cell fates. This situation is distinct from *Shh*$^{-/-}$ embryos, for example, in which ventral neural progenitors do not develop because GliA is never induced by Shh signaling, but GliR is still produced (Chiang et al. 1996). Epistasis experiments show that *Shh*$^{-/-}$;*Gli3*$^{-/-}$ double mutants are still capable of producing some ventral progenitors; the loss of GliR derepresses the Shh pathway under these conditions (Persson et al. 2002). In contrast, mutants that lack cilia are much less sensitive to *Gli3* ablation (Huangfu et al. 2003) (Fig. 2.4).

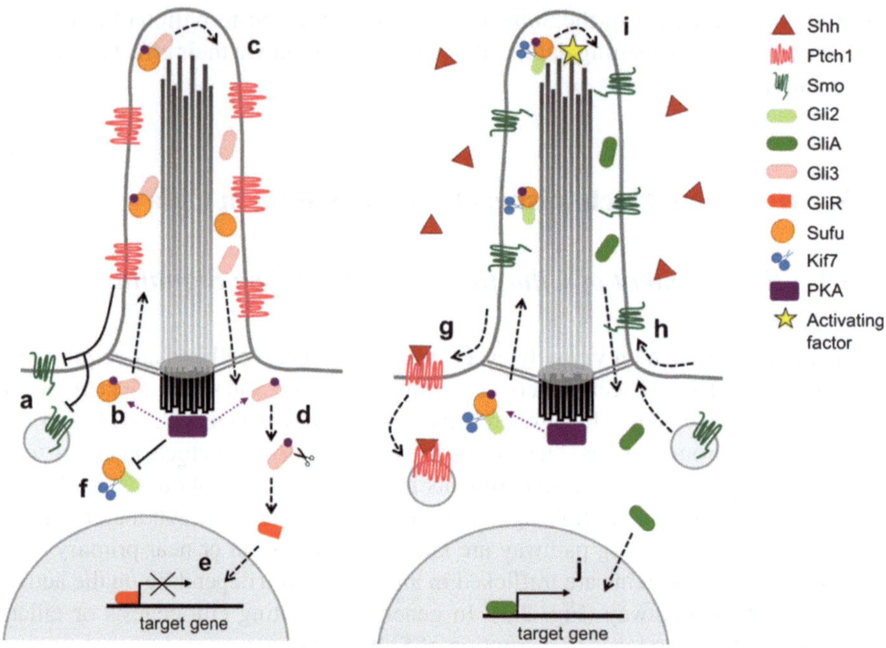

Fig. 2.3 Localization of major Shh pathway proteins within cilia. In the absence of Shh ligand (*left panel*), the Shh receptor Ptch1 is localized to the ciliary membrane, which inhibits the ciliary entry of Smo. Studies have reported that Smo is localized either to the plasma membrane or to intracellular vesicles (or, perhaps, both) in the absence of Shh signaling (**a**). Without Shh stimulation, Gli transcription factors, particularly Gli3, are converted to their repressor form. Gli3 is bound to Sufu, which is regulated via phosphorylation (*purple circles*) by PKA to promote its ciliary localization (**b**). The Sufu/Gli3 complex must be trafficked to the tip of the cilium in order for Gli3 to dissociate from Sufu and promote the formation of GliR (**c**), although it is not known whether this dissociation occurs within the cilium or after the complex exits the cilium. Gli3 itself is also phosphorylated by PKA (*purple circles*) at the base of the cilium to promote its cleavage by the proteasome into GliR (**d**). Processed GliR then translocates to the nucleus, where it represses the transcription of Shh target genes (**e**). At the same time, PKA and Kif7 together inhibit the ciliary localization and activation of Gli2 (**f**). In the presence of Shh (*right panel*), Ptch1 binds to the ligand and is internalized (**g**). The removal of Ptch1 from the ciliary membrane allows Smo to enter the cilium via lateral transport, targeted vesicle fusion, or both (**h**). Activated ciliary Smo promotes the enrichment of Gli proteins at the tip of the cilium, where they are activated to form GliA through a currently unknown mechanism (**i**). GliA then translocates to the nucleus where it initiates transcription of Shh target genes (**j**)

2.3.2 Cilia-Dependent Sonic Hedgehog Effectors

Under baseline conditions, cilia-dependent mechanisms are required to properly regulate the Shh pathway. In the absence of Shh, Ptch1 is enriched in the membrane of the primary cilium (Rohatgi et al. 2007). The presence of Ptch1 serves to inhibit the accumulation of Smo in ciliary membrane, although a small amount of Smo

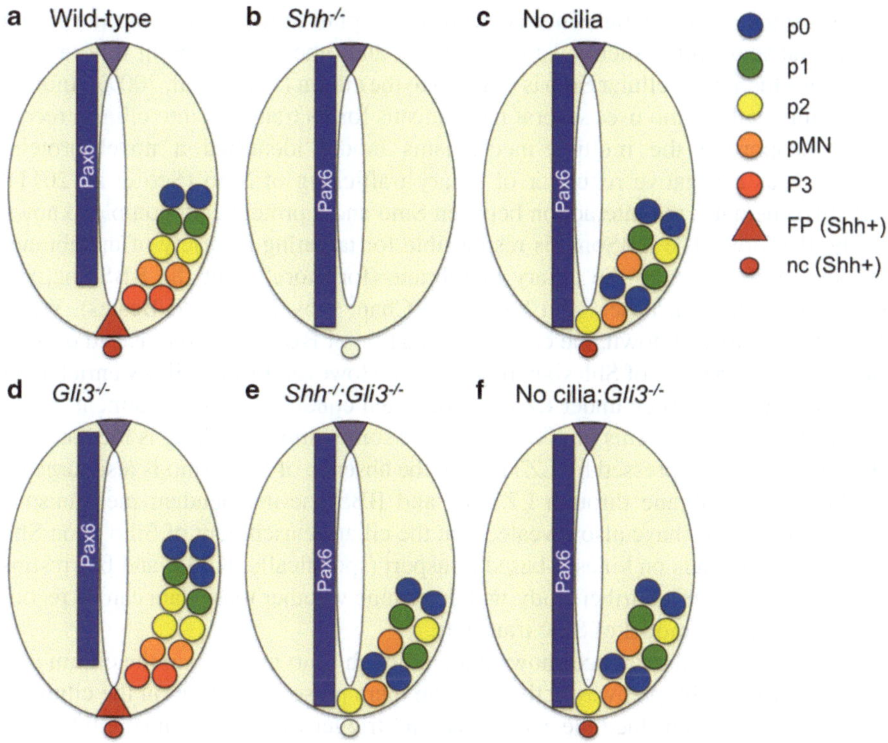

Fig. 2.4 Neural tube patterning requires cilia and Shh. Progenitor domains in the neural tube reveal changes in the balance between GliA and GliR in different mutant embryos. Wild-type embryos (**a**) exhibit six ventral progenitor domains based on graded Shh signaling along the dorsoventral axis. In *Shh*$^{-/-}$ embryos (**b**), no GliA is produced, but GliR still forms, inhibiting all six classes of ventral progenitors and dorsalizing the neural tube. Embryos that lack cilia (**c**) due to mutations in a variety of genes are deficient in both GliA and GliR; the result is that the highest levels of activation and repression cannot be achieved, and the neural tube displays mediolateral progenitor cells throughout the ventral region. These cells are ligand-insensitive; the notochord (nc) continues to produce Shh in cilia mutants even though the floor plate (FP) is not established. Mutations in *Gli3* help reveal more about the mechanism of Shh signaling in various mutants. *Gli3*$^{-/-}$ single mutants (**d**) show a mild phenotype in which mediolateral progenitor domains are slightly expanded and overlapping. Loss of *Gli3* in a *Shh*$^{-/n}$ embryo (**e**) derepresses Shh signaling, allowing mediolateral progenitors to form even in the absence of ligand. In a cilia mutant embryo, however, loss of Gli3 has little effect (**f**) because GliR formation requires cilia

constantly shuttles through the cilium, even in the presence of Ptch1 (Kim et al. 2009; Ocbina and Anderson 2008). Gli2 and Gli3 are also found in cilia in the absence of Shh ligand, and Gli proteins must travel into and out of primary cilia to be properly converted into GliR (Haycraft et al. 2005).

When Shh binds Ptch1, an unknown process causes Ptch1 to exit the cilium and leads to Smo becoming enriched there (Corbit et al. 2005; Rohatgi et al. 2007). The exact mechanism of Smo's movement into cilia remains a matter of debate, although there is some evidence that Smo moves from the plasma membrane into the cilium via lateral transport (Milenkovic et al. 2009), and other studies point to a mechanism by which intracellular Smo is targeted to the cilium (Wang et al. 2009). Another possibility is that Smo uses several mechanisms for its transport into cilia. A recent study supporting the multiple-mechanisms model identified a novel protein, LZTFL1, as a negative regulator of ciliary trafficking of Smo (Seo et al. 2011). LZTFL1 mediates the interaction between Smo and a protein coat complex known as the BBSome. The BBSome is responsible for targeting a variety of membrane-associated receptors to the ciliary membrane (for more about the BBSome, see Chap. 6 by Berbari, Pasek, and Yoder and Chap. 9 by Baker and Beales). When LZTFL1 is knocked down, the ciliary localization of BBSome proteins and of Smo increases, irrespective of Shh signaling activity. However, further ciliary enrichment of Smo can still be seen under LZTFL1-depleted conditions upon treatment with a Shh pathway agonist. This implies that, while some Smo trafficking is mediated by the BBSome and repressed by LZTFL1 in the absence of Shh, Smo is also targeted to the cilia membrane through LZTFL- and BBSome-independent mechanisms. Other experiments have also revealed that the ciliary enrichment of Smo upon Shh stimulation depends on kinesin-based transport (specifically, Kif3a) and β-arrestins (Kovacs et al. 2008). Further study will determine whether these data can be reconciled into a single model of Smo transport.

Pharmacological evidence shows that, although Smo moves into the cilium as a consequence of Shh pathway activation, the mere presence of Smo in the cilium is not sufficient to transduce the Shh signal and trigger the formation of GliA: Smo must also be activated (Wang et al. 2009). The two-step process of Smo-dependent Shh signaling (translocation and activation) has yet to be fully explained. Smo shares many characteristics with G protein-coupled receptors (Ayers and Thérond 2010) and is known to respond to several small molecules in pharmacological screens. Therefore, one appealing model posits that a yet-unknown endogenous molecule – possibly a cholesterol-derived oxysterol (Nachtergaele et al. 2012) – regulates Smo activation, and that Ptch1 inhibits Smo by blocking its access to this molecule, as well as blocking its accumulation in cilia.

The presence of activated Smo in the cilium causes enrichment of Gli proteins at the tip of the cilium, inhibits GliR formation, and triggers the formation of GliA (Chen et al. 2009; Wen et al. 2010). Studies have shown that the processing of Gli proteins into GliA and GliR requires transport of Gli proteins into and out of the primary cilium. It is not known how trafficking through the primary cilium regulates the conversion of full-length Gli proteins into activator or repressor, but experiments thus far have indicated that the process involves the core Shh pathway proteins Sufu, Kif7, and PKA (Chen et al. 2009, 2011; Tukachinsky et al. 2010). Because the balance of GliA and GliR ultimately determines the output of the Shh signaling pathway, proper regulation of Gli proteins is key, and the disruption of Gli processing and/or activation leads to signaling defects that cause abnormal neural patterning in vivo.

Sufu, Kif7, and PKA act at the primary cilium to regulate Gli processing and activation. Sufu binding to Gli proteins inhibits their transcriptional functions.

Sufu can act in a cilia-independent manner to inhibit Gli proteins by sequestering them from the nucleus, but it also moves into and out of cilia with the Glis, becoming enriched in cilia upon Shh stimulation. For GliR or GliA to form, Gli proteins must first dissociate from Sufu. This dissociation requires kinesin-dependent trafficking of the Sufu/Gli complex to primary cilia (Humke et al. 2010; Tukachinsky et al. 2010). In addition to inhibiting the formation of GliR and GliA, Sufu serves to stabilize Gli proteins: genetic ablation of Sufu leads to degradation of Gli2 and Gli3 via the ubiquitin E3 ligase adaptor Spop. Gli3 is more affected by loss of Sufu than Gli2; *Sufu* null embryos exhibit a ventralized neural tube phenotype due to the alteration in the balance of GliA and GliR that results from Gli3 degradation (Wang et al. 2010).

Kif7, an atypical kinesin, normally localizes to the base of the cilium, where it is ideally positioned to regulate the access of Shh pathway proteins to the ciliary compartment. Kif7 serves as a negative regulator of Shh signaling downstream of Smo. Upon activation of the Shh pathway, Kif7 moves to the tip of the cilium concomitantly with Gli proteins, indicating that it may regulate the transport and activation of Gli2, as well as the processing of Gli3 into GliR (Endoh-Yamagami et al. 2009; Liem et al. 2009). One model suggests that, as a motor protein, Kif7 may inhibit Gli activation by moving Gli proteins away from the cilium in the absence of signals from Shh (Liem et al. 2009). Further experiments are needed to test this model. Notably, the role of Kif7 in Shh signaling was initially disputed, as cell-based assays of Kif7 knockdown did not show a significant effect on Shh pathway output (Varjosalo et al. 2006). Thus, in vivo analysis of neural tube patterning proves to be a more sensitive method for examining subtle effects of perturbations to the Shh pathway.

The stepwise nature of the Shh signaling cascade allows for regulation of the pathway by the same effectors at multiple discrete steps. For example, PKA is known to phosphorylate Sufu, thereby promoting Sufu's ciliary localization (Chen et al. 2011). However, analysis of mouse mutants deficient in PKA reveals a more severe neural tube patterning phenotype than $Sufu^{-/-}$ mutants, indicating that PKA is affecting Gli activity and Shh signaling through Sufu-independent mechanisms, in addition to regulating Sufu. Specifically, PKA acts as a negative regulator of the Shh signaling cascade (Epstein et al. 1996; Hammerschmidt et al. 1996; Concordet et al. 1996). PKA is localized at the base of the primary cilium, where it regulates the ciliary entry of Gli proteins, the production of GliR from Gli3, and the activation of Gli2 (Tuson et al. 2011). Although PKA regulates a wide variety of processes in adult vertebrate cells, its most notable function during development seems to be regulation of the Shh pathway in the neural tube as well as in other Shh-responsive tissues (Huang et al. 2002). When a Shh signaling component plays multiple roles in regulating the pathway, as PKA does, it complicates the interpretation of experiments in which this molecule is perturbed. Yet it is not surprising for a single effector to interact with multiple steps of a signaling cascade, especially when all the major signaling components localize to a small organelle like the cilium.

The dynamic regulation of Shh pathway proteins in primary cilia points to the central nature of cilia and ciliary protein trafficking in Shh signaling. In addition to the core Shh pathway proteins described above, a variety of other effectors contribute to the proper function of Shh signaling in cilia and can lead to in vivo neural patterning defects when disrupted.

2.4 Cilia Proteins Essential for Shh Signaling: Lessons from Mutants

2.4.1 Intraflagellar Transport

The link between cilia and Shh signaling was first discovered in a forward genetic screen that identified several mouse mutants with defects in intraflagellar transport (IFT) (Huangfu et al. 2003; Huangfu and Anderson 2005). IFT relies on a highly conserved family of proteins to move cargoes along the microtubules that form the ciliary axoneme. Anterograde IFT relies on the kinesin-2 motor as well as the 11 proteins that constitute the IFTB complex. Retrograde IFT uses the six proteins in the IFTA complex and the cytoplasmic dynein motor (reviewed in: Pedersen and Rosenbaum 2008; Rosenbaum and Witman 2002) (Fig. 2.5). When anterograde IFT is disrupted, ciliogenesis does not occur. (For a more detailed discussion of IFT and ciliogenesis, see Chap. 1, Sect. 4.3). When retrograde IFT is disrupted, IFT cargoes accumulate at the distal end of the cilium and interfere with the normal flow of ciliary protein traffic.

As mentioned before, mutant mouse embryos that lack cilia due to anterograde IFT defects have no Shh signaling; they fail to produce either GliA or GliR (Huangfu et al. 2003; Huangfu and Anderson 2005). In contrast, defects in retrograde IFT can lead to either elevated Shh signaling activity or loss of the Shh response, depending on the causative mutation. Analysis of mouse mutants has revealed the distinct roles of the dynein retrograde motor and the IFTA complex in Shh signaling, as well as several regulatory steps in between them.

Loss of the dynein component Dync2h1 (also called Dnchc2) leads to a reduction in ventral cell types in the neural tube (Huangfu and Anderson 2005; May et al. 2005). These dynein mutant embryos completely lack retrograde IFT and have extreme defects in cilia structure, with very short, almost spherical cilia. Neural tube patterning in *Dync2h1* mutants resembles IFTB mutants, which fail to generate cilia altogether, and loss of dynein also causes reduced levels of cleaved Gli3 compared to wild-type. These phenotypes indicate that the cilia-dependent processes required to generate GliA and GliR are likely to be lost in dynein mutants. Past studies have also proposed that the lack of Shh signaling in *Dync2h1* mutants is due to a failure of Smo to localize to their abnormal cilia (May et al. 2005); however, these interpretations were based on the analysis of motile cilia in the embryonic node, where the role of Shh signaling has not been characterized in as much detail. Later experiments in Shh-responsive cell types have confirmed that Smo does traffic into dynein-deficient primary cilia (Kim et al. 2009; Ocbina and Anderson 2008), so this cannot explain the Shh signaling defects in these mutants.

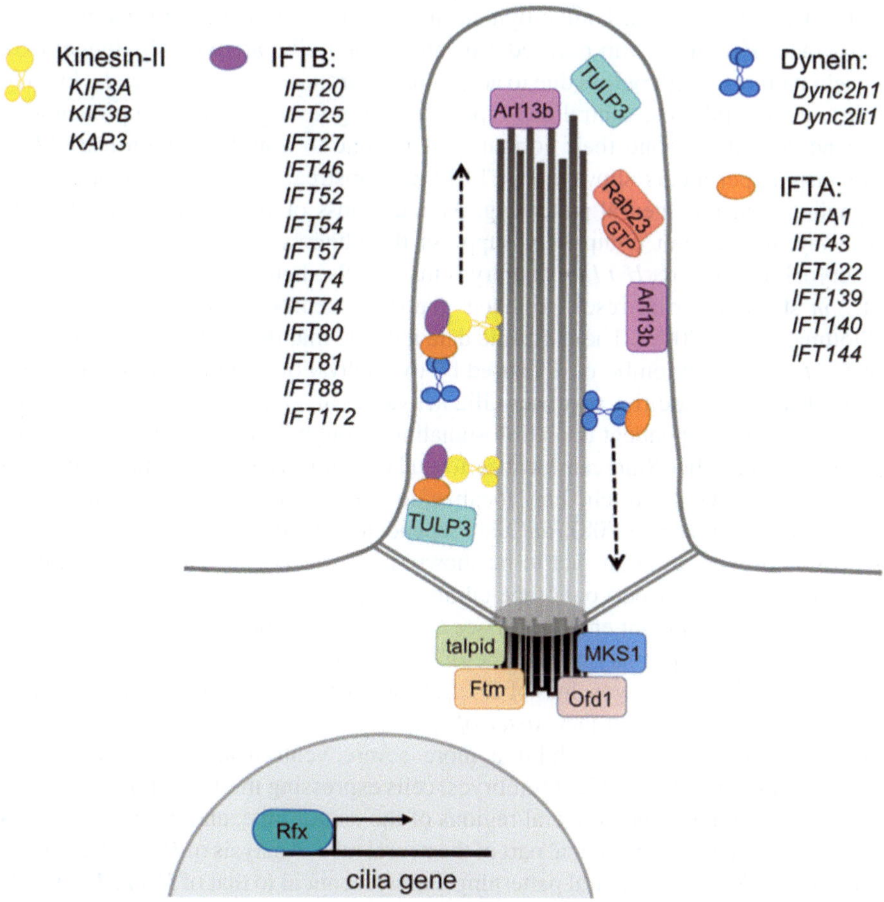

Fig. 2.5 Intraflagellar transport (*IFT*) proteins and other proteins that regulate Shh signaling through ciliogenesis or ciliary protein trafficking. A variety of mutations affecting ciliary proteins have been shown to disrupt Shh signaling. These include proteins associated with intraflagellar transport (the kinesin-II and dynein motors and the IFTA and IFTB complexes) as well as others with functions that are not as well understood. Some of these proteins are found in cilia: TULP3 requires the IFTA complex to enter the cilium and localizes to ciliary tips in the absence of Shh signaling; it is not known whether it associates with the axoneme or the membrane. Rab23 is found in cilia in its GTP-bound state, but not its GDP-bound state. Arl13b is primarily associated with the ciliary membrane but is seen at the base and the tip of the cilium even after the membrane is removed with detergent. Other mutations affecting ciliogenesis and Shh signaling disrupt proteins associated with the basal body: Ofd1, MKS1, Ftm, and talpid are among these. The Rfx transcription factors are not found in cilia, but regulate the transcription of a variety of ciliary genes, thus regulating ciliogenesis and Shh signaling

The neurodevelopmental effects of IFTA complex mutations have been studied using several mouse lines generated in forward genetic screens. In contrast to dynein mutations, the genetic ablation of the IFTA proteins IFT139 (also called THM1 or Ttc21b) or IFT122 results in elevated Shh activity in the neural tube (Qin et al.

2011; Tran et al. 2008). Unlike dynein mutants, IFTA mutants still exhibit some retrograde IFT, albeit with reduced efficiency. Their cilia are typically longer than dynein mutants', with bulges due to accumulated proteins at the distal ends. In *alien (aln)* mutant embryos, ventral cell types, including floor plate and p3 progenitors, expand dorsally beyond their normal domains due to a null mutation in *IFT139*. Epistasis experiments show that IFT139 acts upstream of Gli2 to regulate Shh signaling, and thus neural patterning. Although loss of IFT139 activates the Shh pathway, it does not completely suppress the phenotype of a $Shh^{-/-}$ or $Smo^{-/-}$ embryo. Furthermore, $IFT139^{aln}$ embryos that are also heterozygous for a null allele of *Shh* show a partial rescue of their abnormal neurodevelopmental phenotype (Stottmann et al. 2009). These genetic data indicate that the ventralized neural tube of $IFT139^{aln}$ mutant embryos is caused by overactivation of Gli2 and that IFT139-dependent retrograde IFT regulates Gli2 activation in a manner that is partially, but not completely, dependent upon Shh signaling through Smo. Previous experiments have confirmed that Smo can localize to nodal cilia in $IFT139^{aln}$ mutants and that overexpressed Gli1 protein can localize to cilia in mutant mouse embryonic fibroblasts (Tran et al. 2008); the Shh dependence of Smo and Gli enrichment in $IFT139^{aln}$ primary cilia has yet to be shown, however. Because the IFTA complex mediates protein transport out of the cilium, it is plausible that the overactivation of Gli2 in $IFT139^{aln}$ mutant embryos arises at least in part from Gli proteins' constitutive localization to primary cilia in the $IFT139^{aln}$ mutant.

As in $IFT139^{aln}$ embryos, loss of IFT122 causes an expansion of ventral progenitors in the neural tube. In fact, *sister of open brain (sopb)* mutant embryos with a null mutation in *IFT122* exhibit a more severe ventralization phenotype than $IFT139^{aln}$ mutants. In $IFT122^{sopb}$ embryos, cells expressing markers of the floor plate are found within the mediolateral regions of the neural tube, and pMN progenitors extend into even the most dorsal part of the neural tube. Analysis of $IFT122^{sopb};Shh^{-/-}$ double mutants reveals neural patterning almost identical to that of $IFT122^{sopb}$ single mutants, indicating most of the ventralization phenotype caused by loss of IFT122 is ligand-independent. $IFT122^{sopb};Gli2^{-/-}$ mutants show that, as in $IFT139^{aln}$ mutants, the ventralization of the $IFT122^{sopb}$ neural tube is dependent upon GliA derived primarily from Gli2. Cell biology experiments in $IFT122^{sopb}$ mouse embryonic fibroblasts show that Gli2 and Gli3 are both enriched at the tips of mutant cilia in a ligand-independent manner, but the ligand-dependent ciliary localization of Smo and Sufu are largely unaffected by loss of IFT122 (Qin et al. 2011). It is clear that transport of Gli proteins to the tip of the cilium is required for their activation, and that Glis must dissociate from Sufu to become activated (Haycraft et al. 2005; Liu et al. 2005; Tukachinsky et al. 2010). Perhaps differential regulation of the ciliary trafficking of Sufu and Glis by the IFTA complex contributes to the overactivation of the Shh pathway in these mutants.

Several recent studies examined the interplay between retrograde and anterograde IFT components in ciliogenesis and Shh signaling (Ocbina et al. 2011; Liem et al. 2012). The results suggest that mutations in *IFT172*, which codes for a member of the IFTB complex, are able to rescue defects in *Dync2h1* mutants. Cilia structure and neural tube patterning in $Dync2h^{lln/lln};IFT172^{avc/+}$ double mutants were similar to wild-type. The ability of a mutation in an anterograde trafficking protein to rescue structural defects caused by abnormal retrograde trafficking implies that a balance between anterograde and retrograde IFT regulates cilia formation, and that the Shh signaling defects in *Dync2h1* mutants arise primarily from abnormal cilia structure. Intriguingly, the same study also found that *IFT122^sopb^* can partially rescue neural patterning, cilia structure, and Shh pathway protein trafficking defects in *Dync2h1* mutants and that this rescue is ligand-dependent. This result was interpreted to mean that IFT122 has a role in anterograde IFT as well as retrograde IFT, such that loss of IFT122 can restore the balance between the two that is required for proper ciliogenesis. It also implies that the ligand-independent gain of Shh signaling activity in *IFT122^sopb^* mutants arises from an undescribed IFT-independent function of IFT122, and perhaps the IFTA complex in general, in regulating the Shh pathway (Ocbina et al. 2011). It would be interesting to test whether loss of IFT139 is also able to rescue neural patterning defects in dynein mutants, or whether this additional Shh regulatory function is specific to IFT122.

A subsequent study showed that a severe mutation in the IFTA gene *IFT144* (*IFT144^dmhd^*) can prevent ciliogenesis, further indicating that proteins traditionally associated with retrograde IFT can play a role in anterograde IFT. The hypomorphic allele *IFT144^twt^* did not significantly affect cilia structure, however. Because cilia structure and Shh signaling are so tightly linked, it is perhaps not surprising that the two *IFT144* alleles show distinct neural patterning phenotypes. The severe structural defects in *IFT144^dmhd^* cilia lead to a loss of Shh signaling, while *IFT144^twt^* mutants show ectopic activation of the Shh pathway similar to other IFTA mutants. Furthermore, compound mutants with *IFT144^twt^* and *IFT122^sopb^* alleles showed ciliogenesis and patterning defects as severe as *IFT144^dmhd^*, indicating that the IFTA components IFT144 and IFT122 work together to build cilia. The study also showed that many ciliary membrane proteins, including Smoothened, were mislocalized in IFTA mutant cilia, but that soluble proteins, including Gli2, Sufu, and Kif7, were unaffected by IFTA mutations (Liem et al. 2012). The authors therefore propose that IFTA-dependent trafficking of membrane proteins into cilia may explain the ectopic Shh activation and neural patterning phenotypes in IFTA mutants.

Although both IFTA and IFTB proteins have been linked to ciliogenesis as well as the regulation of Shh signaling, new research on mutant mice indicates that IFT25 (and perhaps IFT27) regulate Shh signaling independently of cilia structure. Mice with a protein null allele of *IFT25* also fail to produce IFT27 and show defects in Shh signaling. Specifically, the mutants show an expansion of mediolateral progenitors at the expense of the most ventral (floor plate) cell types. They also have defects in Ptch1 and Smo localization, Gli processing, and ligand-stimulated enrichment of Gli proteins at the tips of cilia. Despite these defects, their cilia are structurally intact (Keady et al. 2012). These results show that the IFTB complex components IFT25 and IFT27 are required for regulating Shh signaling but are dispensible for ciliogenesis. The exact functions of IFT25 and IFT27 are not known, but they are known to form a complex

(Bhogaraju et al. 2011). As a Rab-like small GTPase, IFT27 may act as a switch to regulate trafficking of Shh pathway components to cilia in a ligand-dependent manner (Qin et al. 2007).

Because IFT complexes and the molecular motors that power IFT are so critical for ciliogenesis and Shh signaling, it is not surprising to learn that factors regulating the expression of IFT genes can affect Shh signaling, as well. The Rfx family of transcription factors is evolutionarily conserved from invertebrates to vertebrates (Dubruille 2002; Swoboda et al. 2000) and is found to regulate the expression of cilia-specific genes, including the dynein component *D2lic* (Bonnafe et al. 2004) and *IFT172* (Ashique et al. 2009). Although genetic manipulations of some Rfx family members, such as *Rfx2* and *Rfx4*, affect both ciliogenesis and Shh signaling (Ashique et al. 2009; Chung et al. 2012), *Rfx3* mutants display no obvious Shh signaling defects, despite having abnormal cilia in the embryonic node and on ependymal cells in the brain (Bonnafe et al. 2004; El Zein et al. 2009). Because Rfx family members are differentially expressed across various tissues, it may be that Rfx3 is not normally expressed in the neural tube, or is expressed redundantly with other Rfx proteins, and is thus dispensable for cilia-dependent neural patterning. Still, it serves as an illustrative example of how factors affecting ciliogenesis and neurodevelopmental Shh signaling may be separable through genetic analysis.

Finally, in addition to the IFT proteins and their motors, a variety of proteins localized to the basal body have been shown to regulate ciliogenesis, such that mutations affecting these proteins leads to phenotypes that resemble anterograde IFT mutants. *Ftm* was discovered as one of six genes deleted in the *fused toes* (*Ft*) mouse mutant, which has defects indicative of reduced Shh signaling activity (Götz et al. 2005; Peters et al. 2002). Targeted knockouts of the individual genes in the *Ft* mutation revealed that *Ftm* is responsible for the majority of defects seen in mutant embryos, including loss of ventral progenitors in the neural tube (Vierkotten et al. 2007). The $Ftm^{-/-}$ phenotype shares many characteristics with anterograde IFT mutants, and indeed, $Ftm^{-/-}$ embryos show reduced numbers of cilia in some tissues such as the node. However, unlike IFTB and kinesin-II mutants, $Ftm^{-/-}$ mutants possess normal motile cilia in some tissues such as the trachea, and $Ftm^{-/-}$ primary fibroblasts can grow cilia in vitro. Thus, Ftm seems to specifically regulate Shh signaling – in particular, the formation of GliA and the processing of GliR – in addition to regulating ciliogenesis and cilia structure in tissues like the node. Ftm is localized to the basal body, placing it in an ideal position from which to regulate the entry of Shh pathway proteins to the cilium. Future experiments are needed to characterize the effects of Ftm ablation on the ligand-dependent ciliary localization of Shh signaling effectors. Other basal body proteins, including Ofd1 (Ferrante et al. 2006; Singla et al. 2010) and talpid (Bangs et al. 2011), affect ciliogenesis and Shh signaling in much the same way as Ftm. Loss of Mks1, another basal body protein, abolishes cilia in most tissues, including the neural tube, but spares some motile cilia. Furthermore, $Mks1^{-/-}$ embryos survive longer than most ciliogenesis mutants and seem to retain a small amount of GliA, suggesting that Mks1 may have functions in Shh signaling outside of ciliogenesis (Weatherbee et al. 2009).

2.4.2 Other Regulators of Shh Signaling in Cilia

Although IFT plays a crucial role in ciliogenesis and Shh signaling, this is not the only mechanism that controls the ciliary localization of Shh pathway proteins or the proper regulation of GliA and GliR transcriptional activity. In recent years, still more effectors of Shh signaling have been identified through genetic manipulations that affect neural tube patterning. Such experiments have revealed positive and negative regulators of Shh signaling that act at multiple distinct steps in the pathway. These Shh signaling effectors come from a variety of different protein families, including some that remain almost completely uncharacterized. In addition, some of these proteins are unique to vertebrates, implying that they may be specific regulators of ciliary Shh signaling (which occurs only in vertebrates, reviewed in: Huangfu and Anderson 2006; Ingham et al. 2011; Varjosalo et al. 2006). This places them in contrast to the IFT complexes, which are more general mediators of ciliogenesis and ciliary maintenance found in organisms ranging from the flagellate alga *Chlamydomonas reinhardtii* to mammals.

While it may be easy to understand why the complete ablation of cilia due to anterograde IFT mutations leads to Shh signaling defects, it is more difficult to conceptualize the complex and sometimes contradictory phenotypes caused by perturbations to retrograde IFT. Recent studies have identified effectors of IFTA that also play a role in regulating Shh signaling (as well as other cilia-dependent signaling pathways; see Chap. 6 by Berbari, Pasek, and Yoder for more details). These proteins are still being characterized, but their further study may shed new light on some of the unsolved mysteries of Shh signaling, including how Smo regulates the formation of GliA.

TULP3, a member of the tubby family of proteins, is one such novel regulator of Shh signaling. Tubby family proteins are conserved across eukaryotes and interact with membrane phosphoinositides to regulate a variety of signaling pathways (reviewed in: Mukhopadhyay and Jackson 2011). Both a targeted null mutation of *TULP3* and the hypomorphic *hitchhiker* mutation result in ventralized neural patterning similar to that seen in IFTA mutants (Norman et al. 2009; Patterson et al. 2009). Epistasis experiments show that TULP3 acts downstream of Shh and Smo, but upstream of Gli2. Loss of TULP3 has no effect on Gli3 processing or Shh-dependent localization of Gli3 to the tips of cilia (Mukhopadhyay et al. 2010; Norman et al. 2009). It is not yet clear, however, how the Shh-dependent ciliary enrichment of Gli2 is affected by mutations in *TULP3*.

TULP3 localizes to cilia and binds to the IFTA complex. The association between TULP3 and IFTA requires some IFTA proteins, including IFT122, but other IFTA proteins in the complex are not required for TULP3 binding (Mukhopadhyay et al. 2010). In *IFT122*^{sopb} mutants, TULP3 is not present in cilia, suggesting that the association between TULP3 and the IFTA complex is required for its ciliary localization. As *IFT122*^{sopb} mutants exhibit more severe neural tube ventralization than *IFT139*^{aln} mutants, it is possible that some of the Shh pathway overactivation in *IFT122*^{sopb} results from mislocalization of TULP3. It seems clear from the current data that TULP3 serves as a negative regulator of Gli2 activation in cilia, but the exact mechanism of its action remains to be elucidated. Because TULP3 is known to be required for the proper ciliary trafficking of several G protein-coupled receptors, one model posits that it may regulate a novel receptor involved in Shh signal transduction (Mukhopadhyay et al. 2010).

The small GTPase Rab23 was identified as a regulator of Shh signaling when the *open brain* (*opb*) mouse line was discovered in a forward genetic screen (Eggenschwiler et al. 2001; Eggenschwiler and Anderson 2000). *Rab23^{opb}* embryos exhibit a ventralized neural tube with expansion of the floor plate, v3, and pMN domains. Double mutant analysis indicates that Rab23 acts as a negative regulator of the Shh pathway downstream of Smo but upstream of Gli2; however, slight differences in patterning between *Rab23^{opb};Smo$^{-/-}$* double mutants and *Rab23^{opb}* single mutants (specifically: double mutants do not show expansion of the floor plate) indicate that Rab23 may also have some Smo-dependent functions (Eggenschwiler et al. 2006).

Cell biology studies of Rab23 present a more complex view of how this GTPase may regulate Shh signaling. Rab family GTPases are involved in vesicle transport to various subcellular compartments (reviewed in Grosshans et al. 2006). Thus, Rab23 was hypothesized to regulate the trafficking of Shh pathway proteins to primary cilia. Recent work indicates that wild-type Rab23 (but not the mutant protein Rab23-S23N, which is a constitutively GDP-bound dominant negative) is localized to cilia and that it regulates ciliary trafficking of Smo (Boehlke et al. 2010). Specifically, Rab23 is thought to act by promoting recycling of Smo that is already localized to the cilium, while inhibiting the ability of other Smo molecules to enter the ciliary compartment. How this proposed mechanism would allow Rab23 to inhibit Gli2 activation is not yet clear. Another recent study showed that Rab23 binds directly to Sufu, promoting Sufu-dependent inhibition of Gli1 transcriptional activity (Chi et al. 2012) by sequestering Gli1 in the cytoplasm. These results provide a more compelling explanation for Rab23's role as a negative regulator of Shh signaling, but the experiments were conducted in cell types that lack primary cilia. It is unclear whether the interaction between Rab23 and Sufu can occur in ciliated cells, or whether ciliary Rab23 has some other function that remains to be described.

Another small GTPase, Arl13b, plays a unique role in the regulation of Shh signaling. The *hennin* (*hnn*) mutant mouse has a protein null mutation in *Arl13b* that leads to loss of the most ventral cell types in the neural tube (floor plate) and an expansion of more mediolateral cell types (pMN). Epistasis experiments show that the *Arl13b^{hnn}* phenotype is caused by abnormal regulation of Gli2 (Caspary et al. 2007). Specifically, the highest levels of GliA in the floor plate are never reached, while a moderate level of GliA drives the expression of Shh target genes throughout most of the neural tube. Gli3, meanwhile, appears to be relatively unaffected by loss of Arl13b. An intact gradient of Gli3-derived GliR regulates the residual patterning in the *Arl13b^{hnn}* neural tube, but *Arl13b^{hnn};Gli3$^{-/-}$* double mutants have essentially no dorsoventral patterning (all cells in the double mutants assume either a v3 or pMN fate).

At the cellular level, Arl13b regulates multiple components of the Shh signaling pathway in cilia. First, *Arl13b^{hnn}* cells have short cilia with defects in the microtubule ultrastructure of the ciliary axoneme (Caspary et al. 2007). Although these defects are not as severe as those seen in IFTB mutants, the abnormal structure of primary cilia could contribute to the misregulation of the Shh pathway in cells lacking Arl13b. Furthermore, many of the proteins that make up the Shh signaling pathway show abnormal localization patterns in *Arl13b*^{hnn} cells (Larkins et al. 2011). Specifically, Ptch1 is found in the cilium even after cells are treated with Shh ligand, and Smo is enriched in cilia even in the absence of Shh. Meanwhile, Sufu, Gli2, and Gli3 fail to become enriched at the tip of the cilium upon Shh treatment in cells lacking Arl13b. Because Gli2 activation requires its transport to the tip of the cilium, the abnormal ligand-dependent trafficking of Shh pathway proteins may explain why GliA function is impaired in *Arl13b*^{hnn} mutants. It is not yet clear whether the abnormal transport of Shh pathway proteins in *Arl13b*^{hnn} cells arises from the structural defects in mutant cilia, or whether Arl13b has a specific function in regulating protein trafficking.

Genetic ablation of Tectonic1, a member of a novel protein family awaiting further characterization, reduces the levels of both GliA and GliR, although the effect seems greater on GliA. Without Tectonic1, the floor plate and v3 progenitors are not specified, indicating a reduction in GliA. Yet, in *Tectonic1*$^{-/-}$;*Shh*$^{-/-}$ double mutants, there are more ventral progenitors than in *Shh*$^{-/-}$ single mutants, implying that less GliR is present to inhibit the specification of these cell types in the double mutant embryos (Reiter and Skarnes 2006). We do not know whether Tectonic regulates Shh signaling from within the cilium, although the related *Tectonic2* gene has been linked to the human ciliopathy Meckel-Gruber syndrome and appears in the cilia proteome database (Shaheen et al. 2011; ciliaproteome.org). (For more on human ciliopathies, see Chap. 6 by Berbari, Pasek, and Yoder and Chap. 9 by Baker and Beales.).

2.4.3 Spatiotemporal Regulation of Shh Signaling

Beyond identifying new effectors of Shh signaling, the analysis of various mouse mutants has revealed unexpected aspects of the mechanism by which Shh signaling patterns the developing neural tube. In particular, many of the mutations described above (*IFT139*^{aln}, *TULP3*$^{-/-}$, *Rab23*^{opb}, *Arl13b*^{hnn}) exhibit patterning defects in the caudal neural tube alone, while the rostral neural tube appears normal despite defects in Shh signal transduction (Caspary et al. 2007; Eggenschwiler et al. 2001; Norman et al. 2009; Tran et al. 2008). Other mutants, like *IFT122*^{sopb} and various alleles of *Dync2h1*$^{-/-}$, show defects throughout the neural tube, but with variable phenotypes along the rostrocaudal axis. One explanation for this puzzling trend may be that the contributions of Gli2 and Gli3 to neural patterning seem to differ along the rostrocaudal axis. Specifically, in the rostral neural tube, ventral cell types still occur in *Gli2*$^{-/-}$ mutants in which Shh signaling has been activated by a *Ptch1* knockout (Motoyama et al. 2003). This indicates that Gli3 can compensate for the loss of Gli2-derived GliA in the rostral neural tube. In the caudal neural tube, however, loss of Gli2 abolishes ventral cell types even when the Shh pathway is maximally activated by the *Ptch1* mutation. Therefore, mutations in Shh effectors that differentially regulate Gli2 versus Gli3 may result in a variable phenotype along the rostrocaudal axis.

Furthermore, the caudal neural tube develops slightly earlier than the rostral neural tube (Papalopulu and Kintner 1996). It is known that the timing of Shh signaling is just as important as the dose of Gli activity received by different spatial domains of the developing tissue (as described earlier in this chapter under Sect. 2.2.1 "Patterning the neural tube"). Therefore, perturbations of the balance between GliA and GliR may have different effects on neural patterning depending on the duration of the Shh-sensitive window of development at different points along the rostrocaudal axis.

2.5 Conclusions and Perspectives

In summary, the regulation of spinal cord development by Shh is completely dependent upon mechanisms that operate within primary cilia. While many aspects of the molecular mechanisms behind this pathway still remain a mystery, over the past 10 years, genetic experiments in mouse models have taught us much about the role of cilia in regulating Shh signaling. Because the mammalian neural tube is so sensitive to subtle shifts in Shh signaling via genetic manipulation, studies using this system have revealed new levels of complexity in the Shh pathway and discovered many unsuspected effectors. The novel, vertebrate-specific regulators of Shh signaling identified most recently (such as TULP3 and Tectonic1) have opened up new avenues of research that will help us solve some of the mysteries in the field, and there are sure to be still more Shh effectors awaiting our discovery in future unbiased genetic screens.

References

Alcedo J, Ayzenzon M, Von Ohlen T et al (1996) The Drosophila smoothened gene encodes a seven-pass membrane protein, a putative receptor for the hedgehog signal. Cell 86:221–232. doi:10.1016/S0092-8674(00)80094-X

Ashique AM, Choe Y, Karlen M et al (2009) The Rfx4 transcription factor modulates Shh signaling by regional control of ciliogenesis. Sci Signal 2:ra70. doi:10.1126/scisignal.2000602

Ayers KL, Thérond PP (2010) Evaluating smoothened as a G-protein-coupled receptor for Hedgehog signalling. Trends Cell Biol 20:287–298. doi:10.1016/j.tcb.2010.02.002

Bai CB, Auerbach W, Lee JS et al (2002) Gli2, but not Gli1, is required for initial Shh signaling and ectopic activation of the Shh pathway. Development 129:4753–4761

Bai CB, Stephen D, Joyner AL (2004) All mouse ventral spinal cord patterning by Hedgehog is Gli dependent and involves an activator function of Gli3. Dev Cell 6:103–115. doi:10.1016/S1534-5807(03)00394-0

Balaskas N, Ribeiro A, Panovska J et al (2012) Gene regulatory logic for reading the Sonic Hedgehog signaling gradient in the vertebrate neural tube. Cell 148:273–284. doi:10.1016/j.cell.2011.10.047

Bangs F, Antonio N, Thongnuek P et al (2011) Generation of mice with functional inactivation of talpid3, a gene first identified in chicken. Development 138:3261–3272. doi:10.1242/dev.063602

Berbari NF, O'Connor AK, Haycraft CJ, Yoder BK (2009) The primary cilium as a complex signaling center. Curr Biol 19:R526–R535. doi:10.1016/j.cub.2009.05.025

Bhogaraju S, Taschner M, Morawetz M et al (2011) Crystal structure of the intraflagellar transport complex 25/27. EMBO J 30:1907–1918. doi:10.1038/emboj.2011.110

Bitgood MJ, Shen L, McMahon AP (1996) Sertoli cell signaling by Desert hedgehog regulates the male germline. Curr Biol 6:298–304. doi:10.1016/S0960-9822(02)00480-3

Boehlke C, Bashkurov M, Buescher A et al (2010) Differential role of Rab proteins in ciliary trafficking: Rab23 regulates smoothened levels. J Cell Sci 123:1460–1467. doi:10.1242/jcs.058883

Bonnafe E, Touka M, AitLounis A et al (2004) The transcription factor RFX3 directs nodal cilium development and left-right asymmetry specification. Mol Cell Biol 24:4417–4427. doi:10.1128/MCB.24.10.4417-4427.2004

Briscoe J (2000) A homeodomain protein code specifies progenitor cell identity and neuronal fate in the ventral neural tube. Cell 101:435–445. doi:10.1016/S0092-8674(00)80853-3

Briscoe J, Ericson J (2001) Specification of neuronal fates in the ventral neural tube. Curr Opin Neurobiol 11:43–49. doi:10.1016/S0959-4388(00)00172-0

Bumcrot DA, Takada R, McMahon AP (1995) Proteolytic processing yields two secreted forms of sonic hedgehog. Mol Cell Biol 15:2294–2303

Caspary T, Anderson KV (2003) Patterning cell types in the dorsal spinal cord: what the mouse mutants say. Nat Rev Neurosci 4:289–297. doi:10.1038/nrn1073

Caspary T, García-García MJ, Huangfu D et al (2002) Mouse dispatched homolog1 is required for long-range, but not juxtacrine, Hh signaling. Curr Biol 12:1628–1632. doi:10.1016/S0960-9822(02)01147-8

Caspary T, Larkins CE, Anderson KV (2007) The graded response to Sonic Hedgehog depends on cilia architecture. Dev Cell 12:767–778. doi:10.1016/j.devcel.2007.03.004

Chamberlain CE, Jeong J, Guo C et al (2008) Notochord-derived Shh concentrates in close association with the apically positioned basal body in neural target cells and forms a dynamic gradient during neural patterning. Development 135:1097–1106. doi:10.1242/dev.013086

Chen M-H, Li Y-J, Kawakami T et al (2004) Palmitoylation is required for the production of a soluble multimeric Hedgehog protein complex and long-range signaling in vertebrates. Genes Dev 18:641–659. doi:10.1101/gad.1185804

Chen M-H, Wilson CW, Li Y-J et al (2009) Cilium-independent regulation of Gli protein function by Sufu in Hedgehog signaling is evolutionarily conserved. Genes Dev 23:1910–1928. doi:10.1101/gad.1794109

Chen Y, Yue S, Xie L et al (2011) Dual phosphorylation of suppressor of fused (Sufu) by PKA and GSK3beta regulates its stability and localization in the primary cilium. J Biol Chem 286:13502–13511. doi:10.1074/jbc.M110.217604

Cheung HO-L, Zhang X, Ribeiro A et al (2009) The kinesin protein Kif7 is a critical regulator of Gli transcription factors in mammalian hedgehog signaling. Sci Signal 2:ra29. doi:10.1126/scisignal.2000405

Chi S, Xie G, Liu H et al (2012) Rab23 negatively regulates Gli1 transcriptional factor in a Su(Fu)-dependent manner. Cell Signal 24:1222–1228. doi:10.1016/j.cellsig.2012.02.004

Chiang C, Litingtung Y, Lee E et al (1996) Cyclopia and defective axial patterning in mice lacking Sonic hedgehog gene function. Nature 383:407–413. doi:10.1038/383407a0

Chizhikov VV, Millen KJ (2005) Roof plate-dependent patterning of the vertebrate dorsal central nervous system. Dev Biol 277:287–295. doi:10.1016/j.ydbio.2004.10.011

Chung M-I, Peyrot SM, LeBoeuf S et al (2012) RFX2 is broadly required for ciliogenesis during vertebrate development. Dev Biol 363:155–165. doi:10.1016/j.ydbio.2011.12.029

Colas JF, Schoenwolf GC (2001) Towards a cellular and molecular understanding of neurulation. Dev Dyn 221:117–145. doi:10.1002/dvdy.1144

Concordet JP, Lewis KE, Moore JW et al (1996) Spatial regulation of a Zebrafish patched homologue reflects the roles of sonic hedgehog and protein kinase A in neural tube and somite patterning. Development 122:2835–2846

Copp AJ, Greene NDE, Murdoch JN (2003) The genetic basis of mammalian neurulation. Nat Rev Genet 4:784–793. doi:10.1038/nrg1181

Corbit KC, Aanstad P, Singla V et al (2005) Vertebrate smoothened functions at the primary cilium. Nature 437:1018–1021. doi:10.1038/nature04117

Denef N, Neubuser D, Perez L, Cohen SM (2000) Hedgehog induces opposite changes in turnover and subcellular localization of patched and smoothened. Cell 102:521–531. doi:10.1016/S0092-8674(00)00056-8

Dessaud E, Yang LL, Hill K et al (2007) Interpretation of the sonic hedgehog morphogen gradient by a temporal adaptation mechanism. Nature 450:717–720. doi:10.1038/nature06347

Dessaud E, McMahon AP, Briscoe J (2008) Pattern formation in the vertebrate neural tube: a sonic hedgehog morphogen-regulated transcriptional network. Development 135:2489–2503. doi:10.1242/dev.009324

Ding Q, Motoyama J, Gasca S et al (1998) Diminished Sonic hedgehog signaling and lack of floor plate differentiation in Gli2 mutant mice. Development 125:2533–2543

Ding Q, Fukami S, Meng X et al (1999) Mouse suppressor of fused is a negative regulator of sonic hedgehog signaling and alters the subcellular distribution of Gli1. Curr Biol 9:1119–1122. doi:10.1016/S0960-9822(99)80482-5

Dubruille R (2002) Drosophila regulatory factor X is necessary for ciliated sensory neuron differentiation. Development 129:5487–5498. doi:10.1242/dev.00148

Echelard Y, Epstein D, St. Jacques B et al (1993) Sonic-hedgehog, a member of a family of putative signaling molecules, is implicated in the regulation of CNS polarity. Cell 75:1417–1430. doi:10.1016/0092-8674(93)90627-3

Eggenschwiler JT, Anderson KV (2000) Dorsal and lateral fates in the mouse neural tube require the cell-autonomous activity of the open brain gene. Dev Biol 227:648–660. doi:10.1006/dbio.2000.9918

Eggenschwiler JT, Anderson KV (2007) Cilia and developmental signaling. Annu Rev Cell Dev Biol 23:345–373. doi:10.1146/annurev.cellbio.23.090506.123249

Eggenschwiler JT, Espinoza E, Anderson KV (2001) Rab23 is an essential negative regulator of the mouse Sonic hedgehog signalling pathway. Nature 412:194–198. doi:10.1038/35084089

Eggenschwiler JT, Bulgakov OV, Qin J et al (2006) Mouse Rab23 regulates hedgehog signaling from smoothened to Gli proteins. Dev Biol 290:1–12. doi:10.1016/j.ydbio.2005.09.022

El Zein L, Ait-Lounis A, Morlé L et al (2009) RFX3 governs growth and beating efficiency of motile cilia in mouse and controls the expression of genes involved in human ciliopathies. J Cell Sci 122:3180–3189. doi:10.1242/jcs.048348

Endoh-Yamagami S, Evangelista M, Wilson D et al (2009) The mammalian Cos2 homolog Kif7 plays an essential role in modulating Hh signal transduction during development. Curr Biol 19:1320–1326. doi:10.1016/j.cub.2009.06.046

Epstein DJ, Marti E, Scott MP, McMahon AP (1996) Antagonizing cAMP-dependent protein kinase A in the dorsal CNS activates a conserved Sonic hedgehog signaling pathway. Development 122:2885–2894

Ericson J, Briscoe J, Rashbass P et al (1997) Graded sonic hedgehog signaling and the specification of cell fate in the ventral neural tube. Cold Spring Harb Symp Quant Biol 62:451–466

Ferrante MI, Zullo A, Barra A et al (2006) Oral-facial-digital type I protein is required for primary cilia formation and left-right axis specification. Nat Genet 38:112–117. doi:10.1038/ng1684

Goodrich LV, Milenkovic L, Higgins KM, Scott MP (1997) Altered neural cell fates and medulloblastoma in mouse patched mutants. Science 277:1109–1113. doi:10.1126/science.277.5329.1109

Götz K, Briscoe J, Rüther U (2005) Homozygous Ft embryos are affected in floor plate maintenance and ventral neural tube patterning. Dev Dyn 233:623–630. doi:10.1002/dvdy.20354

Grosshans BL, Ortiz D, Novick P (2006) Rabs and their effectors: achieving specificity in membrane traffic. Proc Natl Acad Sci USA 103:11821–11827. doi:10.1073/pnas.0601617103

Hammerschmidt M, Bitgood MJ, McMahon AP (1996) Protein kinase A is a common negative regulator of Hedgehog signaling in the vertebrate embryo. Genes Dev 10:647–658. doi:10.1101/gad.10.6.647

Hatten ME, Roussel MF (2011) Development and cancer of the cerebellum. Trends Neurosci 34:134–142. doi:10.1016/j.tins.2011.01.002
Haycraft CJ, Banizs B, Aydin-Son Y et al (2005) Gli2 and Gli3 localize to cilia and require the intraflagellar transport protein polaris for processing and function. PLoS Genet 1:e53. doi:10.1371/journal.pgen.0010053
Huang Y, Roelink H, McKnight GS (2002) Protein kinase A deficiency causes axially localized neural tube defects in mice. J Biol Chem 277:19889–19896. doi:10.1074/jbc.M111412200
Huangfu D, Anderson KV (2005) Cilia and Hedgehog responsiveness in the mouse. Proc Natl Acad Sci USA 102:11325–11330. doi:10.1073/pnas.0505328102
Huangfu D, Anderson KV (2006) Signaling from Smo to Ci/Gli: conservation and divergence of Hedgehog pathways from Drosophila to vertebrates. Development 133:3–14. doi:10.1242/dev.02169
Huangfu D, Liu A, Rakeman AS et al (2003) Hedgehog signalling in the mouse requires intraflagellar transport proteins. Nature 426:83–87. doi:10.1038/nature02061
Humke EW, Dorn KV, Milenkovic L et al (2010) The output of Hedgehog signaling is controlled by the dynamic association between suppressor of fused and the Gli proteins. Genes Dev 24:670–682. doi:10.1101/gad.1902910
Ingham PW, Nakano Y, Seger C (2011) Mechanisms and functions of Hedgehog signalling across the metazoa. Nat Rev Genet 12:393–406. doi:10.1038/nrg2984
Jessell TM (2000) Neuronal specification in the spinal cord: inductive signals and transcriptional codes. Nat Rev Genet 1:20–29. doi:10.1038/35049541
Kawakami T, Kawcak T, Li Y-J et al (2002) Mouse dispatched mutants fail to distribute hedgehog proteins and are defective in hedgehog signaling. Development 129:5753–5765
Keady BT, Samtani R, Tobita K et al (2012) IFT25 Links the signal-dependent movement of Hedgehog components to intraflagellar transport. Dev Cell 22:940–951. doi:10.1016/j.devcel.2012.04.009
Kim J, Kato M, Beachy PA (2009) Gli2 trafficking links Hedgehog-dependent activation of smoothened in the primary cilium to transcriptional activation in the nucleus. Proc Natl Acad Sci USA 106:21666–21671. doi:10.1073/pnas.0912180106
Kogerman P, Grimm T, Kogerman L et al (1999) Mammalian suppressor-of-fused modulates nuclear-cytoplasmic shuttling of GLI-1. Nat Cell Biol 1:312–319. doi:10.1038/13031
Kovacs JJ, Whalen EJ, Liu R et al (2008) Beta-arrestin-mediated localization of smoothened to the primary cilium. Science 320:1777–1781. doi:10.1126/science.1157983
Larkins CE, Aviles GDG, East MP et al (2011) Arl13b regulates ciliogenesis and the dynamic localization of Shh signaling proteins. Mol Biol Cell 22:4694–4703. doi:10.1091/mbc.E10-12-0994
Lee KJ, Jessell TM (1999) The specification of dorsal cell fates in the vertebrate central nervous system. Annu Rev Neurosci 22:261–294. doi:10.1146/annurev.neuro.22.1.261
Lei Q, Zelman AK, Kuang E et al (2004) Transduction of graded Hedgehog signaling by a combination of Gli2 and Gli3 activator functions in the developing spinal cord. Development 131:3593–3604. doi:10.1242/dev.01230
Liem KF, He M, Ocbina PJR, Anderson KV (2009) Mouse Kif7/Costal2 is a cilia-associated protein that regulates Sonic hedgehog signaling. Proc Natl Acad Sci USA 106:13377–13382. doi:10.1073/pnas.0906944106
Liem KF, Ashe A, He M et al (2012) The IFT-A complex regulates Shh signaling through cilia structure and membrane protein trafficking. J Cell Biol 197:789–800. doi:10.1083/jcb.201110049
Litingtung Y, Chiang C (2000) Specification of ventral neuron types is mediated by an antagonistic interaction between Shh and Gli3. Nat Neurosci 3:979–985. doi:10.1038/79916
Liu A, Niswander LA (2005) Bone morphogenetic protein signalling and vertebrate nervous system development. Nat Rev Neurosci 6:945–954. doi:10.1038/nrn1805
Liu A, Wang B, Niswander LA (2005) Mouse intraflagellar transport proteins regulate both the activator and repressor functions of Gli transcription factors. Development 132:3103–3111. doi:10.1242/dev.01894
Lupo G, Harris WA, Lewis KE (2006) Mechanisms of ventral patterning in the vertebrate nervous system. Nat Rev Neurosci 7:103–114. doi:10.1038/nrn1843

Ma Y, Erkner A, Gong R et al (2002) Hedgehog-mediated patterning of the mammalian embryo requires transporter-like function of dispatched. Cell 111:63–75. doi:10.1016/S0092-8674(02)00977-7
Mackie EJ, Tatarczuch L, Mirams M (2011) The skeleton: a multi-functional complex organ: the growth plate chondrocyte and endochondral ossification. J Endocrinol 211:109–121. doi:10.1530/JOE-11-0048
Marigo V, Davey RA, Zuo Y et al (1996) Biochemical evidence that patched is the Hedgehog receptor. Nature 384:176–179. doi:10.1038/384176a0
Martí E, Takada R, Bumcrot DA et al (1995) Distribution of Sonic hedgehog peptides in the developing chick and mouse embryo. Development 121:2537–2547
Matise MP, Epstein DJ, Park HL et al (1998) Gli2 is required for induction of floor plate and adjacent cells, but not most ventral neurons in the mouse central nervous system. Development 125:2759–2770
May SR, Ashique AM, Karlen M et al (2005) Loss of the retrograde motor for IFT disrupts localization of Smo to cilia and prevents the expression of both activator and repressor functions of Gli. Dev Biol 287:378–389. doi:10.1016/j.ydbio.2005.08.050
Milenkovic L, Scott MP, Rohatgi R (2009) Lateral transport of smoothened from the plasma membrane to the membrane of the cilium. J Cell Biol 187:365–374. doi:10.1083/jcb.200907126
Motoyama J (2003) Differential requirement for Gli2 and Gli3 in ventral neural cell fate specification. Dev Biol 259:150–161. doi:10.1016/S0012-1606(03)00159-3
Motoyama J, Milenkovic L, Iwama M et al (2003) Differential requirement for Gli2 and Gli3 in ventral neural cell fate specification. Dev Biol 259:150–161. doi:10.1016/S0012-1606(03)00159-3
Mukhopadhyay S, Jackson PK (2011) The tubby family proteins. Genome Biol 12:225. doi:10.1186/gb-2011-12-6-225
Mukhopadhyay S, Wen X, Chih B et al (2010) TULP3 bridges the IFT-A complex and membrane phosphoinositides to promote trafficking of G protein-coupled receptors into primary cilia. Genes Dev 24:2180–2193. doi:10.1101/gad.1966210
Nachtergaele S, Mydock LK, Krishnan K et al (2012) Oxysterols are allosteric activators of the oncoprotein smoothened. Nat Chem Biol 8:211–220. doi:10.1038/nchembio.765
Norman RX, Ko HW, Huang V et al (2009) Tubby-like protein 3 (TULP3) regulates patterning in the mouse embryo through inhibition of Hedgehog signaling. Hum Mol Genet 18:1740–1754. doi:10.1093/hmg/ddp113
Ocbina PJR, Anderson KV (2008) Intraflagellar transport, cilia, and mammalian Hedgehog signaling: analysis in mouse embryonic fibroblasts. Dev Dyn 237:2030–2038. doi:10.1002/dvdy.21551
Ocbina PJR, Eggenschwiler JT, Moskowitz I, Anderson KV (2011) Complex interactions between genes controlling trafficking in primary cilia. Nat Genet 43:547–553. doi:10.1038/ng.832
Pan Y, Wang C, Wang B (2009) Phosphorylation of Gli2 by protein kinase A is required for Gli2 processing and degradation and the Sonic Hedgehog-regulated mouse development. Dev Biol 326:177–189. doi:10.1016/j.ydbio.2008.11.009
Papalopulu N, Kintner C (1996) A posteriorising factor, retinoic acid, reveals that anteroposterior patterning controls the timing of neuronal differentiation in Xenopus neuroectoderm. Development 122:3409–3418
Park HL, Bai C, Platt KA et al (2000) Mouse Gli1 mutants are viable but have defects in SHH signaling in combination with a Gli2 mutation. Development 127:1593–1605
Patterson VL, Damrau C, Paudyal A et al (2009) Mouse hitchhiker mutants have spina bifida, dorso-ventral patterning defects and polydactyly: identification of Tulp3 as a novel negative regulator of the Sonic hedgehog pathway. Hum Mol Genet 18:1719–1739. doi:10.1093/hmg/ddp075
Pedersen LB, Rosenbaum JL (2008) Intraflagellar transport (IFT) role in ciliary assembly, resorption and signalling. Curr Top Dev Biol 85:23–61. doi:10.1016/S0070-2153(08)00802-8
Pepinsky RB, Zeng C, Wen D et al (1998) Identification of a palmitic acid-modified form of human Sonic hedgehog. J Biol Chem 273:14037–14045. doi:10.1074/jbc.273.22.14037
Persson M, Stamataki D, te Welscher P et al (2002) Dorsal-ventral patterning of the spinal cord requires Gli3 transcriptional repressor activity. Genes Dev 16:2865–2878. doi:10.1101/gad.243402

Peters T, Ausmeier K, Dildrop R, Rüther U (2002) The mouse fused toes (Ft) mutation is the result of a 1.6-Mb deletion including the entire Iroquois B gene cluster. Mamm Genome 13:186–188. doi:10.1007/s00335-001-2142-7

Porter JA, Young KE, Beachy PA (1996) Cholesterol modification of hedgehog signaling proteins in animal development. Science 274:255–259. doi:10.1126/science.274.5285.255

Qin H, Wang Z, Diener D, Rosenbaum J (2007) Intraflagellar transport protein 27 is a small G protein involved in cell-cycle control. Curr Biol 17:193–202. doi:10.1016/j.cub.2006.12.040

Qin J, Lin Y, Norman RX et al (2011) Intraflagellar transport protein 122 antagonizes Sonic Hedgehog signaling and controls ciliary localization of pathway components. Proc Natl Acad Sci USA 108:1456–1461. doi:10.1073/pnas.1011410108

Reiter JF, Skarnes WC (2006) Tectonic, a novel regulator of the Hedgehog pathway required for both activation and inhibition. Genes Dev 20:22–27. doi:10.1101/gad.1363606

Roelink H, Porter JA, Chiang C et al (1995) Floor plate and motor neuron induction by different concentrations of the amino-terminal cleavage product of sonic hedgehog autoproteolysis. Cell 81:445–455. doi:10.1016/0092-8674(95)90397-6

Rohatgi R, Milenkovic L, Scott MP (2007) Patched1 regulates hedgehog signaling at the primary cilium. Science 317:372–376. doi:10.1126/science.1139740

Rosenbaum JL, Witman GB (2002) Intraflagellar transport. Nat Rev Mol Cell Biol 3:813–825. doi:10.1038/nrm952

Sánchez-Camacho C, Bovolenta P (2009) Emerging mechanisms in morphogen-mediated axon guidance. Bioessays 31:1013–1025. doi:10.1002/bies.200900063

Sasaki H, Nishizaki Y, Hui C et al (1999) Regulation of Gli2 and Gli3 activities by an amino-terminal repression domain: implication of Gli2 and Gli3 as primary mediators of Shh signaling. Development 126:3915–3924

Seo S, Zhang Q, Bugge K et al (2011) A novel protein LZTFL1 regulates ciliary trafficking of the BBSome and smoothened. PLoS Genet 7:e1002358. doi:10.1371/journal.pgen.1002358

Shaheen R, Faqeih E, Seidahmed MZ et al (2011) A TCTN2 mutation defines a novel Meckel Gruber syndrome locus. Hum Mutat 32:573–578. doi:10.1002/humu.21507

Singla V, Romaguera-Ros M, Garcia-Verdugo JM, Reiter JF (2010) Ofd1, a human disease gene, regulates the length and distal structure of centrioles. Dev Cell 18:410–424. doi:10.1016/j.devcel.2009.12.022

Stone DM, Hynes M, Armanini M et al (1996) The tumour-suppressor gene patched encodes a candidate receptor for Sonic hedgehog. Nature 384:129–134. doi:10.1038/384129a0

Stottmann RW, Tran PV, Turbe-Doan A, Beier DR (2009) Ttc21b is required to restrict sonic hedgehog activity in the developing mouse forebrain. Dev Biol 335:166–178. doi:10.1016/j.ydbio.2009.08.023

Svärd J, Heby-Henricson K, Henricson KH et al (2006) Genetic elimination of suppressor of fused reveals an essential repressor function in the mammalian Hedgehog signaling pathway. Dev Cell 10:187–197. doi:10.1016/j.devcel.2005.12.013

Swoboda P, Adler HT, Thomas JH (2000) The RFX-type transcription factor DAF-19 regulates sensory neuron cilium formation in C. elegans. Mol Cell 5:411–421. doi:10.1016/S1097-2765(00)80436-0

Taipale J, Cooper MK, Maiti T, Beachy PA (2002) Patched acts catalytically to suppress the activity of smoothened. Nature 418:892–897. doi:10.1038/nature00989

Tian H, Jeong J, Harfe BD et al (2005) Mouse Disp1 is required in sonic hedgehog-expressing cells for paracrine activity of the cholesterol-modified ligand. Development 132:133–142. doi:10.1242/dev.01563

Tran PV, Haycraft CJ, Besschetnova TY et al (2008) THM1 negatively modulates mouse sonic hedgehog signal transduction and affects retrograde intraflagellar transport in cilia. Nat Genet 40:403–410. doi:10.1038/ng.105

Tukachinsky H, Lopez LV, Salic A (2010) A mechanism for vertebrate Hedgehog signaling: recruitment to cilia and dissociation of SuFu-Gli protein complexes. J Cell Biol 191:415–428. doi:10.1083/jcb.201004108

Tuson M, He M, Anderson KV (2011) Protein kinase A acts at the basal body of the primary cilium to prevent Gli2 activation and ventralization of the mouse neural tube. Development 138:4921–4930. doi:10.1242/dev.070805 van den Heuvel M, Ingham P (1996) Smoothened encodes a serpentine protein required for hedgehog signalling. Nature 382:547–551

Varjosalo M, Li S-P, Taipale J (2006) Divergence of hedgehog signal transduction mechanism between Drosophila and mammals. Dev Cell 10:177–186. doi:10.1016/j.devcel.2005.12.014

Vierkotten J, Dildrop R, Peters T et al (2007) Ftm is a novel basal body protein of cilia involved in Shh signalling. Development 134:2569–2577. doi:10.1242/dev.003715

Wang B, Fallon JF, Beachy PA (2000) Hedgehog-regulated processing of Gli3 produces an anterior/posterior repressor gradient in the developing vertebrate limb. Cell 100:423–434. doi:10.1016/S0092-8674(00)80678-9

Wang Y, Zhou Z, Walsh CT, McMahon AP (2009) Selective translocation of intracellular smoothened to the primary cilium in response to Hedgehog pathway modulation. Proc Natl Acad Sci USA 106:2623–2628. doi:10.1073/pnas.0812110106

Wang C, Pan Y, Wang B (2010) Suppressor of fused and Spop regulate the stability, processing and function of Gli2 and Gli3 full-length activators but not their repressors. Development 137:2001–2009. doi:10.1242/dev.052126

Weatherbee SD, Niswander LA, Anderson KV (2009) A mouse model for Meckel syndrome reveals Mks1 is required for ciliogenesis and Hedgehog signaling. Hum Mol Genet 18:4565–4575. doi:10.1093/hmg/ddp422

Wen X, Lai CK, Evangelista M et al (2010) Kinetics of hedgehog-dependent full-length Gli3 accumulation in primary cilia and subsequent degradation. Mol Cell Biol 30:1910–1922. doi:10.1128/MCB.01089-09

Whitfield JF (2008) The solitary (primary) cilium–a mechanosensory toggle switch in bone and cartilage cells. Cell Signal 20:1019–1024. doi:10.1016/j.cellsig.2007.12.001

Wijgerde M, McMahon JA, Rule M, McMahon AP (2002) A direct requirement for Hedgehog signaling for normal specification of all ventral progenitor domains in the presumptive mammalian spinal cord. Genes Dev 16:2849–2864. doi:10.1101/gad.1025702

Wong SY, Reiter JF (2008) The primary cilium: at the crossroads of mammalian Hedgehog signaling. Curr Top Dev Biol 85:225–260. doi:10.1016/S0070-2153(08)00809-0

Chapter 3
Primary Cilia and Brain Development

Evangelia Tasouri, Marc August Willaredt, and Kerry L. Tucker

Contents

Abstract The primary cilium is a small microtubule-based organelle projecting from the plasma membrane of practically all cells in the mammalian body. In the past 8 years a flurry of papers has indicated a crucial role of this long-neglected structure in the development of a wide variety of organs. We summarize here the function that primary cilia have in the development of the brain, with especial attention paid to the morphogenesis of the embryonic cerebral cortex and cerebellum. Cilia-dependent signaling of Sonic hedgehog to the transcription factor Gli3 is analyzed, and potential roles of Wnt signaling are discussed.

E. Tasouri • M.A. Willaredt • K.L. Tucker (✉)
Interdisciplinary Center for Neurosciences, Institute of Anatomy and Cell Biology,
University of Heidelberg, 69120 Heidelberg, Germany
e-mail: Kerry.Tucker@urz.uni-heidelberg.de

K.L. Tucker and T. Caspary (eds.), *Cilia and Nervous System Development and Function*,
DOI 10.1007/978-94-007-5808-7_3,

Keywords Primary cilia • Cerebral cortex • Forebrain • Cerebellum • Diencephalon • Sonic hedgehog • Gli3 • Telencephalon • Thombenecephalon • Wnt • Joubert syndrome

3.1 Introduction

In the third week of human embryonic development, the central nervous system arises from a uniform sheet of neuroectodermal tissue. Through the action of proteins that move the cells together in a coordinated vertical fashion, the central portion of the neural plate folds downwards to form a neural groove, while the lateral edges fold upwards to ultimately meet and fuse at the dorsal aspect. Thereby the neural tube is formed, starting in the center of the cranial-caudal body axis and proceeding in both directions simultaneously. A special population of cells found directly lateral to the neural plate gives rise to the migratory neural crest, which in turn generates the peripheral nervous system, including sympathetic and parasympathetic ganglia, dorsal root ganglia, Schwann cells, and the adrenal medulla (Le Douarin and Kalcheim 1999). The neural tube initially consists of a single uniform layer of pseudo-stratified neural epithelium that surrounds the ventricular space formed by the fusion of the neural groove edges. The former will give rise in turn to a marvelously complex arrangement of central nervous system organs, whereas the latter will give rise to the fluid-filled ventricular system deep within the adult brain. The caudal portions of the neural tube will give rise to the spinal cord, while the cranial portions will give rise to the brain including its largest components the cerebral cortex and the cerebellum, in terms of volume and cell number, respectively.

Between the 4th and 5th weeks of human gestation, the cranial end of the neural tube expands greatly in size, and initially three compartments can be recognized, with the prosencephalon at the cranial end followed by the mesencephalon and the rhombencephalon. By the 7th gestational week, the prosencephalon goes on to divide into two further subsegments: the telencephalon generates the cerebral cortex and some of the basal ganglia, while the diencephalon will generate the thalamus, epithalamus, subthalamus, and hypothalamus. The rhombencephalon is in its turn also subdivided into two elements, the metencephalon generating the pons and the cerebellum, while the myencephalon generates the medulla oblongata. The purpose of this chapter is to illustrate how the tiny organelle the primary cilium has been demonstrated in just a few years to be essential for the development of many of the structures mentioned above, with a special focus on the development of the cerebral cortex and the cerebellum. Other chapters will detail the role of primary cilia in the development of the spinal cord (Chap. 2 by Tamara Caspary) and the ventricular system (Chap. 8 by Nathalie Spassky). For a detailed description of the ultrastructure of primary cilia, please refer to Chap. 1.

3.2 Primary Cilia and Prosencephalon Development

3.2.1 Primary Cilia and Cerebral Cortex Development

3.2.1.1 Primary Cilia and Shh Signaling in Embryonic Cerebral Cortex

Intraflagellar transport (IFT) is the system in which protein cargo is transported along the axoneme of cilia and is very important not just for the genesis and the maintenance of the organelle but also for signal transduction cascades such as vertebrate Hh (Scholey and Anderson 2006). Anterograde transport is accomplished by the 11 IFT members of the IFTB complex, which form a scaffold to interact between cargo and the transport motor kinesin-II. Retrograde transport is accomplished by the 6 IFT proteins of the IFTA complex in association with dynein. One of the best-studied developmental pathways that are implicated in primary cilia involves the IFT of Hh signaling components. The three vertebrate ligands for Hh signalling, Sonic (Shh), Indian, and Desert, all bind to the specific receptor for Hh-family ligands Patched (Ptch1). In the absence of any ligand, Ptch1 prevents Smoothened (Smo), a G-coupled transmembrane protein, to associate with the membrane, thus repressing signal transduction. In the presence of a ligand, Ptch1 moves out of the cilium, thereby relieving the inhibition of Smo activity (Corbit et al. 2005; Rohatgi et al. 2007). Smo moves into the primary cilium (Corbit et al. 2005) and transduces the Hh signal through its effects upon the transcription factors Gli2 and Gli3, which also localize to primary cilia (Haycraft et al. 2005). Both transcription factors are activated upon association with cilia, while Gli3 is proteolytically converted from a full-length form to a shorter repressor form (Gli3R) in a cilia-dependent fashion.

Through the use of forward and reverse mouse genetics, several mutants in IFT genes have been discovered and engineered, respectively. These mutants have been one of the primary reagents for investigating the role of primary cilia in brain development. The first report to examine in detail the effects that cilia have upon forebrain morphogenesis employed a hypomorphic allele of the IFT complex B *Ift88* gene called *cobblestone* (*cbs*), which had been generated by *N*-ethyl-*N*-nitrosourea mutagenesis followed by positional cloning (Willaredt et al. 2008). In this mutant, the open reading frame of the *Ift88* cDNA remains normal but Ift88 mRNA and protein levels decrease by 70–80%. As seen in many other ciliary mutants (e.g. Gorivodsky et al. 2009), this leads to exencephaly in the homozygous state in a high fraction of embryos (Fig. 3.1), which of course hinders analysis of cranial developmental processes. However, in the *cbs* mutant, exencephaly proved to be specific to the genetic background, and so preservation of the strain on a C57BL/6 background reduced levels of exencephaly to 10% and thus allowed for analysis of forebrain defects, illustrating one of the most common themes to run through phenotypes in ciliary mutants in many different organs, namely both a highly variable penetrance and expressivity in the phenotypes.

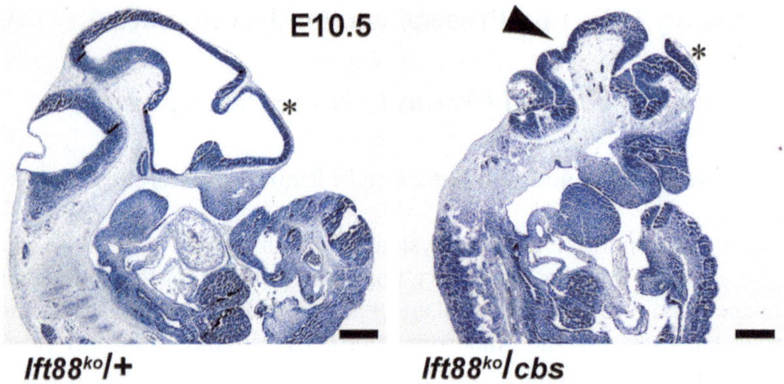

Fig. 3.1 Loss of *Ift88* expression leads to exencephaly. Hematoxylin stain of parasagittal sections of E10.5 embryos from wildtype (+/+) and compound heterozygotes (*Ift88*^{ko}/*cbs*) for the *cbs* allele and a full deletion of the *Ift88* gene. Arrowhead indicates exencephaly in the caudal forebrain and midbrain of the *Ift88* mutant. * Cerebral cortex. Dorsal is to the *left*, cranial to the *top of each panel*. *Scale bar*: 0.3 mm

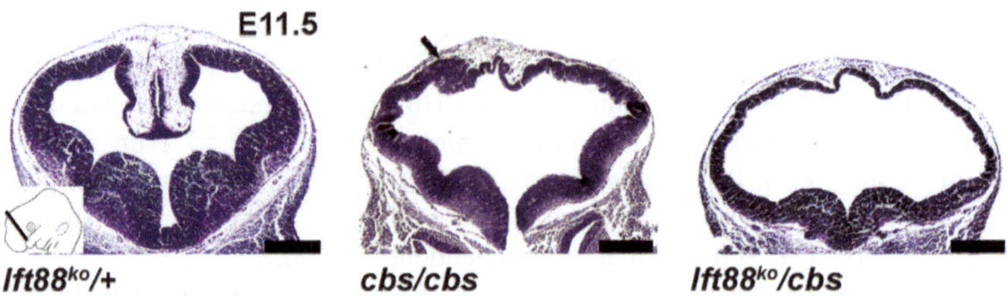

Fig. 3.2 Loss of *Ift88* expression results in a profound disarray in the morphogenesis of the dorsal telencephalon. Hematoxylin-stained coronal sections of a E11.5 *Ift88*^{ko}/+ embryo, a *cbs*/*cbs* embryo, and a *Ift88*^{ko}/*cbs* compound heterozygote littermate. Inset (*lower left*) indicates the plane of section. *Arrow* indicates heterotopias. Dorsal is to the *top in each panel*. *Scale bar*: 0.3 mm (Figure adapted from a publication by Willaredt et al. 2008)

Nonexencephalic *cbs* mutants displayed severe malformations of medial dorsal structures in the telencephalon, including the hippocampus, choroid plexus and cortical hem, in addition to subpial ectopic rosettes of proliferating precursors surrounded by a shell of newborn neurons which resembled small neural tubes in cross section, giving rise to the mutant's name (Fig. 3.2). The dismorphology was even more pronounced upon performance of a complementation analysis, crossing the *cbs* mutant to a mouse line in which exons 4–6 of the *Ift88* gene have been ablated through genetic recombination (Haycraft et al. 2007). In the double heterozygotes, invagination of the dorsal midline barely occurred, leading to a single rostral vesicle, a situation similar to that seen in holoprosencephaly (Fig. 3.2). The pallial-subpallial boundary did not form properly in *cbs* mutants, and examination of the dorsal border between the telencephalon and the caudally-situated diencephalon revealed similar boundary identity problems, in that the caudal telencephalon displayed patches of tissue expressing markers of both telencephalic as well as thalamic tissue (Willaredt et al. 2008).

Most of the above-mentioned phenotypes are comparable to the ones observed in the spontaneous *Gli3* deletion mutant *extra toes-Jackson* (Xt^J) (Fotaki et al. 2006; Theil et al. 1999; Tole et al. 2000), and given the importance of IFT for Shh signaling it proved fruitful to compare the two mutants and examine Shh signaling in the *cbs* mutant. In the spinal cord, the hallmark of defective Shh signaling is a disruption in the dorsal-ventral patterning scheme, such that neurons normally specified in the ventral region, where Shh concentrations are at their highest, are lost at the expense of more dorsally-located subtypes (see Chap. 2 for a thorough discussion), and this ventralization phenotype can be seen in the forebrain, where Shh is also secreted from ventrally-localized structures. Surprisingly, examination of the *cbs* mutant revealed no strong differences in dorsal-ventral patterning, in that markers of the cerebral cortex and the ventrally-located striatum were clearly expressed where they should be (Willaredt et al. 2008). This stands in contrast to the situation seen in other ciliary mutants, where quite strong shifts in pallial-subpallial domains have been observed (Besse et al. 2011; Stottmann et al. 2009, see below). Examination of *Shh* expression revealed no major changes in the *cbs* mutant (Fig. 3.3), which must partially account for the lack of ventralization in the telencephalon, in contrast to the situation in the spinal cord, where ventralization can clearly be observed accompanying the loss of *Shh* expression from the floorplate. In situ and quantitative RT-PCR analysis of Shh signalling targets, such as *Ptch1* and *Gli1*, showed in fact a surprising increase in their expression in the *cbs* mutants (Willaredt et al. 2008). As the effects of Shh on its downstream transcriptional targets are known to work through the Gli family of transcription factors, expression of Gli3 was examined and found at the mRNA level to be normal (Fig. 3.4a, Willaredt et al. 2008), as were the expression of two downstream targets of Gli3 signaling *Emx1* (Fig. 3.4b) and *Emx2* (Fig. 3.4c), whose expression is eliminated and downregulated, respectively, in the Xt^J mutant (Theil et al. 1999; Tole et al. 2000). However, Western blot analysis indicated an increase in the relative levels of the full-length Gli3 isoform. The shift of Gli3 proteolytic processing is taken as a reflection of a breakdown of normal ciliary function, and it is speculated that the recorded phenotypes reflect either a loss of Gli3 repressor isoform function or a dominant negative effect of the increased levels of the full-length Gli3 isoform. Most surprising, when compared to other ciliary mutants, both transmission and scanning electron microscopy showed that primary cilia were still present and morphologically and ultrastructurally normal (Willaredt et al. 2008). Finally, examination of changes in Wnt signaling also revealed an upregulation of canonical Wnt signaling, as reflected in the increased and ectopic expression of *Wnt7b* (Fig. 3.5a), *Wnt8b* (Fig. 3.5b), and *Axin2* (Willaredt et al. 2008), but whether this reflects a direct effect of primary cilia upon Wnt signaling or a secondary disturbance attendant with abnormal Shh signaling remains untested.

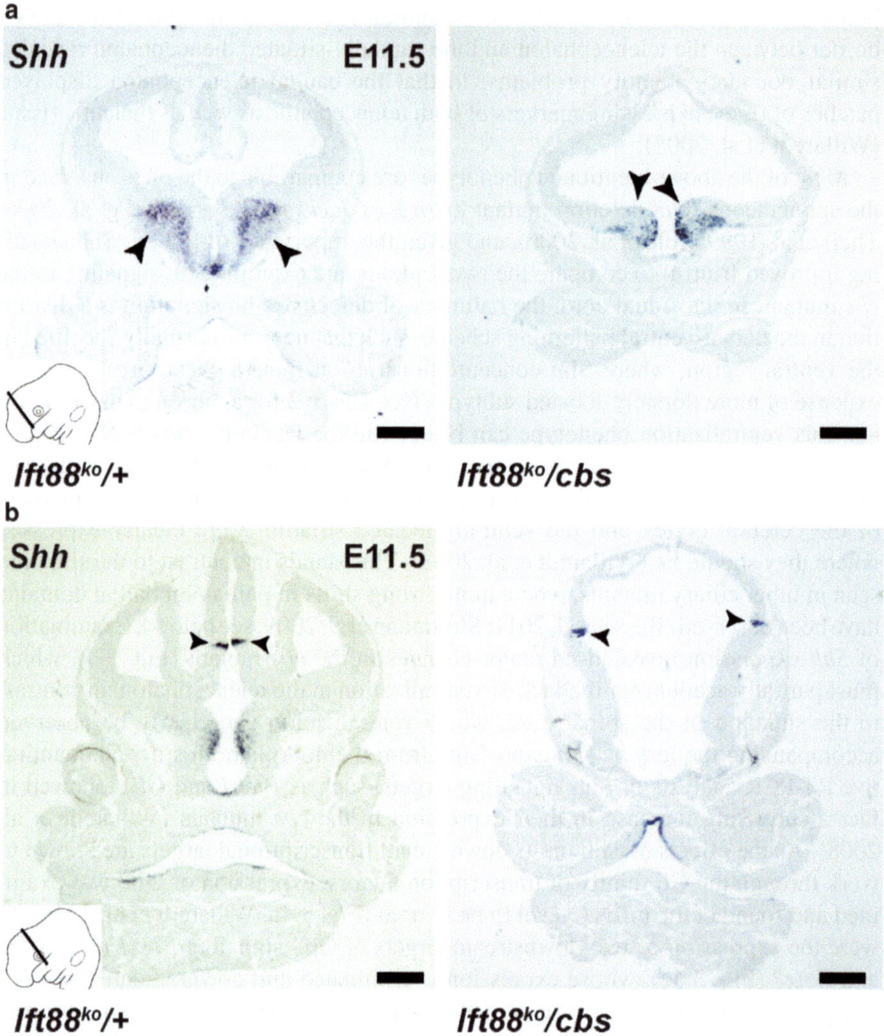

Fig. 3.3 The expression domains of *Shh* in the forebrain of *Ift88* mutants remain essentially intact. In situ hybridization on coronal sections of E11.5 embryos from wildtype (+/+) and compound heterozygotes (*Ift88*^{ko}/*cbs*) for the *cbs* allele and a full deletion of the *Ift88* gene. **a**, **b**, *Shh*. **a**, Rostral sections indicating *Shh* expression in the ventral subpallium (*arrowheads*). **b**, Caudal sections indicating *Shh* expression in the zona limitans intrathalamica (ZLI, *arrowheads*). *Inset* (*lower left*) indicates the plane of section. Dorsal is to the *top in each panel*. *Scale bar*: 0.3 mm

A study from Gorivodsky et al. (2009) examined loss of another gene encoding a ciliary protein, *Ift172*, which resulted in severe cranio-facial malformations, failure to close the cranial neural tube, and holoprosencephaly. Their work showed that in these *Ift172* mutants the expression of *nodal* is practically abolished by embryonic

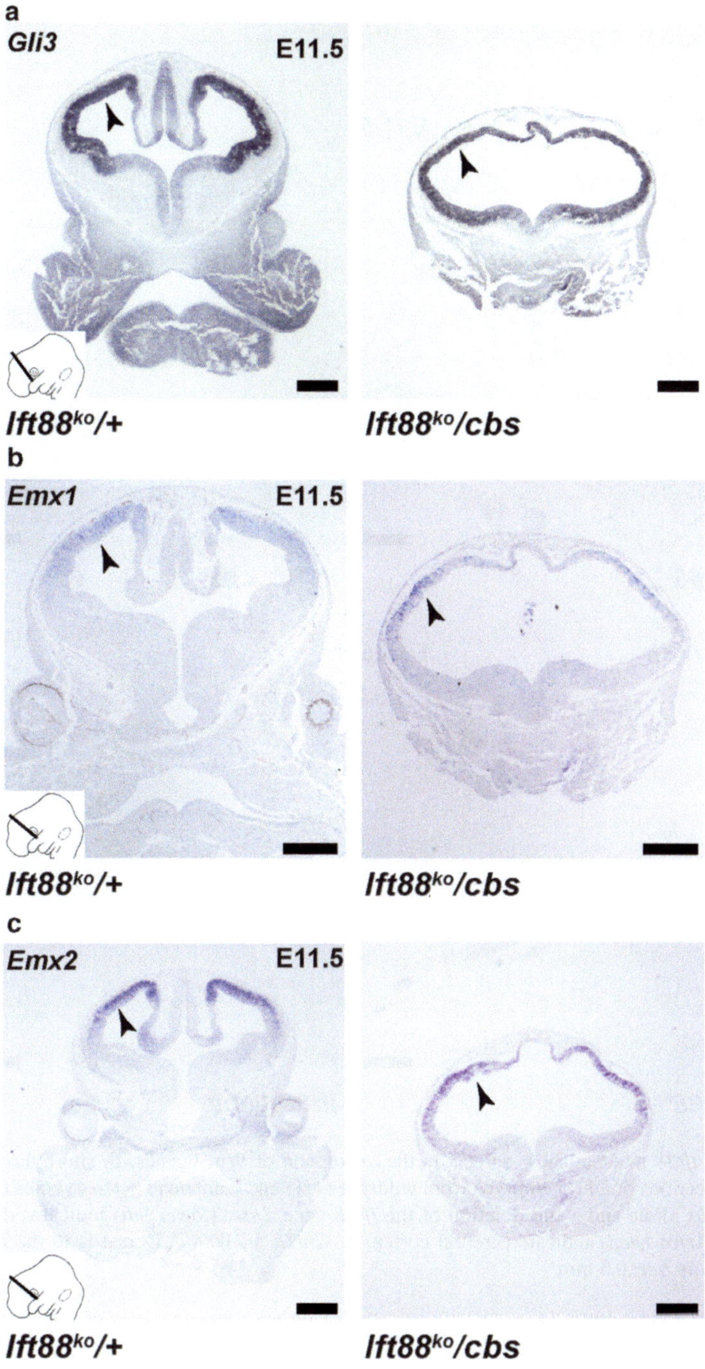

Fig. 3.4 *Ift88* mutants show no defects in the expression of *Gli3* or its downstream targets *Emx1* and *Emx2*. In situ hybridization on coronal sections of E11.5 embryos from wildtype (+/+) and compound heterozygotes (*Ift88*^{ko}/*cbs*) for the *cbs* allele and a full deletion of the *Ift88* gene. *Inset* (*lower left*) indicates the plane of section. *Arrowheads* indicate cerebral cortex. **a**, *Gli3*. **b**, *Emx1*. **c**, *Emx2*. Dorsal is to the *top in each panel*. *Scale bar*: 0.3 mm

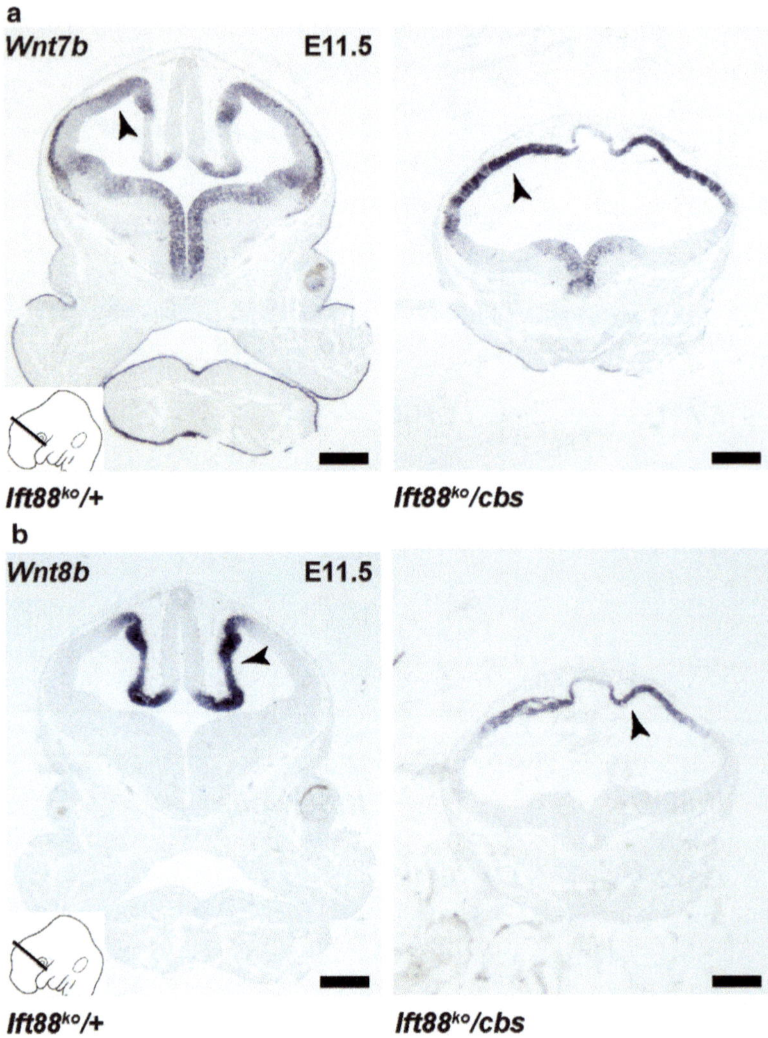

Fig. 3.5 *Ift88* mutants show defects in the expression of Wnt ligands. In situ hybridization on coronal sections of E11.5 embryos from wildtype (+/+) and compound heterozygotes (*Ift88*^{ko}/*cbs*) for the *cbs* allele and a full deletion of the *Ift88* gene. *Inset* (*lower left*) indicates the plane of section. *Arrowheads* indicate cerebral cortex. **a**, *Wnt7b*. **b**, *Wnt8b*. Dorsal is to the *top in each panel*. *Scale bar*: 0.3 mm

day (E) 7.5 (Gorivodsky et al. 2009). This could partially explain the forebrain phenotypes seen in other ciliary mutants, as nodal is fundamental for the induction of the anterior mesendoderm that later expresses *Shh*, *Foxa2*, and *Gsc* (Vincent et al. 2003). Both the *Ift88* and *Ift172* mutations occurred in genes encoding IFTB complex proteins that are associated with anterograde IFT, and the simplest interpretation of their phenotypes is to conclude that in these mutants both the full-length and short isoforms of Gli3 are not being produced and activated in a normal fashion, thereby leading to a decrease in Shh signaling.

Stottmann et al. examined mice mutant for a ciliary gene, tetratricopeptide repeat domain 21B (*Ttc21b*), which encodes the ciliary protein IFT139 involved in retrograde IFT (Stottmann et al. 2009; Tran et al. 2008). Its absence led to a defect in retrograde IFT, resulting in the accumulation of IFT proteins (Tran et al. 2008). In contrast to ciliary mutants in other studies where Shh signalling is decreased (Besse et al. 2011; Gorivodsky et al. 2009; Wilson et al. 2011), *Ttcb21* mutants showed an elevation of Shh signalling since the anterograde transport is intact and components of the pathway, like Smo, can enter the cilium. In the *Ttcb21* mutants, elevated Shh signalling was also responsible for the dramatic ventralization observed, and naturally downstream targets such as *Ptch1* and *Foxa2* also showed increased expression (Stottmann et al. 2009). Strong evidence for an increase in Shh signaling in this mutant comes from the observation that when Shh expression was genetically reduced, the *Ttcb21* phenotype was altered, suggesting that Ttcb21 is responsible for inhibiting Shh expression in the dorsal mesendoderm and midbrain, ensuring appropriate patterning of midbrain and forebrain (Stottmann et al. 2009). Analysis of Wnt signalling in the *Ttcb21* mutants revealed a downregulation of a Wnt-sensitive lacZ transgene in the forebrain although no major changes in levels of beta-catenin were detected.

Besse et al. (2011) examined a mouse mutant with a targeted deletion of the *Ftm* gene, whose human ortholog RPGRIP1L has been found mutated in a subset of patients with the ciliopathies Meckel or Joubert syndrome (Arts et al. 2007; Delous et al. 2007). In *Ftm* mutants the protein was not localized to the base of primary cilia at the apical side of E12.5 telencephalic neuroepithelia, as is the case in the wildtype. As might be expected, the protein was instead absent in the mutants. At the ventricular surface, *Ftm* mutant forebrain exhibited short, small, button-shaped cilium-like structures with basal bodies of normal appearance. Interestingly, ventricular tight junctions seemed unaffected, which speaks for a normal apical-basal polarity of the neuroepithelial cells. Examination of patterning markers in the telencephalon of these mutants revealed that the dorsal expression of *Ngn2* and *Pax6* was severely decreased in the anterior telencephalon, whereas the expression of *Dlx2* and *Gsx 2* was enlarged dorsally. Interestingly, the dorsal expansion of ventral structures was limited to the most anterior part of the telencephalon. As seen in many other ciliary mutants, the ratio of full-length Gli3 to the processed isoform Gli3R was increased in the *Ftm* mutant. This last result was in accordance with a decrease in Gli3 activity, because the extent of the dorsomedial structures were diminished and deformed, and matches that seen in the forebrain of the *cobblestone* mutant (Willaredt et al. 2008) and in many other ciliary mutants in other developing organs (Haycraft et al. 2005; Huangfu and Anderson 2005; Liu et al. 2005; May et al. 2005; Tran et al. 2008). Interestingly, the phenotype did not seem to be explained by increased Shh signalling. The expression pattern of the three target genes *Shh*, *Gli1* and *Nkx2.1* of the Hh signalling pathway were more or less unaffected in the ventral telencephalon of E9.5 to E12.5 *Ftm* mutants. Strikingly, the authors were able to rescue the ventralization of the telencephalon by crossing the *Ftm* mutants to mice expressing an engineered mutant allele of *Gli3* that specifically produces a truncated form of Gli3 that should act as a constitutive Gli3R isoform (Böse et al. 2002). Even more impressively, the double mutants did not display primary cilia at the ventricle, demonstrating that the *Ftm* phenotype was caused by defective Gli3R activity in the cortex and that ciliary defects can be rescued by expression of the appropriate downstream signaling components (Besse et al. 2011).

To summarize these findings, all of these initial studies clearly showed a need for primary cilia in the Shh-determined dorsal-ventral identity of embryonic telencephalic structures, but interestingly the identity of the ciliary gene that was mutated affected the nature of the dorsal-ventral change observed in the mutation. In mutations in IFTB complex proteins, the observed effects could be explained by a lack of both full-length and the short isoform of Gli3 (Gorivodsky et al. 2009; Willaredt et al. 2008). In the mutation in *Ttc21b*, a component of the IFTA complex, Shh signaling was found to be increased and thus a ventralization of the forebrain was observed (Stottmann et al. 2009). Finally, in a mutation in the gene encoding the basal body protein Ftm, the effects could be explained by a loss of Gli3R activity.

3.2.1.2 Primary Cilia and the Control of Proliferation in the Cerebral Cortex

All of the aforementioned mouse mutants suffer from the fact that they are constitutive, and thus the relative contribution that primary cilia may make to earlier and later stages of telencephalic development cannot be discerned. The only study to address this potential problem utilized an inducible mutation of the *Kif3a* gene (Marszalek et al. 1999), a kinesin protein responsible for anterograde IFT in cilia. The authors employed a nestin::Cre line that is expressed in the central nervous system from E10.5 (Petersen et al. 2002) to specifically eliminate *Kif3a* from developing telencephalon at the beginning of cortical neurogenesis. Already at E12.5, primary cilia were reported to be lost from the ventricular zone of the cortex (Wilson et al. 2011). Mutant mice survived until birth and displayed markedly larger cerebral cortices and ectopic rosettes similar to those seen in *Gli3* (Fotaki et al. 2006; Theil et al. 1999) and *cobblestone* (Willaredt et al. 2008) mutants. The morphology, apical-basal polarity, and junctional contacts of neuroepithelial cells were unaffected in *Kif3a* mutants. The increase of cortical volume in the mutants could not be explained by a decrease in cell death, which was analyzed by TUNEL staining on E13.5 tissue. *Kif3a* mutants actually exhibited nearly twice the number of apoptotic cells as heterozygous littermates. In addition, the rate of neurogenesis was actually shown to increase in *Kif3a* mutants, which would also argue against a larger cortical volume. Careful analysis of the cell cycle indicated that precursors were speeding up in the G1 phase, resulting in a shorter cell cycle that could lead to the overproduction of cortical cells seen at birth.

To explain these changes the authors examined several pathways known to influence proliferation in the cortex and discovered that Gli3 processing is specifically altered to favor production of the full-length Gli3 isoform, as have countless other studies (e.g. Haycraft et al. 2005; Huangfu and Anderson 2005; Liu et al. 2005; May et al. 2005; Tran et al. 2008; Willaredt et al. 2008). Several downstream targets of Shh signaling, including *Ptch1*, *Fgf15*, and *cyclinD1*, were also shown to be upregulated. By examination of a *Gli3* mutant (Wang et al. 2007) lacking a 68-amino acid domain containing the proteolytic processing site to convert Gli3 to Gli3R, the authors could observe changes in cell cycle dynamics similar to that seen in the *Kif3a* mutants. The results suggest that the ratio of Gli3 to Gli3R, which decreases with time during development (Wilson et al. 2011), could explain the changes in proliferation rates seen in the *Kif3a* mutants. However, the *Gli3* mutants analyzed did not display such a dramatic overproliferation phenotype as the *Kif3a* mutant, suggesting possible alternative roles for Kif3a in this respect. Also in contrast to *Gli3* mutants and all other reported ciliary mutants discussed so far, the *Kif3a* mutants retained their normal dorsal-ventral cortical identities and boundaries, as shown by an unaffected expression pattern of *Pax6*, *Emx1* and *Emx2*, whereas *Mash1*, *Nkx2.2* and *Dlx2* were normally expressed in the ventral forebrain. Finally, to identify a downstream target of altered Gli3:Gli3R isoform ratios, the authors looked at *Fgf15*. Fgf15 was overexpressed by in utero electoporation in the cortices of wildtype embryos at E12.0 and analyzed 2 days later with BrdU labeling. Areas of the cortex electroporated with an *Fgf15* expression construct did show a shortened cell cycle, demonstrating that this is sufficient to change cell cycle dynamics like that seen in the *Kif3a* and *Gli3* mutants.

Neuroepithelial cells that constitute the initial progenitor population of the developing cerebral cortex bear a primary cilium at their apical surface that juts out into the ventricle (Cohen and Meininger 1987) and which is dysfunctional or missing in ciliary mutants (Besse et al. 2011; Stottmann et al. 2009; Willaredt et al. 2008; Wilson et al. 2011). A recent study carefully examined the location of primary cilia on newly-generated precursors that are in the process of delamination and subsequent migration away from the ventricular surface to become so-called "basal progenitors" (Wilsch-Brauninger et al. 2012). Cells destined to become basal progenitors, as identified by expression of the gene *Tbr2*, were found to establish primary cilia facing away from the ventricle and localized at the basolateral plasma membrane, despite the retention of adherens junctions with neighboring cells. The number of cells bearing such cilia increased with embryonic age, correlating with the increase in cortical neurogenesis seen from midgestation. Interestingly, the number of cells demonstrating basolaterally-oriented cilia increased upon overexpression of the transcription factor insulinoma-associated 1, which promotes the production of basal progenitors from their apical progenitor mothers (Farkas et al. 2008). This finding has major implications for the control of the differentiation of basal progenitors, as their primary cilia are no longer exposed to the ventricular milieu with its particular mixture of growth factors, but are exposed instead to the extracellular matrix of the developing cortex.

In the majority of cilia-bearing cells, primary cilia are present during the G0/G1 phase of the cell cycle, and they are resorbed when the cells prepare to reenter the cell cycle (Tucker et al. 1979a, b). Evidence has surfaced showing that cilia place a brake upon cell-cycle reentry, in that their forced resorption can push cells into S-phase. One group (Li et al. 2011) focused upon the dynein light chain subunit Tctex-1, which dissociates from the dynein complex upon phosphorylation of residue Thr94 (Chuang et al. 2005). The authors first performed in vitro experiments with the retinal pigmented epithelial (RPE-1) cell line, in which ciliogenesis occurs after serum withdrawal. shRNA-based knockdown of *Tctex-1* resulted in cell cycle block upstream of S-phase entry. This block could be rescued by a phosphomimetic mutant of the protein, demonstrating the importance of Tctex-1 phosphorylation for the G1 to S transition. The phosphorylated form of Tctex-1 was found to localize to the transition zone, a specialized structure between the basal body and the base of the ciliary axoneme. Using mutant versions of Tctex-1, the authors could show that ciliary resorption is dependent upon the relocalization of phosphorylated Tctex-1 to the transition zone, and thus entry into S-phase. Tctex-1 is enriched in radial glia of the developing cerebral cortex, and so the authors switched to in utero electroporation to perform in vivo experimentation. Examination of murine cerebral cortex from E11-E17 with an anti-Tctex-1 antibody revealed that the protein localizes to the transition zone between the basal body and the primary cilium at the apical pole of radial glia. The authors then performed complementary loss- and gain-of-function studies to demonstrate a similar role for Tctex-1 in control of radial glial proliferation. *Tctex-1* knockdown resulted in excessive neuronal differentiation and a depletion of radial glial cells. In contrast, overexpression of a phosphomimetic mutant of Tctex-1 increased the mitotic index and shortened the length of G1 phase of the cell cycle. Unfortunately, neither of these experiments examined a direct effect of *Tctex-1* knockdown/overexpression upon the number and identity of cells bearing cilia, but the authors argue that the conclusions derived from the cell culture experiments are directly applicable to the in vivo situation. These data are nicely complementary to the results seen in the *Kif3a* conditional mutant, in which G1 phase shortening was observed upon loss of primary cilia (Wilson et al. 2011). Similar results were found by manipulation of the centrosomal protein Nde1 in RPE-1 cells and in Zebrafish embryos (Kim et al. 2011).

To summarize, the most recent publications examining the control of cortical development by primary cilia have focused upon the action at the ventricular zone and the control of proliferation of the radial glia that give rise, either directly or indirectly through the generation of basal neuronal progenitors, to most of the neurons in the cerebral cortex. Primary cilia borne at the apical edge of neuroepithelia respond in a Gli3-dependent fashion to Shh signals and contribute to a lengthening of the G1 phase of the cell cycle and thereby constrain proliferation at the ventricular zone. The net effect of this activity is not only a morphological one in the sense that cilia thereby sculpt the shape of the developing cortex. Ultimately cilia thereby maintain a normal equilibrium between neural precursor and neuronal populations. The finding that basally-located precursors orient their cilia away from the ventricle is another provocative finding that will surely affect our understanding of how this tiny organelle affects the construction of one of the most complicated organs in the body.

3.2.2 Primary Cilia and Hippocampus Development

Shh is not necessary for the embryonic specification of the hippocampus (Machold et al. 2003), but in the *Gli3* mutant Xt^J, the formation of the hippocampus is abrogated in that the dorsal telencephalic midline does not invaginate (Theil et al. 1999) and hippocampal markers such as *Ephb1* and KA1 are not expressed in the remaining dorsal tissue (Tole et al. 2000). In keeping with these findings, primary cilia are necessary for the proper morphological development of the hippocampus, but they do not seem to be necessary for the determination of this structure. In a hypomorphic allele of the *Ift88* gene, the embryonic hippocampus shows a disorganized morphology and reduced levels of both *EphB1* and *Lhx2* expression, the latter of which is important for normal hippocampal development (Willaredt et al. 2008). In the postnatal mammal, and throughout life, granule neuron precursors of the dentate gyrus in the hippocampus are specified and migrate into the dentate gyrus, where they proliferate and finally reside in the subgranular zone, thereby generating new neurons throughout adult life. Shh signalling is crucial for both proliferation and maintenance of postnatal neural stem cells (Lai et al. 2003; Machold et al. 2003). Several recent studies have demonstrated a role for primary cilia in adult neurogenesis (Breunig et al. 2008; Han et al. 2008; Town et al. 2008), and these reports are dissected in Chap. 4.

3.2.3 Primary Cilia and Olfactory Bulb Development

Although the *Shh* mutant (Chiang et al. 1996) and the *Gli3* deletion mutant Xt^J (Johnson 1967) both lack olfactory bulbs, olfactory epithelia does develop in the Xt^J mutant (Sullivan et al. 1995) and also in the medial proboscis of the *Shh* mutant (Balmer and LaMantia 2004). Thus it would be of interest to see if cilia have anything to do with the development of this structure, as was discussed in a publication by Besse et al. (2011). A previous study by Delous et al. showed that *Ftm* mouse mutants exhibit apparent olfactory bulb agenesis (Delous et al. 2007). Tbx21, a marker for mitral cells of the olfactory bulb, was used to identify an ectopically-located primordial olfactory bulb in the pallium of *Ftm* mutants. Even though this mislocalized olfactory bulb-like structure is properly specified at E13.5, it is localized dorsolaterally. Additionally, it exhibited failure to contact olfactory neurons followed by an inability to laminate normally. The authors demonstrated that normal olfactory bulb development can be achieved by crossing *Ftm* mutants to the *Gli3* mutant (Böse et al. 2002) described above, in which a truncated form of Gli3 that should act as a constitutive Gli3R isoform is expressed at the Gli3 locus.

3.2.4 Primary Cilia and Diencephalon Development

One publication has investigated the role of primary cilia in the development of diencephalic structures such as the thalamus, hypothalamus, subthalamus and epithalamus.

Willaredt et al. showed that *cbs* mutants, described above as bearing a hypomorphic allele of *Ift88*, exhibit a breakdown in the boundary between the telencephalon and diencephalon along with high numbers of rosettes containing newborn neurons (Willaredt et al. 2008). Another report demonstrated similar observations in the *Gli3* deletion mutants Xt^J (Fotaki et al. 2006). Even though in ciliary mutants nothing has been shown regarding the determination of diencephalic substructures, there is a legitimate hypothesis that tissues such as the hypothalamic and thalamic nuclei will be affected, as their differentiation has been shown to be dependent upon Shh and Gli3 (Alvarez-Bolado et al. 2012; Haddad-Tovolli et al. 2012). In addition, numerous reports where ciliary mutants exhibit vaguely-described eye defects (Caspary et al. 2007; Cortellino et al. 2009; Gorivodsky et al. 2009; Qin et al. 2011; Tran et al. 2008; Vierkotten et al. 2007), including ocular coloboma in hypomorphic *Ift88* mutants (Willaredt et al. 2008), and a subset of patients with Joubert (Satran et al. 1999) or Meckel Syndrome (Thompson and Baraitser 1986) displaying coloboma, have shown that there is a link between cilia and eye development; however no study has been published yet investigating this matter.

3.3 Primary Cilia and Mesencephalon Development

A study in 2009 investigating a mutation on one of the IFT proteins involved in the IFTA complex, *Ift172*, showed that loss of Ift172 results in severe craniofacial malformations in addition to failure in cranial neural tube closure and holoprosencephaly (Gorivodsky et al. 2009). Along with these phenotypes, Ift172 mutants demonstrate defective midbrain-hindbrain patterning. While rhombomeres 2–7 seem to develop normally, rhombomere 1 does not, failing to express the marker *Gbx2*. Furthermore, even though midbrain-hindbrain markers were properly expressed at the beginning, they soon crossed over their respective boundaries. The authors have showed that the isthmic organizer was still positioned in the correct manner (Gorivodsky et al. 2009), despite the loss of *Gbx2* expression in rhombomere 1, which is thought to establish the midbrain/hindbrain boundary in cooperation with *Otx2* (Liu and Joyner 2001). The establishment of roughly normal boundaries but with substantial mixing of marker populations is reminiscent of the breakdown of pallial-subpallial and telencephalic-diencephalic boundaries in the *cbs* and Xt^J mutants. Whether this reflects a misspecification or a subsequent mixing of determined populations remains unclear. Somewhat surprisingly, no other publications have addressed the development of known Shh-dependent structures in the midbrain such as the substantia nigra or serotonergic neuronal populations, but such studies are certain to be published soon.

3.4 Primary Cilia and Rhombencephalon Development

All of the reports so far linking primary cilia to rhombencephalon development have dissected the important role that primary cilia play in the morphogenesis of the cerebellum. Genes involved in ciliogenesis, cilia function or maintenance have been linked with malformations of the cerebellum in humans. The best known disorder characterized by cerebellar malformations is Joubert syndrome (Joubert et al. 1969), in which cerebellar vermal hypoplasia is revealed by a distinct "molar tooth" sign upon MRI imaging (Quisling et al. 1999). Mutations in 14 different genes encoding proteins that localize to primary cilia have been found in Joubert syndrome patients (Sattar and Gleeson 2011), including *AHI1* (Dixon-Salazar et al. 2004; Ferland et al. 2004), *NPHP1* (Nephronophthisis 1) (Parisi et al. 2004), and *CEP290* (Centrosomal protein 290 Da) (Valente et al. 2006). Cerebellar hypoplasia is a phenotype found also in numerous other cilia-related disorders, such as Meckel-Gruber syndrome (MKS1 and MKS3) (Cincinnati et al. 2000). In the syndromes mentioned above, the molecular and the developmental basis of cerebellar hypoplasia are poorly understood. Chizhikov et al. (2007) investigated the central nervous system-specific inactivation of two ciliary genes, *Ift88* and *Kif3a*, in order to decipher the role of cilia in developmental disorders (Chizhikov et al. 2007). Previous findings, along with data showing that *Ift88* is broadly expressed in the cerebellum, have led to the clarification that both Purkinje cell and granule cell progenitors bear primary cilia. Chizhikov and his colleagues showed that lack of either *Ift88* or *Kif3a* gene expression in the central nervous system led to severe cerebellar hypoplasia but also in abnormalities in foliation. These manifestations due to loss of either of these two ciliary genes were attributed to the inability of the neonatal granule cell progenitor population to expand. However, the determination of cerebellar granule neurons (CGN) was not affected, in full agreement with previous data generated from pharmacological blockade of Shh (Dahmane and Ruiz i Altaba 1999; Wallace 1999; Wechsler-Reya and Scott 1999) and also from conditional ablation of genes in the Hedgehog signalling, *Shh*, *Smo*, *Gli1*, and *Gli2* (Corrales et al. 2004, 2006; Lewis et al. 2004). Taking into account that Ift88 is a critical player in Shh transduction in multiple cell layers, Chizhikov's et al. findings suggest that loss of responsiveness to Shh signalling is the cause of the loss of granule cell progenitor proliferation. Their data demonstrate the significance that both *Ift88* and *Kif3a* genes have in ciliary function, since in both mutants the cerebellar malformations were phenotypically similar.

In the study by Chizhikov et al. (2007) Purkinje cells were morphologically disorganized in the stratum purkinjense and showed abnormal dendritic aborization. A potential explanation for this phenotype could be the lower numbers of mature granule cells since Purkinje cells do not express the hGFAP::Cre transgene used to delete *Ift88*. In order to further investigate the direct result of cilia on the determination of Purkinje cells the authors studied a conditional ablation of *Ift88* promoted by the nestin::CRE transgene, which is known to be expressed in the majority of cerebellar progenitors. The authors reported that in these *Ift88* mutants Purkinje neurons are present. Spassky et al. (2008) supported and extended these findings with similar results using a conditional deletion of *Kif3a*. In this study, direct evidence that Shh signaling is involved was provided by examination of a conditional mutant *Smo* allele under control of the hGFAP::CRE deleter strain, the same one used in the Chizhikov study (Chizhikov et al. 2007). The *Smo* mutant mice also displayed a hypoplastic cerebellum and a more drastic foliation defect than the *Kif3a* conditional mutant. More importantly, analysis of a *Kif3a*; *Smo* double mutant indicated a milder phenotype, more comparable to the *Kif3a* single knockout. Thus, *Kif3a* was found to be epistatic to (i.e. downstream of) *Smo*, and thus provides direct genetic evidence that Shh signaling is acting through primary cilia. Finally, Moyer et al. (1994) reported extensive CGN heterotopias in *orpk*, a hypomorphic *Ift88* allele, and CGN heterotopias have also been observed in a mouse model in which a constitutively-active form of Gli2 is specifically expressed in CGN precursors (Han et al. 2009).

The involvement of primary cilia in Wnt signaling has proven to be one of the most controversial aspects in the field. Although in vitro analysis indicated that primary cilia constrain canonical Wnt signaling (Corbit et al. 2008), the evidence from in vivo studies is contradictory. Analysis in the mouse and in the zebrafish *Danio rerio* of Wnt-dependent developmental processes revealed no role for primary cilia (Huang and Schier 2009; Ocbina et al. 2009), but perturbations in Wnt signaling have been recorded in the later development of many organ systems, including the forebrain, cerebellum, kidney, pancreas, mammary gland, and long bones (reviewed in Tasouri and Tucker 2011). These studies have not clarified whether perturbations in Wnt signaling are a primary or secondary effect of disturbances in ciliary function.

To more directly address the issue of Wnt signaling in primary cilia-dependent development, Lancaster et al. (2011) used a mouse model where the gene *Ahi1*, the first gene linked to Joubert syndrome, was deleted (Louie et al. 2010). The authors demonstrated that generally the brain morphology of the *Ahi1* mutants was normal apart from a hypoplastic vermis showing a slightly defective foliation pattern and a cerebellum whose size was decreased by nearly 40% and reduced in total folia number. Examination of cerebellar anatomy during embryogenesis revealed a midline fusion defect at the cerebellar vermis primordium, and as early as E12.5 the mutants showed a lengthened and widened rhombic roof plate. *Ahi1* mutants compared to control brains also showed a fusion at the site of the cerebellar hemispheres. Similar phenotypes were observed upon examination of a mouse line with a targeted deletion of *Cep290*, another ciliary gene shown to be defective in patients with Joubert syndrome (Valente et al. 2006). The authors used MRI to examine human fetuses later diagnosed with Joubert syndrome, and amazingly were able to record fusion defects between gestational weeks 21 and 25, at which time point fusion has already occurred in normal fetuses.

Immunohistofluorescent staining of Jbn in CGNs revealed localization of the protein to the basal body. The authors employed tagged constructs expressing jouberin (Jbn), the protein encoded by the *Ahi1* locus that is known to localize to primary cilia, and containing three distinct mutations seen in Joubert syndrome patients. All three mutant Jbn proteins failed to localize at the primary cilium. In the *Ahi1* mutant mice there was no difference in the number or morphology of cilia in the CGN precursor population, suggesting that Jbn is not necessary for ciliogenesis but rather important for ciliary-mediated signalling. β-Catenin colocalisation experiment showed that β-catenin is found with Jbn at the base of the cilium, pointing towards a potential role for the primary cilium in the modulatory role that Jbn has for Wnt signalling. To examine the role of Jbn in Wnt signaling, luciferase assays were used to assess whether these mutant proteins were defective in modulating Wnt signalling. Overexpression of wild type Jbn in cells treated with Wnt3a showed a very modest increase in Wnt report activity compared to the vector control, but overexpression of the mutant constructs did not show any increase in Wnt activity. In addition, *Ahi1* mutant mice were tested for defective Wnt signalling by crossing them to Wnt reporter BATgal transgenic mice. Whole-mount staining at E13.5, when midline fusion begins, showed a downregulation of Wnt signalling at the cerebellar midline in *Ahi1* mutants. BrdU labelling to assess proliferation of precursors surrounding the site of midline fusion revealed significantly lower numbers of labelled cells in the fusion site in mutant *Ahi1* embryos. To establish a mechanistic explanation, the Wnt-signaling agonist lithium was injected intraperitoneally into pregnant dams at E12.5 and E13.5. Most excitingly, *Ahi1* mutant embryos from lithium-injected females were rescued, in that they showed expanded cerebellar hemisphere tissue with a reduced interhemispheric separation, showing no significant difference between mutants and control littermates. The proliferation marker Ki67 showed a significant reduction of proliferating cells at the site of fusion in untreated *Ahi1* mutant embryos, whereas lithium had rescued this phenotype in the *Ahi1* mutants. In conclusion, Lancaster et al. (2011) suggest that Joubert syndrome seems to be associated more with a specific defect in cilia function or signalling rather than a more general defect in ciliogenesis, as the *Ahi1* and *Cep290* mutants show milder phenotypes in the cerebellum compared to the *Ift88* and *Kif3a* knockouts, in which ciliogenesis is abrogated (Chizhikov et al. 2007; Spassky et al. 2008).

Surprisingly, the *Ahi1* mutants had distinct differences from the symptoms seen in patients with Joubert syndrome. They showed neither the molar tooth sign nor did they display fragmentation of deep cerebellar nuclei, anatomical abnormalities of the descending trigeminal tract, brainstem nuclei, and pyramidal tract decussation. Despite the cerebellar defects, the cellular organization, neuronal morphology, and layering of the cerebellar cortex was not affected. BrdU labeling from E16.5 and until postnatal day 5 revealed a reduced proliferation in the vermis at prenatal but not at postnatal timepoints. Since the Shh signalling pathway that has a key role in CGN proliferation is mostly active after birth (Corrales et al. 2004), these results suggested that it is not implicated in the prenatal proliferation defects. To show this, the authors investigated Shh signaling using the downstream Shh target N-myc and showed no difference in *Ahi1* mutant mice when compared to their littermate controls in the protein abundance of the N-myc in whole cerebellar extracts. Nor were any differences seen in *Ptch1* or *Gli1* expression patterns or levels. Therefore the authors concluded that postnatal Shh signalling is not affected in *Ahi1*-mutant cerebella.

To summarize the studies of primary cilia in the development of the cerebellum, it is quite clear that both Wnt and Shh signaling are responsible for the control of proliferation in cerebellar precursor populations in both embryonic and postnatal stages, respectively. Primary cilia are however not responsible for determination of the two most important neuronal types in the cerebellar cortex, the CGNs and the Purkinje neurons. Prenatal defects in proliferation of the roof plate was accounted for by defects in Wnt signaling, contributing to a midline fusion defect of the vermis (Lancaster et al. 2011). Postnatally, the proliferation of CGN precursors in the foliae of the cerebral hemispheres was reduced in IFTB complex mutants, reflecting defects in Shh signal transduction caused by a loss of cilia (Chizhikov et al. 2007; Spassky et al. 2008). It still remains to be resolved what is causing the classic "molar tooth" abnormality in the vermis of Joubert patients, as well as several other anatomical abnormalities detected in these patients, but other mouse models in the dozen other genes implicated in Joubert syndrome may shed light on this matter.

3.5 Conclusions and Outlook

Although the number of studies dealing with primary cilia and the brain is dwarfed by those examining primary cilia and spinal cord development, a mechanistic picture is slowly becoming clear, at least with respect to the role of Shh in the forebrain. Primary cilia at the apical pole of neuroepithelial progenitor cells respond to Shh signal by lengthening the G1 phase of the cell cycle in a Gli3-dependent fashion. Loss of ciliary function, through impaired IFT or a loss of the cilia themselves, leads to an overproliferation of progenitors that results in overgrowth within the cortex. However, many issues remain unclear. What do cilia have to do with rosette formation in both the cortex and the cerebellum? Even within the category of mutants in IFTB components (e.g. the hypomorphic *Ift88* allele *cbs* (Willaredt et al. 2008) and the targeted disruption of *Kif3a* (Wilson et al. 2011)), discrepancies need to be resolved in the control of boundary formation between the pallium and the subpallium in the embryonal cortex. What are the different roles that primary cilia are playing in basal progenitors, with their cilia projecting basolaterally instead of into the ventricle? What other signaling pathways beside Shh and Wnt are utilizing primary cilia in the context of neural development? Are cilia controlling the migration of the neurons away from the ventricular zone? These questions are of current research interest. Topics that will be addressed in the future include a thorough analysis of cilia-dependent diencephalic development, mechanistic experiments to definitively determine a role for Wnt signaling in forebrain development (or not), medium throughput analysis of other signaling cascades that are developmentally relevant and may be cilia dependent, and, crucially, a closer investigation of patients with ciliopathies for defects in forebrain development and a subsequent modeling of these defects with appropriate model organisms. For example, in certain patients with Joubert or Oral-Facial-Digital Type I syndromes, periventricular heterotopias, polymicrogyria, and cerebral cysts have been observed (outlined in great detail in Chap. 9), which could reflect the postnatal outcome of the embryonic rosettes observed in the cortex of some ciliary mouse mutants (Willaredt et al. 2008; Wilson et al. 2011). Although the prospect for healing neurodevelopmental aspects of ciliopathies may seem at first glance like an impossible task, the groundbreaking work of Lancaster et al. (2011) shows that in fact, at least in animal models, it is conceivable that prenatal exposure to drugs that mimic events downstream of ciliary function may offer hope for certain situations, at least in those families in which genetic analysis has already indicated a predisposition to a ciliopathy.

Acknowledgements The authors were funded by the German Research Society (DFG SFB 488, Teilprojekt B9) and would like to thank Joachim Kirsch for generous scientific support and Tamara Caspary for thoughtful commentary to the text.

References

Alvarez-Bolado G, Paul FA, Blaess S (2012) Sonic hedgehog lineage in the mouse hypothalamus: from progenitor domains to hypothalamic regions. Neural Dev 7:4

Arts HH, Doherty D, van Beersum SE, Parisi MA, Letteboer SJ, Gorden NT, Peters TA, Marker T, Voesenek K, Kartono A, Ozyurek H, Farin FM, Kroes HY, Wolfrum U, Brunner HG, Cremers FP, Glass IA, Knoers NV, Roepman R (2007) Mutations in the gene encoding the basal body protein RPGRIP1L, a nephrocystin-4 interactor, cause Joubert syndrome. Nat Genet 39:882–888

Balmer CW, LaMantia AS (2004) Loss of Gli3 and Shh function disrupts olfactory axon trajectories. J Comp Neurol 472:292–307

Besse L, Neti M, Anselme I, Gerhardt C, Ruther U, Laclef C, Schneider-Maunoury S (2011) Primary cilia control telencephalic patterning and morphogenesis via Gli3 proteolytic processing. Development 138:2079–2088

Böse J, Grotewold L, Ruther U (2002) Pallister-Hall syndrome phenotype in mice mutant for Gli3. Hum Mol Genet 11:1129–1135

Breunig JJ, Sarkisian MR, Arellano JI, Morozov YM, Ayoub AE, Sojitra S, Wang B, Flavell RA, Rakic P, Town T (2008) Primary cilia regulate hippocampal neurogenesis by mediating sonic hedgehog signaling. Proc Natl Acad Sci USA 105:13127–13132

Caspary T, Larkins CE, Anderson KV (2007) The graded response to Sonic Hedgehog depends on cilia architecture. Dev Cell 12:767–778

Chiang C, Litingtung Y, Lee E, Young KE, Corden JL, Westphal H, Beachy PA (1996) Cyclopia and defective axial patterning in mice lacking Sonic hedgehog gene function. Nature 383:407–413

Chizhikov VV, Davenport J, Zhang Q, Shih EK, Cabello OA, Fuchs JL, Yoder BK, Millen KJ (2007) Cilia proteins control cerebellar morphogenesis by promoting expansion of the granule progenitor pool. J Neurosci 27:9780–9789

Chuang JZ, Yeh TY, Bollati F, Conde C, Canavosio F, Caceres A, Sung CH (2005) The dynein light chain Tctex-1 has a dynein-independent role in actin remodeling during neurite outgrowth. Dev Cell 9:75–86

Cincinnati P, Neri ME, Valentini A (2000) Dandy-Walker anomaly in Meckel-Gruber syndrome. Clin Dysmorphol 9:35–38

Cohen E, Meininger V (1987) Ultrastructural analysis of primary cilium in the embryonic nervous tissue of mouse. Int J Dev Neurosci 5:43–51

Corbit KC, Aanstad P, Singla V, Norman AR, Stainier DY, Reiter JF (2005) Vertebrate Smoothened functions at the primary cilium. Nature 437:1018–1021

Corbit KC, Shyer AE, Dowdle WE, Gaulden J, Singla V, Chen MH, Chuang PT, Reiter JF (2008) Kif3a constrains beta-catenin-dependent Wnt signalling through dual ciliary and non-ciliary mechanisms. Nat Cell Biol 10:70–76

Corrales JD, Rocco GL, Blaess S, Guo Q, Joyner AL (2004) Spatial pattern of sonic hedgehog signaling through Gli genes during cerebellum development. Development 131:5581–5590

Corrales JD, Blaess S, Mahoney EM, Joyner AL (2006) The level of sonic hedgehog signaling regulates the complexity of cerebellar foliation. Development 133:1811–1821

Cortellino S, Wang C, Wang B, Bassi MR, Caretti E, Champeval D, Calmont A, Jarnik M, Burch J, Zaret KS, Larue L, Bellacosa A (2009) Defective ciliogenesis, embryonic lethality and severe impairment of the Sonic Hedgehog pathway caused by inactivation of the mouse complex A intraflagellar transport gene Ift122/Wdr10, partially overlapping with the DNA repair gene Med1/Mbd4. Dev Biol 325:225–237

Dahmane N, Ruiz i Altaba A (1999) Sonic hedgehog regulates the growth and patterning of the cerebellum. Development 126:3089–3100

Delous M, Baala L, Salomon R, Laclef C, Vierkotten J, Tory K, Golzio C, Lacoste T, Besse L, Ozilou C, Moutkine I, Hellman NE, Anselme I, Silbermann F, Vesque C, Gerhardt C, Rattenberry E, Wolf MT, Gubler MC, Martinovic J, Encha-Razavi F, Boddaert N, Gonzales M, Macher MA, Nivet H, Champion G, Bertheleme JP, Niaudet P, McDonald F, Hildebrandt F, Johnson CA, Vekemans M, Antignac C, Ruther U, Schneider-Maunoury S, Attie-Bitach T, Saunier S (2007) The ciliary gene RPGRIP1L is mutated in cerebello-oculo-renal syndrome (Joubert syndrome type B) and Meckel syndrome. Nat Genet 39:875–881

Dixon-Salazar T, Silhavy JL, Marsh SE, Louie CM, Scott LC, Gururaj A, Al-Gazali L, Al-Tawari AA, Kayserili H, Sztriha L, Gleeson JG (2004) Mutations in the AHI1 gene, encoding jouberin, cause Joubert syndrome with cortical polymicrogyria. Am J Hum Genet 75:979–987

Farkas LM, Haffner C, Giger T, Khaitovich P, Nowick K, Birchmeier C, Paabo S, Huttner WB (2008) Insulinoma-associated 1 has a panneurogenic role and promotes the generation and expansion of basal progenitors in the developing mouse neocortex. Neuron 60:40–55

Ferland RJ, Eyaid W, Collura RV, Tully LD, Hill RS, Al-Nouri D, Al-Rumayyan A, Topcu M, Gascon G, Bodell A, Shugart YY, Ruvolo M, Walsh CA (2004) Abnormal cerebellar development and axonal decussation due to mutations in AHI1 in Joubert syndrome. Nat Genet 36:1008–1013

Fotaki V, Yu T, Zaki PA, Mason JO, Price DJ (2006) Abnormal positioning of diencephalic cell types in neocortical tissue in the dorsal telencephalon of mice lacking functional Gli3. J Neurosci 26:9282–9292

Gorivodsky M, Mukhopadhyay M, Wilsch-Braeuninger M, Phillips M, Teufel A, Kim C, Malik N, Huttner W, Westphal H (2009) Intraflagellar transport protein 172 is essential for primary cilia formation and plays a vital role in patterning the mammalian brain. Dev Biol 325:24–32

Haddad-Tovolli R, Heide M, Zhou X, Blaess S, Alvarez-Bolado G (2012) Mouse thalamic differentiation: gli-dependent pattern and gli-independent prepattern. Front Neurosci 6:27

Han YG, Spassky N, Romaguera-Ros M, Garcia-Verdugo JM, Aguilar A, Schneider-Maunoury S, Alvarez-Buylla A (2008) Hedgehog signaling and primary cilia are required for the formation of adult neural stem cells. Nat Neurosci 11:277–284

Han YG, Kim HJ, Dlugosz AA, Ellison DW, Gilbertson RJ, Alvarez-Buylla A (2009) Dual and opposing roles of primary cilia in medulloblastoma development. Nat Med 15:1062–1065

Haycraft CJ, Banizs B, Aydin-Son Y, Zhang Q, Michaud EJ, Yoder BK (2005) Gli2 and Gli3 localize to cilia and require the intraflagellar transport protein polaris for processing and function. PLoS Genet 1:e53

Haycraft CJ, Zhang Q, Song B, Jackson WS, Detloff PJ, Serra R, Yoder BK (2007) Intraflagellar transport is essential for endochondral bone formation. Development 134:307–316

Huang P, Schier AF (2009) Dampened Hedgehog signaling but normal Wnt signaling in zebrafish without cilia. Development 136:3089–3098

Huangfu D, Anderson KV (2005) Cilia and Hedgehog responsiveness in the mouse. Proc Natl Acad Sci USA 102:11325–11330

Johnson DR (1967) Extra-toes: a new mutant gene causing multiple abnormalities in the mouse. J Embryol Exp Morphol 17:543–581

Joubert M, Eisenring JJ, Robb JP, Andermann F (1969) Familial agenesis of the cerebellar vermis. A syndrome of episodic hyperpnea, abnormal eye movements, ataxia, and retardation. Neurology 19:813–825

Kim S, Zaghloul NA, Bubenshchikova E, Oh EC, Rankin S, Katsanis N, Obara T, Tsiokas L (2011) Nde1-mediated inhibition of ciliogenesis affects cell cycle re-entry. Nat Cell Biol 13:351–360

Lai K, Kaspar BK, Gage FH, Schaffer DV (2003) Sonic hedgehog regulates adult neural progenitor proliferation in vitro and in vivo. Nat Neurosci 6:21–27

Lancaster MA, Gopal DJ, Kim J, Saleem SN, Silhavy JL, Louie CM, Thacker BE, Williams Y, Zaki MS, Gleeson JG (2011) Defective Wnt-dependent cerebellar midline fusion in a mouse model of Joubert syndrome. Nat Med 17:726–731

Le Douarin NM, Kalcheim C (1999) In: Bard JBL, Barlow PW, Kirk DL (eds) The neural crest, 2 edn. Developmental and cell biology. Cambridge University Press, Cambridge

Lewis PM, Gritli-Linde A, Smeyne R, Kottmann A, McMahon AP (2004) Sonic hedgehog signaling is required for expansion of granule neuron precursors and patterning of the mouse cerebellum. Dev Biol 270:393–410

Li A, Saito M, Chuang JZ, Tseng YY, Dedesma C, Tomizawa K, Kaitsuka T, Sung CH (2011) Ciliary transition zone activation of phosphorylated Tctex-1 controls ciliary resorption, S-phase entry and fate of neural progenitors. Nat Cell Biol 13:402–411

Liu A, Joyner AL (2001) Early anterior/posterior patterning of the midbrain and cerebellum. Annu Rev Neurosci 24:869–896

Liu A, Wang B, Niswander LA (2005) Mouse intraflagellar transport proteins regulate both the activator and repressor functions of Gli transcription factors. Development 132:3103–3111

Louie CM, Caridi G, Lopes VS, Brancati F, Kispert A, Lancaster MA, Schlossman AM, Otto EA, Leitges M, Grone HJ, Lopez I, Gudiseva HV, O'Toole JF, Vallespin E, Ayyagari R, Ayuso C, Cremers FP, den Hollander AI, Koenekoop RK, Dallapiccola B, Ghiggeri GM, Hildebrandt F, Valente EM, Williams DS, Gleeson JG (2010) AHI1 is required for photoreceptor outer segment development and is a modifier for retinal degeneration in nephronophthisis. Nat Genet 42:175–180

Machold R, Hayashi S, Rutlin M, Muzumdar MD, Nery S, Corbin JG, Gritli-Linde A, Dellovade T, Porter JA, Rubin LL, Dudek H, McMahon AP, Fishell G (2003) Sonic hedgehog is required for progenitor cell maintenance in telencephalic stem cell niches. Neuron 39:937–950

Marszalek JR, Ruiz-Lozano P, Roberts E, Chien KR, Goldstein LS (1999) Situs inversus and embryonic ciliary morphogenesis defects in mouse mutants lacking the KIF3A subunit of kinesin-II. Proc Natl Acad Sci USA 96:5043–5048

May SR, Ashique AM, Karlen M, Wang B, Shen Y, Zarbalis K, Reiter J, Ericson J, Peterson AS (2005) Loss of the retrograde motor for IFT disrupts localization of Smo to cilia and prevents the expression of both activator and repressor functions of Gli. Dev Biol 287:378–389

Moyer JH, Lee-Tischler MJ, Kwon HY, Schrick JJ, Avner ED, Sweeney WE, Godfrey VL, Cacheiro NL, Wilkinson JE, Woychik RP (1994) Candidate gene associated with a mutation causing recessive polycystic kidney disease in mice. Science 264:1329–1333

Ocbina PJ, Tuson M, Anderson KV (2009) Primary cilia are not required for normal canonical Wnt signaling in the mouse embryo. PLoS One 4:e6839

Parisi MA, Bennett CL, Eckert ML, Dobyns WB, Gleeson JG, Shaw DW, McDonald R, Eddy A, Chance PF, Glass IA (2004) The NPHP1 gene deletion associated with juvenile nephronophthisis is present in a subset of individuals with Joubert syndrome. Am J Hum Genet 75:82–91

Petersen PH, Zou K, Hwang JK, Jan YN, Zhong W (2002) Progenitor cell maintenance requires numb and numblike during mouse neurogenesis. Nature 419:929–934

Qin J, Lin Y, Norman RX, Ko HW, Eggenschwiler JT (2011) Intraflagellar transport protein 122 antagonizes Sonic Hedgehog signaling and controls ciliary localization of pathway components. Proc Natl Acad Sci USA 108:1456–1461

Quisling RG, Barkovich AJ, Maria BL (1999) Magnetic resonance imaging features and classification of central nervous system malformations in Joubert syndrome. J Child Neurol 14:628–635, discussion 669–72

Rohatgi R, Milenkovic L, Scott MP (2007) Patched1 regulates hedgehog signaling at the primary cilium. Science 317:372–376

Satran D, Pierpont ME, Dobyns WB (1999) Cerebello-oculo-renal syndromes including Arima, Senior-Loken and COACH syndromes: more than just variants of Joubert syndrome. Am J Med Genet 86:459–469

Sattar S, Gleeson JG (2011) The ciliopathies in neuronal development: a clinical approach to investigation of Joubert syndrome and Joubert syndrome-related disorders. Dev Med Child Neurol 53:793–798

Scholey JM, Anderson KV (2006) Intraflagellar transport and cilium-based signaling. Cell 125:439–442

Spassky N, Han YG, Aguilar A, Strehl L, Besse L, Laclef C, Ros MR, Garcia-Verdugo JM, Alvarez-Buylla A (2008) Primary cilia are required for cerebellar development and Shh-dependent expansion of progenitor pool. Dev Biol 317:246–259

Stottmann RW, Tran PV, Turbe-Doan A, Beier DR (2009) Ttc21b is required to restrict sonic hedgehog activity in the developing mouse forebrain. Dev Biol 335:166–178

Sullivan SL, Bohm S, Ressler KJ, Horowitz LF, Buck LB (1995) Target-independent pattern specification in the olfactory epithelium. Neuron 15:779–789

Tasouri E, Tucker KL (2011) Primary cilia and organogenesis: is Hedgehog the only sculptor? Cell Tissue Res 345:21–40

Theil T, Alvarez-Bolado G, Walter A, Ruther U (1999) Gli3 is required for Emx gene expression during dorsal telencephalon development. Development 126:3561–3571

Thompson E, Baraitser M (1986) An autosomal recessive mental retardation syndrome with hepatic fibrosis and renal cysts. Am J Med Genet 24:151–158

Tole S, Ragsdale CW, Grove EA (2000) Dorsoventral patterning of the telencephalon is disrupted in the mouse mutant extra-toes(J). Dev Biol 217:254–265

Town T, Breunig JJ, Sarkisian MR, Spilianakis C, Ayoub AE, Liu X, Ferrandino AF, Gallagher AR, Li MO, Rakic P, Flavell RA (2008) The stumpy gene is required for mammalian ciliogenesis. Proc Natl Acad Sci USA 105:2853–2858

Tran PV, Haycraft CJ, Besschetnova TY, Turbe-Doan A, Stottmann RW, Herron BJ, Chesebro AL, Qiu H, Scherz PJ, Shah JV, Yoder BK, Beier DR (2008) THM1 negatively modulates mouse sonic hedgehog signal transduction and affects retrograde intraflagellar transport in cilia. Nat Genet 40:403–410

Tucker RW, Pardee AB, Fujiwara K (1979a) Centriole ciliation is related to quiescence and DNA synthesis in 3T3 cells. Cell 17:527–535

Tucker RW, Scher CD, Stiles CD (1979b) Centriole deciliation associated with the early response of 3T3 cells to growth factors but not to SV40. Cell 18:1065–1072

Valente EM, Silhavy JL, Brancati F, Barrano G, Krishnaswami SR, Castori M, Lancaster MA, Boltshauser E, Boccone L, Al-Gazali L, Fazzi E, Signorini S, Louie CM, Bellacchio E, Bertini E, Dallapiccola B, Gleeson JG (2006) Mutations in CEP290, which encodes a centrosomal protein, cause pleiotropic forms of Joubert syndrome. Nat Genet 38:623–625

Vierkotten J, Dildrop R, Peters T, Wang B, Ruther U (2007) Ftm is a novel basal body protein of cilia involved in Shh signalling. Development 134:2569–2577

Vincent SD, Dunn NR, Hayashi S, Norris DP, Robertson EJ (2003) Cell fate decisions within the mouse organizer are governed by graded nodal signals. Genes Dev 17:1646–1662

Wallace VA (1999) Purkinje-cell-derived Sonic hedgehog regulates granule neuron precursor cell proliferation in the developing mouse cerebellum. Curr Biol 9:445–448

Wang C, Pan Y, Wang B (2007) A hypermorphic mouse Gli3 allele results in a polydactylous limb phenotype. Dev Dyn 236:769–776

Wechsler-Reya RJ, Scott MP (1999) Control of neuronal precursor proliferation in the cerebellum by Sonic Hedgehog. Neuron 22:103–114

Willaredt MA, Hasenpusch-Theil K, Gardner HARG, Kitanovic I, Hirschfeld-Warneken VC, Gojak CP, Gorgas K, Bradford CL, Spatz J, Wolfl S, Theil T, Tucker KL (2008) A crucial role for primary cilia in cortical morphogenesis. J Neurosci 28:12887–12900

Wilsch-Brauninger M, Peters J, Paridaen JT, Huttner WB (2012) Basolateral rather than apical primary cilia on neuroepithelial cells committed to delamination. Development 139:95–105

Wilson SL, Wilson JP, Wang C, Wang B, McConnell SK (2011) Primary cilia and Gli3 activity regulate cerebral cortical size. Dev Neurobiol 72:1196–1212

Chapter 4
Primary Cilia in Cerebral Cortex: Growth and Functions on Neuronal and Non-neuronal Cells

Matthew R. Sarkisian, Jon I. Arellano, and Joshua J. Breunig

Contents

M.R. Sarkisian (✉)
Department of Neuroscience, McKnight Brain Institute, University of Florida College of Medicine, Gainesville, FL 32610, USA
e-mail: msarkisian@ufl.edu

J.I. Arellano
Department of Neurobiology, Yale University, New Haven, CT, USA
e-mail: jon.arellano@yale.edu

J.J. Breunig
Department of Biomedical Sciences, Regenerative Medicine Institute, Cedars-Sinai Medical Center, Los Angeles, CA 90048, USA
e-mail: Joshua.Breunig@cshs.org

K.L. Tucker and T. Caspary (eds.), *Cilia and Nervous System Development and Function*, DOI 10.1007/978-94-007-5808-7_4,

Abstract The prevailing view until very recently was that primary neuronal cilia, which were first described in electron microscopic studies of the central nervous system (CNS) approximately 50 years ago, were likely vestigial. This was due in large part to their lost motility during the course of evolution. For decades, further investigation into these structures was hampered by the lack of methods to specifically label cilia and the paucity of information about their growth and function in the CNS. In this chapter, we review the unexpected roles that primary cilia have in shaping the CNS and in particular the generation and maturation of cells in the postnatal cerebral cortex. We discuss newly available research tools for detecting cilia and manipulating ciliogenesis. Focusing on the mammalian cerebral cortex, this chapter reviews the patterns of growth of neuronal cilia, signaling profiles and putative functions of neuronal and non-neuronal cilia, and potential consequences of abnormal ciliogenesis in these cell types.

Keywords Cerebral cortex • Hippocampus • Primary cilium • Axoneme • Basal body • Ciliogenesis • Excitatory • Inhibitory • Glia • G protein-coupled receptor • Sonic hedgehog

4.1 Introduction

Showcased in virtually every neuroscience textbook are the exquisite drawings of Golgi-stained neurons and glial cells by Santiago Ramón y Cajal, arguably one of history's most famous neuroanatomists. His meticulous studies of the brain laid foundation to textbook views of cell morphology, circuitry and communication between neurons (Swanson and Swanson 1995). His drawings illustrate in precise detail the enormous 'square-footage' of cortical pyramidal neurons manifested by elaborate dendrites and long distance axonal projections. But as pointed out by others (Fuchs and Schwark 2004), he did not portray a cellular process common to virtually every neuron: the primary cilium, a hair-like structure that extends a few microns off the cell body. Though the first description of primary cilia in neurons is about 50 years old (Dahl 1963), they had been essentially dismissed for many decades because they were viewed as vestigial cellular appendages, in part because it seemed they lost their ability to be motile during the course of evolution (Fuchs and Schwark 2004; Louvi and Grove 2011; Whitfield 2004) and because it was not clear that they were present in all neurons. So why we are only now focusing on these clearly sophisticated organelles?

The importance of neural primary cilia in the brain is an emerging field of neuroscience, lagging significantly behind our knowledge of cilia biology in both other mammalian organ systems (e.g. kidney, retina) and model organisms (e.g. *C. elegans* or *Chlamydomynas*) (Vincensini et al. 2011). Assisting in the rise of primary cilia to prominence in the field is the identification of genes both necessary for cilia formation and whose mutation disrupts CNS development both in rodent and human (Bennouna-Greene et al. 2011; Breunig et al. 2008; Cantagrel et al. 2008; Davenport et al. 2007; Davenport and Yoder 2005;

Rooryck et al. 2007; Willaredt et al. 2008). In addition, the availability of cilia markers allowing detection of neural cilia (Berbari et al. 2007; Bishop et al. 2007; Brailov et al. 2000; Domire et al. 2011; Domire and Mykytyn 2009; Handel et al. 1999) allows studying their distribution, molecular composition and addressing basic questions about the significance of cilia throughout the brain. For example, what is the potential role of cilia during neurogenesis? When do cilia emerge from different neural cell types? Are the molecular constituents that enable cilia formation and function preserved from region to region, cell type to cell type? What do neural cilia sense and how do those signals get translated in the cell? Do neuronal cilia serve similar functions as glial cilia? Why are there so many different neurological manifestations associated with ciliopathies? In this chapter, we will discuss recent work that begins to address some of these questions. Even though cilia are found throughout the brain (Bishop et al. 2007), we will focus on the cerebral cortex. Our goal is to review the developmental profiles of both neuronal and non-neuronal cilia in cerebral cortex, and discuss some of the consequences of aberrant cilia development on cortical cell types.

4.2 Identification of Cilia: Specific Markers and Distribution in the Brain

Over the past decade there have been several excellent review articles describing the historical and regional descriptions of neuronal cilia throughout the CNS (Fuchs and Schwark 2004; Green and Mykytyn 2010; Louvi and Grove 2011; Whitfield 2004). Briefly, the earliest reports of cilia on neurons in the brain date back roughly 50 years ago (Dahl 1963). These observations were made using electron microscopy (EM) that was the only way to identify cilia at the time, and still is the gold standard for identifying and describing cilia architecture. Those first cilia were noted on granular neurons of the dentate gyrus in the hippocampus of adult rats (Dahl 1963), and were followed by other reports recognizing primary cilia in other brain regions including Purkinje cells of the rat cerebellum (Del Cerro and Snider 1969) and pyramidal neurons of human neocortex (Mandl and Megele 1989). The electron microscopy data suggested that both neurons and glial cells had a single primary cilium (Dahl 1963; Del Cerro and Snider 1969; Karlsson 1966). However the limited number of cells analyzed precluded drawing definitive conclusions. More recently, the study of neural cilia has gone through a renaissance, owing to the characterization of a host of antibodies that detect protein enrichment in the cilium. More specifically, cilia are specialized organelles isolated from the cell cytoplasm and they exhibit differential composition compared to the rest of the cell. Antibodies to proteins enriched in cilia have allowed for a more quantitative perspective on CNS ciliogenesis—notably highlighting their ubiquity—and, in addition, have greatly increased our understanding of the differential localization of signaling pathways therein. In the following section we will summarize the cytoskeletal and signaling markers that are enriched in the cilium (also reviewed by Domire and Mykytyn 2009).

4.2.1 Cytoskeletal Markers

In most cell types, cilia axonemes are enriched in acetylated-alpha tubulin, a cytoskeletal protein that is also present in the mitotic spindle and mid-body tubules of cells in mitosis, but also more generally in cytoplasmic microtubules (Piperno and Fuller 1985; also see Wheatley et al. 1996). Tufts of acetylated alpha tubulin positive cilia are found in the motile ependymal cells lining the lateral ventricles; however most cortical neuronal cilia appear to have generally lost or evolved a different posttranslationally modified form of tubulin (Berbari et al. 2007; Stanic et al. 2009; Wheatley et al. 1996; MRS unpublished observation). The significance or mechanisms underlying this change are unclear, but could be due to evolved changes in tubulin acetyltransferase expression (Leroux 2010; Shida et al. 2010). The basal bodies of cilia are typically enriched in either gamma-tubulin or pericentrin (Fig. 4.1a) (Anastas et al. 2011; Arellano et al. 2012; Miyoshi et al. 2006, 2009b).

4.2.2 Signaling Markers

The cilia membrane is enriched with several receptors largely involved in neuromodulation. Interestingly, known receptors enriched in the cilia membrane are mostly subtypes of G-protein coupled receptors (GPCRs) including somatostatin receptor 3 (SSTR3) (Fig. 4.1b–d) (Handel et al. 1999; Stanic et al. 2009), serotonin receptor 6 (5HT6 (Fig. 4.1e, f) Brailov et al. 2000), melanin concentrating hormone receptor 1 (MCH1R; Berbari et al. 2008a, b), and dopamine receptor 1 (D1, and possibly D2 and 5; Domire et al. 2011; Marley and von Zastrow 2010). Also, neurotrophin receptor $p75^{NTR}$ has been described in cilia (Chakravarthy et al. 2010). In addition, other molecules particularly enriched in neural cilia include Arl13b, a member of a family of ADP-ribosylation factor (Arf) proteins which have regulatory roles in membrane trafficking and cytoskeletal dynamics (Cantagrel et al. 2008; D'Souza-Schorey and Chavrier 2006; Zhou et al. 2006); $G\alpha 11$, a G-protein that bridges neurotransmitter receptors and second messenger signaling (Fuchs and Schwark 2004) and adenylyl cyclase 3 (ACIII) (Fig. 4.1a), a mediator of G-protein and cAMP signaling that labels many cilia in the brain (Anastas et al. 2011; Arellano et al. 2012; Bishop et al. 2007).

Neuronal cilia GPCRs such as SSTR3 are extensively distributed on cilia in numerous brain regions, with apparent differences in the levels of expression (Fig. 4.1b, d). Cilia staining was demonstrated at the EM level (Fig. 4.1c), and was particularly strong in the dentate gyrus, piriform and retrosplenial cortex, the internal plexiform and granular layers of the olfactory bulb and in the transition cortex-amygdala, and was absent in some regions including the striatum or the cerebellar cortex (Handel et al. 1999). Antibodies against 5HT6 showed labeling on cilia in restricted brain regions including the nucleus accumbens, the caudate-putamen, the olfactory tubercles and the islands of Calleja (Hamon et al. 1999), and subsequently

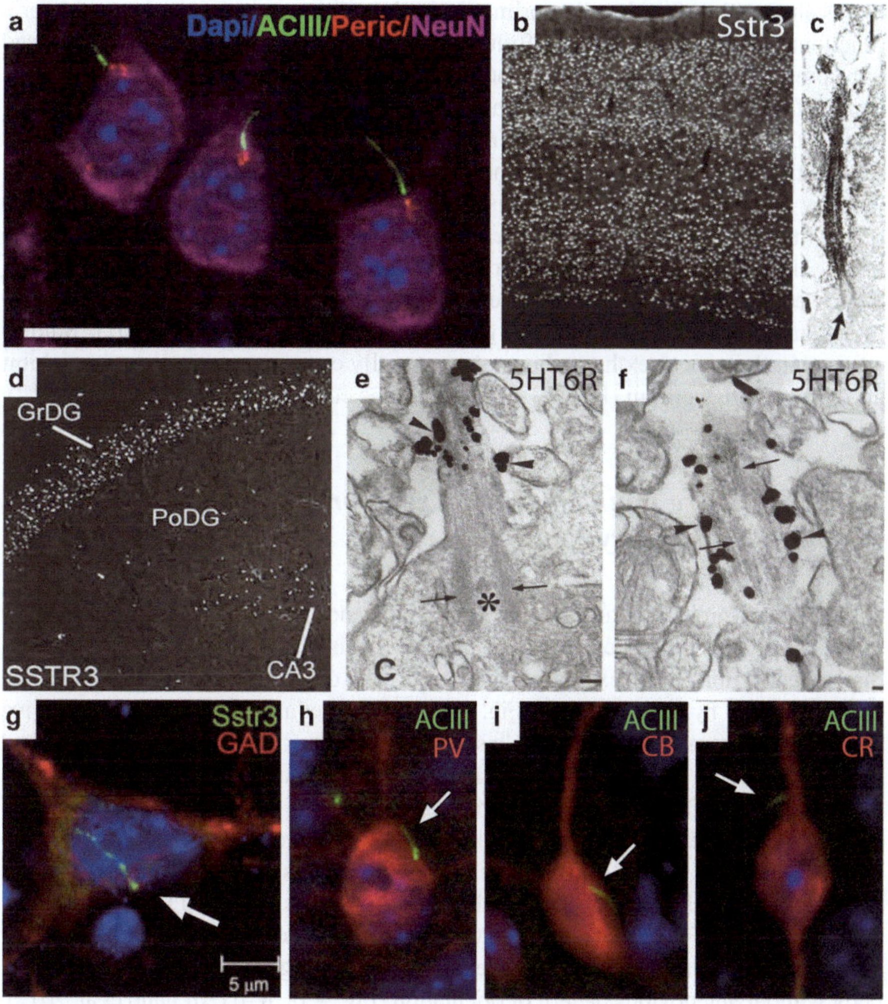

Fig. 4.1 Cilia markers. (**a**) ACIII is specifically enriched in cilia of pyramidal cells stained with NeuN and DAPI in the adult mouse neocortex, while pericentrin (Peric) is a specific marker of basal bodies. (**b**) SSTR3 is also enriched in neuronal cilia in the neocortex of adult mice. (**c**) Electron microscope image showing immuno detection of SSTR3 in neuronal cilia. (**d**) SSTR3 is enriched in neuronal cilia in the granular and polymorphic layers of the dentate gyrus (GrDG and PoDG respectively) and hippocampal CA3 fields. Note that SSTR3 is only expressed in neuronal layers, indicating lack of this marker in cilia from non-neuronal cells. (**e, f**) Electron microscope images showing 5HT6 receptor immunostaining mostly in the cilium membrane (*arrowhead*). The staining does not seem associated with axonemal microtubules (*arrows*). (**g**) GABAergic cells in culture expressing glutamic acid decarboxylase (GAD) exhibit Sstr3 positive cilia (*arrow*). (*h–j*), ACIII positive cilia (*arrows*) are present in neocortical GABAergic cells expressing parvalbumin (PV) (**h**), calbindin (CB) (**i**) and calretinin (CR) (**j**) in vivo. Panels *a*, *h–j* are from (Arellano et al. 2012). Panels *b–f* were reproduced with permission by ELSEVIER BV (Handel et al. 1999; Stanic et al. 2009; Brailov et al. 2000). Panel *g* was reproduced with permission by JOHN/WILEY & SONS, INC (Berbari et al. 2007)

it was shown that they localize primarily in the cilia membrane, as expected for a functional receptor (Fig. 4.1e, f) (Brailov et al. 2000). Similarly, Mchr1 was reported to be present in cilia only in certain brain regions: the hippocampus, nucleus accumbens, olfactory bulb, and hypothalamus (Berbari et al. 2008a). D1 receptor distribution in the brain has not been yet reported, and available data only shows cilia localization in cultured cells from the amygdala (Domire et al. 2011) and striatum (Marley and von Zastrow 2010). Bishop and colleagues used ACIII antibodies to report that this marker was expressed by cilia in most regions of the adult mouse brain, although differences in expression could be found: ACIII was strongly expressed in sensory and limbic areas, while there was paucity of expression in motor areas or in the thalamus or globus pallidus (Bishop et al. 2007). Studies focused in the neocortex have shown that virtually every neocortical neuron exhibit an ACIII positive cilium (Fig. 4.1a, h–j) (Anastas et al. 2011; Arellano et al. 2012).

4.2.3 Cilia Are Found on Both Excitatory and Inhibitory Neurons

Cilia appear to grow from all neuronal subtypes. In addition to the large evidence pointing to presence of cilia in excitatory neurons, analysis of hippocampal neurons in culture revealed that cells expressing glutamic acid decarboxylase (GAD) had cilia immunostained by ACIII or Sstr3 (Fig. 4.1g) (Berbari et al. 2007), and studies in vivo in the dentate gyrus and olfactory bulb showed that cells positive for GAD67 exhibited Sstr3 positive cilia (Stanic et al. 2009). In the neocortex, Arellano et al. (2012) showed that in all layers GABAergic interneurons, as defined by expression of parvalbumin, calbindin and calretinin, showed typically an ACIII positive cilium, similar in length to pyramidal neurons (Fig. 4.1h–j). Unlike pyramidal neurons whereby the cilium extends consistently near the base of the apical dendrite (Anastas et al. 2011; Arellano et al. 2012; Dahl 1963; Fuchs and Schwark 2004), interneurons do not appear to have a clear polarity to their cilium growth. It is often found elongating from the middle of the cell body without any obvious pattern of orientation (Fig. 4.1a, h–j) (Arellano et al. 2012). Thus, we now know that both excitatory and inhibitory neurons have a primary cilium (although olfactory epithelial neurons can grow ~ 11/neuron (Fuchs and Schwark 2004)). Whether the mechanisms of growth or signaling properties of interneurons are similar to those of excitatory neurons awaits further studies.

4.3 Overview of Neuronal Cilia Development: Focus on Cerebral Cortex

As discussed in Chap. 3, many studies have reported that progenitors in the ventricular zone typically exhibit a single primary cilium facing the ventricle (Caspary et al. 2007; Li et al. 2011; Willaredt et al. 2008; Wilson et al. 2012). We also know that most if not all neurons in adult animals exhibit cilia. However the process of neuronal ciliogenesis is not well known, and has only recently been addressed in neocortical neurons (Arellano et al. 2012) and in postnatally generated granule cells in the dentate gyrus (Kumamoto et al. 2012). Here we discuss the developmental appearance of neuronal cilia, factors regulating receptor localization to them and consequences of neuronal cilia structure/function.

4.3.1 Protracted Process of Neuronal Ciliogenesis

In the mouse neocortex, ciliogenesis was studied using ACIII antibodies and ultrastructural serial section analysis at different time points from late embryonic to postnatal animals (Arellano et al. 2012). ACIII immunoblotting showed very little expression early embryonically that increased at E18.5 leading to prominent postnatal expression, suggesting that ciliogenesis occurs mostly postnatally (Fig. 4.2a). However, ACIII immunostaining of the cortical plate at P0 and P4 revealed mostly a punctate pattern, suggesting that neuronal cilia are not elongated yet (Fig. 4.2c, d). Analysis at P8 and P14 revealed increasingly longer ACIII positive cilia, with slightly different rate of elongation depending on laminar position. Interestingly, ACIII positive neuronal cilia will not reach maximum length (4–6 μm on average) until P60-P90, showing a slow elongation process in all neocortical layers (Fig. 4.2d). Analysis in animals 6 months and 1 year old revealed shorter ACIII positive cilia, raising the possibility that cilia length homeostasis may change with age (Arellano et al. 2012).

The ultrastructural analysis of the developing cilia confirmed the findings obtained with ACIII, and analysis at embryonic day 16.5 revealed that neurons with migratory morphology did not have a cilium and the centrioles were observed free in the cytoplasm (Fig. 4.3a, f). The mother centriole frequently exhibited an attached vesicle, but no cilia were detected (Arellano et al. 2012). In neurons with non-migratory morphology that likely had reached their final position in the cortical plate, the mother centriole was typically attached to the cell membrane configuring the basal body (Fig. 4.3b, f). Furthermore, the basal body sometimes exhibited a membranous expansion resembling a short cilium (0.2–0.5 μm) typically filled with vesicles, tubular structures and diffuse content, but lacking clear axonemal organization. Thus, the authors named this structure a procilium (Fig. 4.3c, f). Procilia were reported as typically surrounded by vesicles and with the Golgi apparatus in close proximity, supporting the possibility of intense molecular trafficking between the cytoplasm and the cilium. Those procilia resembled the developing cilia described in diverse avian and mammalian tissues by Sorokin (1962, 1968a, b) and in the chick developing neural tube by Sotelo and Trujillo-Cenoz (1958). Also, they resemble the cilia buds frequently observed in neurons with compromised intraflagellar transport (Han et al. 2008). Electron microscopy analysis at P0 and P4 revealed that most neurons in the neocortex still exhibited a slightly longer procilium (0.5–2 μm) (Fig. 4.3c, f) without a developed axoneme except in rare cells in the subplate (P0) and in deep layers (P4) which showed some organized microtubular

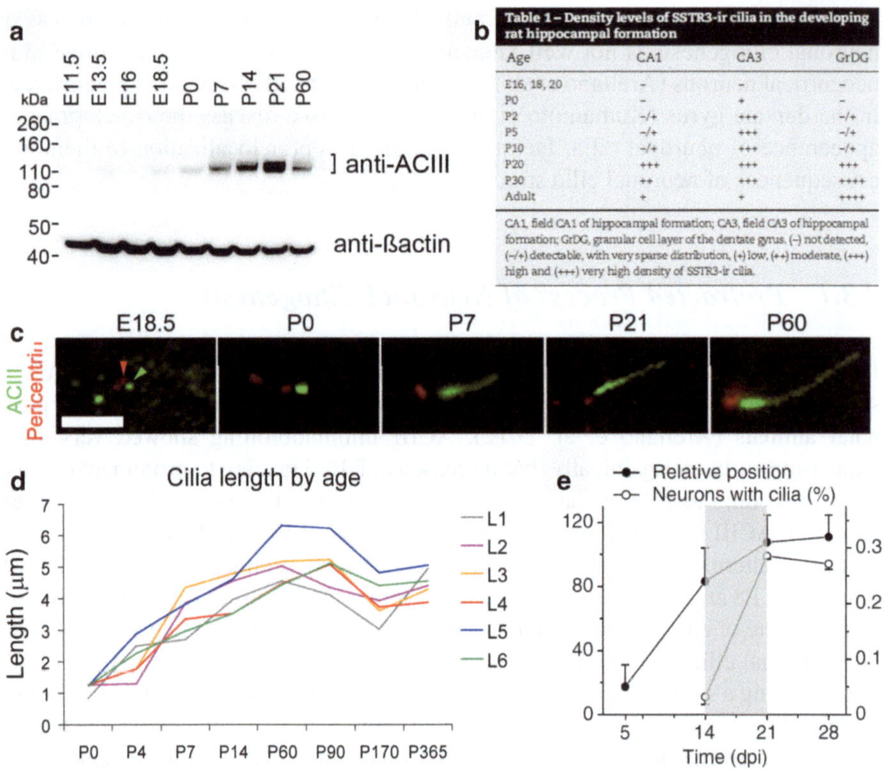

Table 1 – Density levels of SSTR3-ir cilia in the developing rat hippocampal formation

Age	CA1	CA3	GrDG
E16, 18, 20	–	–	–
P0	–	+	–
P2	–/+	+++	–/+
P5	–/+	+++	–/+
P10	–/+	+++	–/+
P20	+++	+++	+++
P30	++	+++	++++
Adult	+	+	++++

CA1, field CA1 of hippocampal formation; CA3, field CA3 of hippocampal formation; GrDG, granular cell layer of the dentate gyrus. (–) not detected, (–/+) detectable, with very sparse distribution, (+) low, (++) moderate, (+++) high and (++++) very high density of SSTR3-ir cilia.

Fig. 4.2 Development of neuronal cilia. (**a**) Western blot analysis of ACIII expression in mouse cortical lysates from embryonic day 11.5 (E11.5) to young adult (~P60) showing a band close to the predicted molecular weight of unglycosylated ACIII (~125 kDa). Although detectable at very low levels from E13.5, expression increases notably after E18.5, and especially postnatally between P0 and P21. At P60, there is a decrease in the intensity of the ACIII signal. β-Actin (*lower blot*) was a loading control. (**b**) Developmental distribution of SSTR3 positive cilia in the mouse hippocampus. Note the heterogeneity in the onset of SSTR3 expression between fields, and the decrease of expression in pyramidal fields in adult animals. (**c**) Examples of ACIII positive cilia (*green arrowhead*) at different developmental stages in mouse neocortical neurons illustrating the relative delay in elongation of cilia that only occurs postnatally. Pericentrin is used to label the basal body (*red arrowhead*). (**d**) Graph illustrating layer differences in elongation of ACIII positive cilia. Elongation is a slow process, with cilia reaching maximum length at P60-P90. Note the apparent shortening in animals 6 months and 1 year old compared to P60-P90 animals. (**e**) Analysis of the appearance of cilia in postnatally generated neurons in the dentate gyrus. The relative position of retrovirally labeled granule cells (*black dots*) and the percentage of ciliated neurons (*white dots*) was recorded at different time points after infection. The graph shows that cilia appear between 2 and 3 weeks after cells are generated (*shaded area*), when migration is terminating, suggesting that cilia start growing when migration is completed. Panels **a**, **c**, **d** are from (Arellano et al. 2012); Panel **b** was reproduced with permission of ELSEVIER BV (Stanic et al. 2009). Panel **e** was from (Kumamoto et al. 2012) (Panel **e**)

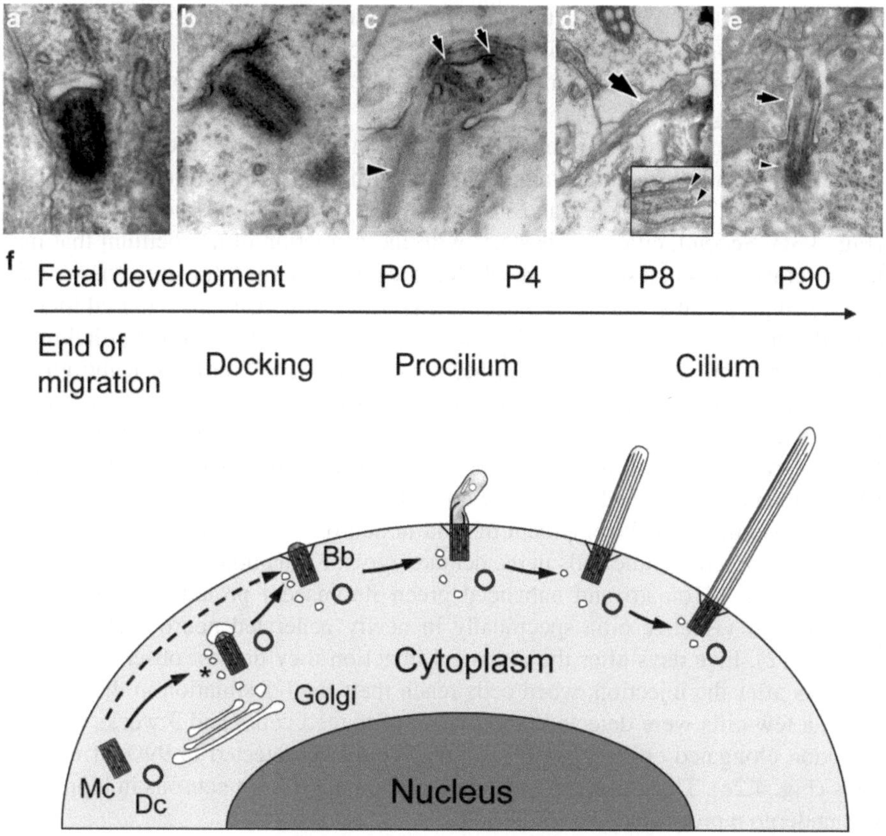

Fig. 4.3 Model of ciliogenesis in mouse neocortical neurons according to Arellano et al. 2012. Migrating neurons have their mother centriole (Mc) and daughter centriole (Dc) free in the cytoplasm. Once cells terminate migration and reach their appropriate lamina, the mother centriole attaches a vesicle (**a, f**), likely from the Golgi apparatus, buds a very short procilium and docks to the plasma membrane (alternatively the mother centriole docks directly to the plasma membrane without vesicle attachment as indicated by the *discontinuous arrow*). Docking to the membrane involves developing specific structures such as transition filaments and the mother centriole will become a basal body (Bb), frequently surrounded by vesicles (*asterisk*) (**b, f**). The basal body grows a procilium: a membranous expansion about 0.5–2 μm in length, lacking a proper axoneme and typically containing vesicles, short and disorganized tubular structures and electrondense diffuse content (**c**). This procilium does not display typical axonemal characteristics until ~P8 (**d**) although axonemal growth seem to start ~P0 in some subplate cells and could start ~P4 in some populations of neurons that showed early elongation of cilia (e.g. some layer five neurons). Overall, cilia will take weeks to fully elongate towards a peak ~P60-P90 (**e**), with some differences between layers. Arrows indicate procilia/cilia; arrowheads indicate basal bodies except in inset in D, which indicate microtubules (All panels from Arellano et al. 2012)

content. Analysis at P8 revealed elongated cilia with axonemal organization in neurons (Fig. 4.3d, f), suggesting that axoneme elongation could exhibit different onset in different cortical layers between P4 and P8, in agreement with the ACIII immunostaining data (Arellano et al. 2012).

The conclusions of this analysis in developing rodent neocortex are fourfold: first, that centrosome docking and basal body formation in immature neurons occurs after migration is completed and neurons reach their final position in the cortex (Fig. 4.3f). Second, ciliogenesis starts with the formation of a procilium that may last several days, suggesting that newly formed cilia undergo a fairly extensive organizational period, potentially packing the procilium with materials required for cilia growth and function. The molecular constituents contained within the procilium in different neuron subtypes await further characterization. Third, axoneme growth starts between P4-P8 in the neocortex and seems to drive cilia elongation. Fourth, cilia elongation is a protracted process that lasts about 2–3 months postnatally, and both the rate of growth and the maximal length are layer dependent (Fig. 4.2d) (Arellano et al. 2012; Fuchs and Schwark 2004).

The postmigratory development of cilia in neurons has been confirmed in postnatally generated granule cells in the dentate gyrus. Kumamoto et al. used a retrovirus encoding tdTomato and enhanced green fluorescent protein (EGFP) tagged Centrin-2 to visualize cilia specifically in newly generated neurons (Kumamoto et al. 2012). Five days after the retroviral injection they did not observe cilia, but 2 weeks after the injection, when cells reach their final destination in the granular layer, a few cilia were detected (about 10% of tagged cells) and 3 weeks after the injection elongated cilia (about 3 μm long) could be detected in 99% of infected cells (Fig. 4.2e). Thus, like the neocortex, cilia in adult-born neurons in hippocampus undergo a protracted growth period.

This slow ciliogenesis in cortical neurons is in sharp contrast with reports on cultured fibroblasts that can grow cilia in a matter of hours (Tucker et al. 1979a, b). Similarly, cultured neurons derived from fetal/perinatal tissues from a variety of brain regions including hippocampus, striatum, amygdala, cerebellum, and spinal cord can grow cilia in days (Barzi et al. 2010; Belgacem and Borodinsky 2011; Berbari et al. 2007; Domire et al. 2011; Miyoshi et al. 2009a). These differences in the timing of ciliogenesis could be partially explained by the migration process that neurons need to complete in vivo. However the protracted procilia stage in settled neurons in the neocortex does not seem to occur in cultured neurons or this process is accelerated. In addition, different brain regions exhibit marked differences in ciliogenesis. For example, cilia immunostaining and EM analysis reveals that limbic regions such as the hippocampus and piriform cortex exhibit elongated cilia even at PO (Han et al. 2008; Stanic et al. 2009; JIA unpublished observation), suggesting that procilia stage could be shorter and/or axoneme growth could be faster in those regions. The factors that control the rate of cilia elongation are not clear, but are likely attributable to numerous extracellular and intracellular factors (e.g., levels of soluble tubulin and polymerized actin (Sharma et al. 2011)). Similarly, Ou and colleagues showed that inhibition of ACIII produced significant elongation of cilia in a variety of cultured cells including human astrocytes, and proposed that cilia elongation could be regulated by the ACIII-cAMP pathway (Ou et al. 2009). They also showed that lithium has even larger effect on cilia elongation in culture, an effect also demonstrated in vitro and in vivo in mice by Miyoshi et al. (2009a). They reported that chronic ingestion of lithium (Li_2CO_3) in mice produced significant elongation of cilia in the dorsal striatum and nucleus accumbens. However the mechanism of action of lithium is not clear. Considering microtubule turnover in cilia is constantly being added and broken down (Marshall and Rosenbaum 2001), it is feasible that a final cilia length for neurons is not fixed, but rather may fluctuate with age.

4.3.2 Appearance and Targeting of Specific GPCRs to Neuronal Cilia

The developmental appearance of neuronal cilia GPCRs has not been extensively characterized. However we know that at least in hippocampus and neocortex, Sstr3 is detectable in cilia around birth (Anastas et al. 2011; Miyoshi et al. 2006; Stanic et al. 2009). Stanic et al. analyzed SSTR3 immunostaining in the developing hippocampus from embryonic stages to adult animals. This study revealed that Sstr3 is first detectable postnatally in cilia in the hippocampal fields and dentate gyrus (DG). However Sstr3 may not be a consistent marker of cilia as revealed by the low density of cilia immunostained in 5 months old animals in several hippocampal fields (Fig. 4.2b), and the lack of immunostaining in granular cells from the dentate gyrus at P0, when EM analysis has revealed the presence of elongated cilia with a well-developed axoneme (Han et al. 2008; Stanic et al. 2009). This suggests that there is not a correlation between the expression of signal transduction molecules and membrane receptors (e.g. ACIII, Sstr3) and the presence of a mature axoneme.

Why and how do these very specific receptor subtypes localize to the plasma membrane of neuronal cilia? One mechanism appears to be the conserved ciliary localization domains encoded within the protein amino acid sequence (Berbari et al. 2008a; Domire et al. 2011). For example, in 5HT6, SSTR3 and Mchr1 receptors, a sequence is found in the 3rd intracellular loop of the protein partially or largely responsible for the localization to cilia (Berbari et al. 2008a). Tagging these sequences onto other non-cilia localizing receptors can cause aberrant GPCR trafficking into cilia (Fig. 4.4). The entry of Sstr3 and Mchr1 into neuronal cilia is also coordinated by Tubby-like protein 3 (TULP3), a protein bridges IFT-A proteins and membrane phosphoinosides to allow GPCRs (e.g. Sstr3) to enter into cilia (Mukhopadhyay et al. 2010). Although not yet demonstrated for neurons, most GPCRs likely go through a conserved pericentrosomal preciliary complex, a recently identified vesiculotubular structure that stores and recycles ciliary GPCRs in an endosome manner (Kim et al. 2010). Moreover, very recent findings suggest that a transition zone complex is essential for cilia formation and acts as a two-way diffusion barrier, effectively preventing non-ciliary membrane proteins entering the cilium while retaining specific ciliary proteins inside the cilium (Chih et al. 2012). This complex includes the transmembrane component TMEM231, and the intracellular proteins Sept2, CC2D2A, and B9D1. These components likely

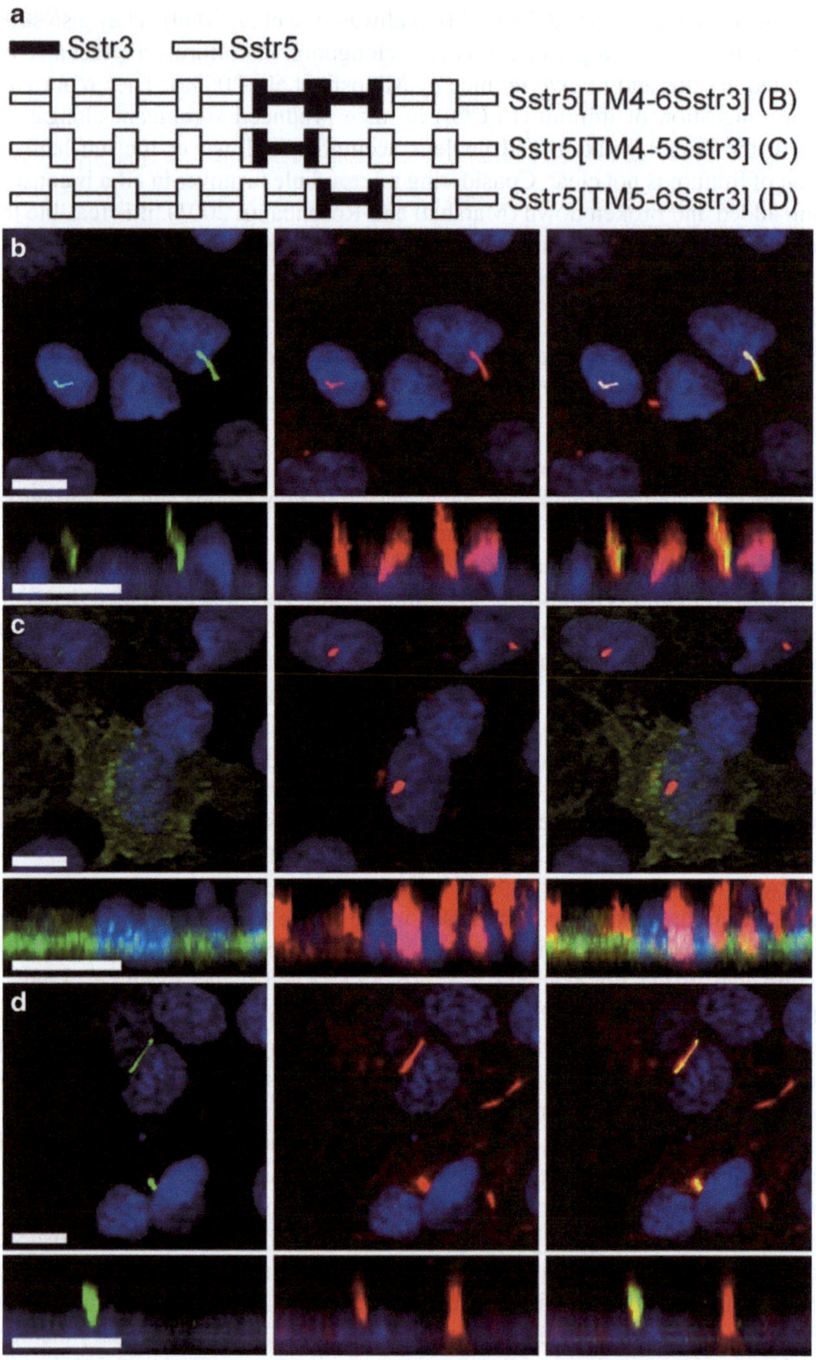
a
Sstr3
Sstr5
Sstr5[TM4-6Sstr3] (B)
Sstr5[TM4-5Sstr3] (C)
Sstr5[TM5-6Sstr3] (D)
b
c
d

form a small part of a larger ciliopathy-associated complex, including the other B9-C2 domain containing proteins and a host of others given the similar phenotypes seen upon loss of these proteins (Dowdle et al. 2011; Garcia-Gonzalo et al. 2011; Town et al. 2008; Weatherbee et al. 2009).

Other factors that direct GPCRs to neuronal cilia are Bardet Biedl Syndrome (BBS) proteins which form a large complex of proteins (called BBSome complex) (Berbari et al. 2008b; Jin et al. 2010; Ou et al. 2005). In the absence of BBS2 and BBS4, there is a loss of Sstr3 and Mchr1 localization in hippocampal neuronal cilia (Berbari et al. 2008b). This group found that by adding back BBS2 protein into BBS2 knockout neurons, Mchr1 localization into cilia is restored. In other areas of the brain (e.g., striatum, olfactory tubercle and amygdyla), loss of BBS4 can result in an abnormal influx of the D1 receptor into amygdyla neurons (Domire et al. 2011). Considering at least 14 *BBS* genes have been identified (Billingsley et al. 2010; Domire et al. 2011), the control of neuronal cilia GPCR targeting is likely highly regulated and dependent on various combinations of BBSome proteins. The roles for BBS proteins have expanded beyond the cilium as recent findings suggest they may also regulate gene transcription (Gascue et al. 2012). Thus it is tempting to speculate BBS proteins could act as a signaling bridge between the cilia and the nucleus.

4.3.3 Consequences of Mutations that Disrupt Neuronal Cilia Structure/Signaling: Role in Learning and Memory?

A host of ciliopathies in human (e.g. Bardet-Biedl Syndrome and Joubert Syndrome discussed further in Chap. 9) are associated with developmental delay, intellectual disability, cognitive dysfunction, autism spectrum and mood disorders (Bachmann-Gagescu et al. 2012; Bennouna-Greene et al. 2011; Green et al. 1989; for review see: Lee and Gleeson 2011; Louvi and Grove 2011). This begs the question: what is the consequence of ablating cilia/cilia function in the brain? In 2007, Davenport and colleagues analyzed IFT88 ^{Tg737} mice hypomorphic for the IFT-B complex protein and Kif3a conditional mice deficient for an anterograde motor required for ciliogenesis (Davenport et al. 2007). Surprisingly there was no catastrophic consequence on brain development or premature neurodegeneration (Davenport et al. 2007; Louvi and Grove 2011). However, these authors found that loss of cilia function associated

Fig. 4.4 The third intracellular (i3) loop of SSTR3 is sufficient to localize SSTR5 to cilia. (**a**) Schematic of chimeric receptors containing portions of SSTR3 (*black lines*) and SSTR5 (*white lines*) fused to EGFP. Transmembrane domains are depicted as *boxes*. (**b–d**) Images of IMCD cells expressing the indicated chimeric receptors. *Left*, EGFP (*green*); *middle*, acetylated α-tubulin (*red*); *right*, merged images. Chimeric receptors SSTR5[TM4-6SSTR3] (**b**) and SSTR5[TM5-6SSTR3] (**d**) localize to cilia, whereas chimeric receptor Sstr5[TM4-5Sstr3] (**c**) does not localize to cilia, suggesting that the i3 loop of SSTR3 contains ciliary localization sequences. Nuclei are *blue* (Figure and modified legend reproduced with permission by JOHN/WILEY & SONS, Inc. Berbari et al. 2008a)

with the hypothalamic area of the brain led to an obese phenotype. This phenotype is observed by other cilia mutations which would presumably affect hypothalamus (e.g. ACIII Wang et al. 2009). The mechanisms and functions of cilia in these regions are discussed in greater detail in later chapters of this book.

The lack of a catastrophic brain phenotype in mice lacking functional cilia does not completely suggest the absence of significant functional/network disruptions. In 2010, Einstein and colleagues found that mice lacking SSTR3 (or wildtype mice injected with an SSTR3 antagonist) displayed deficits in discriminating novel objects and recalling familiar objects (Einstein et al. 2010). These mice also exhibit deficits in adenylyl cyclase/cAMP-induced long-term potentiation in the hippocampus. Similarly, learning and memory defects are observed in mice deficient for MchR1 and ACIII (Adamantidis et al. 2005; Wang et al. 2011). If true, how could a loss of signaling from cilia lead to a change neural network function? Potentially shedding light on this is very recent work revealing that the loss of cilia on adult born hippocampal neurons can result in a slow change in dendritic morphology concurrently altering synaptic integration. In hippocampus, newborn neurons that are deprived from forming a cilium fail to extend normal dendritic arbors and receive weaker input from the entorhinal cortex (primary afferent input to DG) (Fig. 4.5a–f). These authors proposed that failed ciliogenesis may lead to abnormally high Wnt and ß catenin levels which has been observed in other types of cilia mutations (Corbit et al. 2008; Kumamoto et al. 2012). Interestingly, in contrast to blunt neuronal cilia, abnormally long cilia are also linked to ectopic dendrite outgrowth (Massinen et al. 2011). Collectively, the loss of signaling from cilia during cortical neuronal maturation may set the stage for network maturity and may contribute to the etiology of neurodevelopmental manifestations associated with ciliopathies.

4.4 Identification and Functions of Cilia on Non-neuronal Cells

As mentioned earlier, some cilia markers, like SSTR3 or 5HT6, have so far only be detected in neurons, and in general there is little information regarding the distribution of cilia in non-neuronal cells. However, it has been reported about 50% of GFAP positive cells in culture exhibited ACIII positive cilia (Berbari et al. 2007). Interestingly, Bishop et al. (2007) reported only rare ACIII positive cilia in cells immunostained for GFAP in tissue sections. These differences could be due to methodological issues or age difference in both studies: cultured P7 hippocampal cells in one case (Berbari et al. 2007) and tissue from adult mouse in the other (15), but could also indicate that cilia might have different features in vitro and in vivo. Moreover, neural progenitors in the brain exhibit astroglial features such as GFAP and nestin expression, and they contain specific receptors that are not present in neuronal cilia. We will discuss them in the next section.

Regarding other types of glia, Bishop and colleagues indicated that ACIII was not detected in oligodendrocytes or microglia (Bishop et al. 2007). However, the lack of

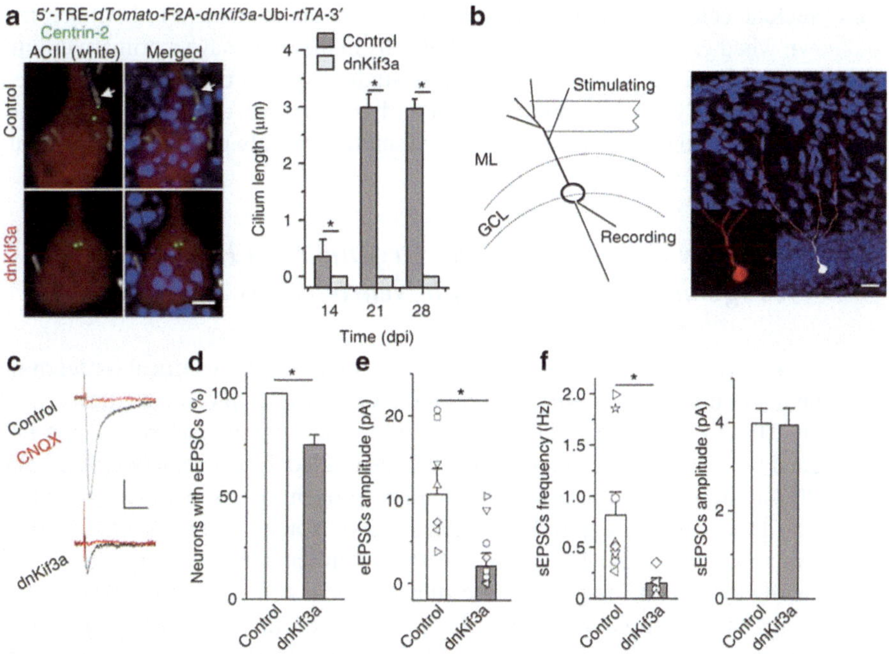

Fig. 4.5 Primary cilia depletion in adult born neurons is associated with impaired glutamatergic synapse formation. (**a**) *Top*, inducible retroviral vector used to ectopically express dominant-negative Kif3a (dnKif3a) in adult-born dentate granule cells (DGCs) to deplete primary cilia. *Bottom*, typical adult-born DGCs labeled with ACIII, centrioles and adult-born DGCs (*red*, dTomato). Note the ACIII positive axoneme (*arrow*) is absent in dnKif3a expressing DGCs. *Right*, cilia are blunt in dnKif3a-positive adult-born DGCs compared to control. (**b**) Electrophysiological recordings of newborn DGC filled with biocytin. The stimulating electrode was placed in the outer molecular layer (ML) to principally excite the entorhinal cortical projections. (**c–f**) Glutamatergic synaptic transmission recorded from control and dnKif3a-positive adult-born DGCs at 21 days post infection (dpi). Sample traces of glutamatergic synaptic transmission in the presence of 5 µM bicuculline are shown (**c**). DGCs lacking cilia on average showed lower amplitude of evoked excitatory postsynaptic currents (eEPSCs)(**e**) and lower frequency of spontaneous EPSCs (sEPSCs) (**f**) compared to control (Figure and modified legend from Kumamoto et al. 2012)

ACIII does not necessarily imply the lack of cilium, and studies in the peripheral nervous system indicate that pre-myelinating oligodendrocytes may have cilia that are lost during myelination (Yoshimura and Takeda 2012). It will be interesting to ascertain whether this is the case in the CNS though it will be a complex identification process as in CNS the oligodendrocyte lineage is convoluted. While many cells in this lineage become mature oligodendrocytes, others remain as polydendrocytes (NG2+ macroglia) and are the most proliferative cell type in the adult brain (Nishiyama et al. 2009). We are not aware of reports examining cilia expression in these cell types.

It is also unknown whether cortical microglia have primary cilia. However microglia are among the most dynamic and motile cell types in the brain (Nimmerjahn et al. 2005), and microtubule dynamics due to migration and/or changes in morphology may preclude centriole docking and the extension of an axoneme. Furthermore, it is unknown whether pericytes and endothelial cells in the brain have a primary cilium. We suspect if these cells do bear cilia, future studies will not only characterize their molecular profiles but also conditionally perturb their growth and functions to better understand their contributions to cortical, and more generally to brain development.

4.4.1 Primary Cilia as Signaling Organelles in Postnatal Progenitors: Roles in Adult Neurogenesis

It is well known that in postnatal hippocampus, a brain structure critical for learning and memory, there is ongoing proliferation of neuronal precursors throughout lifespan. These proliferating cells are situated in the subgranular layer of the DG (Ming and Song 2005). The chief precursors (also called Type 1 cells (Kempermann et al. 2004)) are positive for astrocyte/glial and progenitor cell markers, including glial fibrillary acid protein (GFAP), Nestin and Notch1 (Breunig et al. 2007; Kempermann et al. 2004; Lugert et al. 2010). Similar GFAP+neural stem cells reside along the lateral ventricles in the subventricular zone (SVZ), another area of continuing neurogenesis. However, the presence of this neuroneogenic SVZ-RMS-OB stream appears to diminish rapidly in humans and is likely absent in adults (Sanai et al. 2011; reviewed in Breunig et al. 2011). Here we discuss the results of studies that suggest these cells are responsive to Shh via their cilium that mediates the genesis of new neurons in these brain regions.

The cilia of Type 1 cells or astrocyte-like neural precursors (ALNPs) in mouse hippocampus lies within an invagination of the cell soma and close to the nucleus (Fig. 4.6), a characteristic feature of cilia also reported in both rat and human astrocytes (Moser et al. 2009; Yoshimura et al. 2011). Like neurons, the cilia of these GFAP-positive cells also enrich ACIII along the length of their cilia, but they are typically short (~1–2 μm) compared to neighboring neuronal cilia and are recessed substantially within the ciliary pocket. Unlike neighboring neuronal cilia, these cells appear to concentrate molecules important for Shh signaling. These cells can traffic Smo and Gli1 to their cilium, indicative of active Shh signaling in the cilium (Breunig

Fig. 4.6 Primary cilia are found on Type 1 neural precursor cells in the dentate gyrus. (**a**) Inset: Inducible Gfap-CreER^{T2}-mediated EGFP labeling of SGZ neural precursor cells. Mice were pulsed with tamoxifen for 3 days and sacrificed on the fifth day to label early precursor cells. (**b**) EGFP positive radial precursor (*green*) in the SGZ of the dentate gyrus. Note the ACIII positive cilia in the granule cell layer and SGZ. (**b'**) Enlarged view of outlined cell in **b**, demonstrating the small primary cilium (ACIII positive axoneme [*red*] extending from basal body [*blue*; gamma tubulin]) of this cell. (**c–c'**) 3D reconstruction from confocal z-stacks of cell in **b**. Note that this cilium is likely recessed in the membrane but this is hard to discriminate do to the limits of confocal microscopy. (**d–e$_3$**) Double immunoEM demonstrating localization of an ACIII positive cilium within a GFAP positive SGZ precursor cell body. (**f**) 3D reconstruction of cilium in d–e$_3$ showing that this cilium is recessed within the membrane of this cell (Figure and modified legend from Breunig et al. 2008)

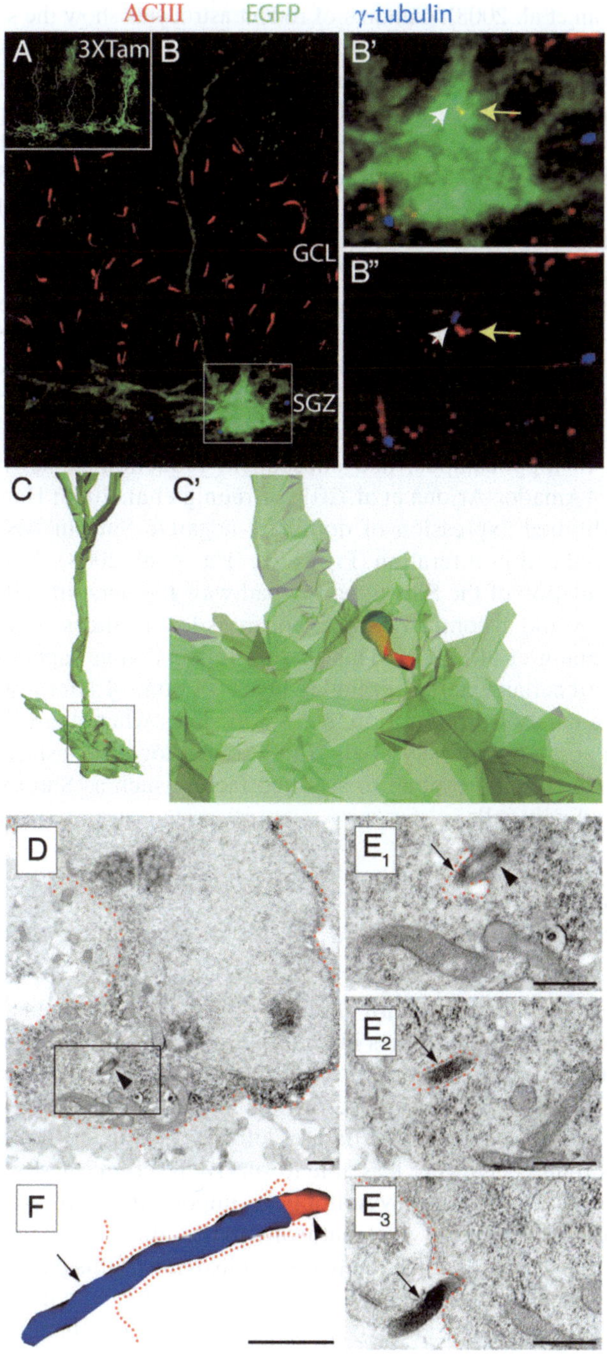
ACIII
EGFP
γ-tubulin
A
3XTam
B
B'
B"
GCL
SGZ
C
C'
D
E1
E2
E3
F

et al. 2008; Han et al. 2008). Cultures of rodent astrocytes show the same ability to traffic Shh signaling components (Yoshimura et al. 2011). Interestingly, SVZ neural stem cells contact the lumen of the lateral ventricles, forming a pinwheel structure with surrounding ependymal cells (Mirzadeh et al. 2008). At their apical domain, these GFAP+neural stem cells project a cilium into the ventricle. These cells are known to be Shh-responsive (Balordi and Fishell 2007; Palma et al. 2005) but whether Shh machinery localizes to the cilium in these cells is unknown. Furthermore, it is currently unknown whether transit-amplifying neuronal subtypes born from GFAP positive precursors in the DG and SVZ harbor cilia. It may be that their proliferative nature precludes this. In any event, the cilium in ALNPs is poised to respond to Shh to contribute to the continued production of neurons in the postnatal brain.

Multiple mouse mutants that disrupt ALNP proliferation show dramatic defects in cell proliferation and Shh signaling in hippocampus. For example, loss or dysfunction of *Kif3a*, *Ift88*, *Ift20* and *Stumpy* result in failed or aberrant genesis of cilia in hippocampus, and subsequent reduction in newborn neurons (Fig. 4.7a, b) (Amador-Arjona et al. 2011; Breunig et al. 2008; Han et al. 2008). Further, conditional expression of dominant-negative Smo in ANLP cilia also leads to reduced cell proliferation (Fig. 4.7c) (Han et al. 2008). Not only is there severe dysregulation of the Shh signaling pathway, the burst in cell proliferation that normally would accompany Shh exposure is lost in slices containing abnormal cilia (Breunig et al. 2008). This finding reflects what happens in slices of developing cerebellar external granular layer. In control slices, application of Shh induces a burst of proliferation, but not in slices where there is loss of cilia due to Kif3a absence (Spassky et al. 2008). Taken together, astrocyte-like cells extend cilia which may respond to secreted factors such as Shh to promote the proliferation of new cells.

4.4.2 Signaling via Glial Cilia: Roles in Cell Survival

Hedgehog-mediated signaling thru glial cell cilia may also influence a cell's ability to survive under stress. Studies in cultured rat astrocytes have shown that cell death (induced by serum starvation) could be mitigated if cells are maintained in the presence of hedgehog pathway activator and Smo-agonist (SAG) (Yoshimura et al. 2011). Further when glial cilia are deformed using shRNA against Ift20, these authors found cells were less able to survive a serum starvation. Thus, signaling via the glial cell cilia may not only be important for stimulating cell proliferation, but activating survival pathways when the cell is presented harsh extracellular conditions. The mechanisms downstream of this pathway need to be elucidated, but could involve survival factors such as Akt which can be recruited to primary cilia (Michaud and Yoder 2006) and mediate hedgehog signaling (Riobo et al. 2006).

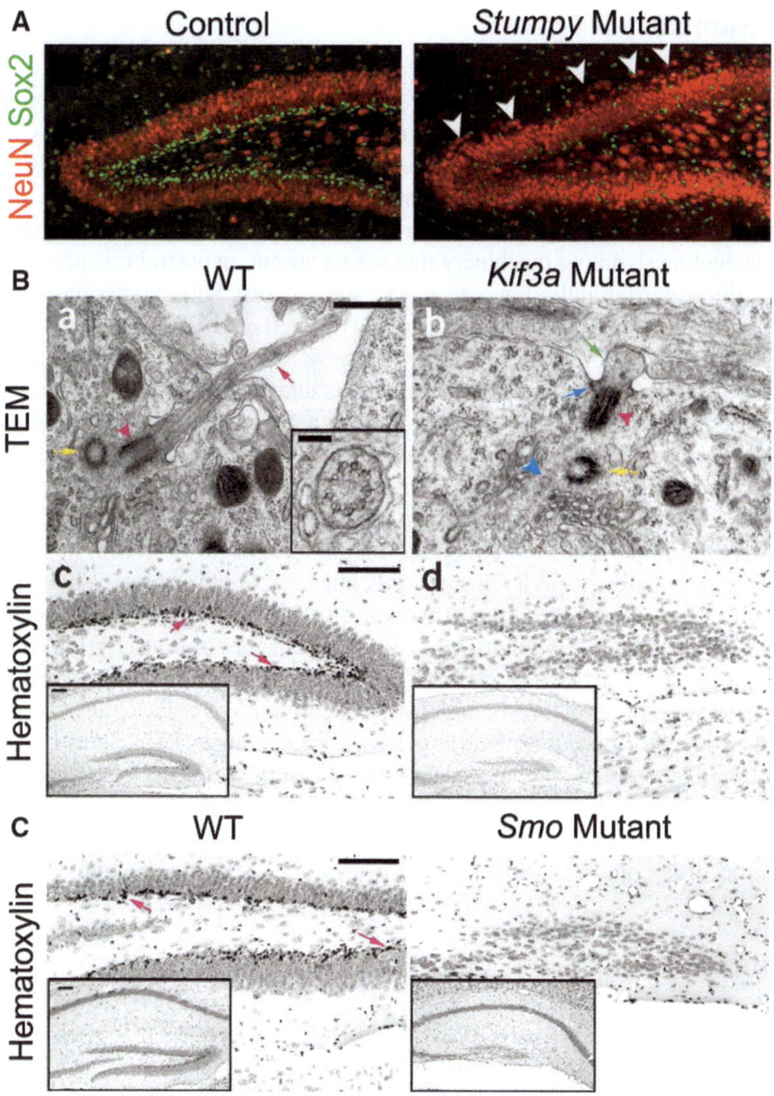

Fig. 4.7 Dentate gyrus defects observed in ciliary and Shh mutants. (**A**) Comparison of control and *Stumpy* conditional mutant mouse immunolabeled with NeuN (*red*; mature neurons) and Sox2 (*green*; astroglia and precursor cells), displaying the lack of SGZ Sox2 positive precursors and ectopic neurons above the granule cell layer (*arrowheads*). (**B**) TEM (*a–b*) demonstrating the presence and defective formation of cilia in the *WT* and *Kif3a* mutant mouse, respectively. (*c–d*) Hematoxylin staining of the dentate gyrus of control and *Kif3a* mutants. Note the hypoplasia of the Kif3a granule cell layer. (**C**) The hypoplasia of the Kif3a GCL is phenocopied in mice lacking *Smo*, the Shh pathway effector which localizes to precursor cilia. Panels A and modified legend from Breunig et al. (2008). Panels B and C and modified legend from Han et al. (2008)

4.5 Summary

Given these studies, neuronal cilia have shed their status as vestigial structures. It is clear that not all cilia are built equally in neurons and glia. Based on cell type, cilia differentially enrich various types of neuromodulatory receptors. Similarly, ALNP cilia appear to be capable of controlling continued genesis of neurons in the hippocampus and other types of non-neuronal cilia potentially regulate myelination, using hedgehog signaling machinery that seems absent on postmitotic neuronal cilia in the cortex. Intriguingly, a recent report showed that Shh expression persists in layer 5 pyramidal neurons in the adult brain (Harwell et al. 2012), and it is tempting to speculate that Shh release from these neurons could act on a variety of nearby cilia. The potential importance of neural cilia is underscored by the various types of learning and memory deficits seen in both mutant mice and human ciliopathies, which may in part be attributable to a function for cilia in coordinating dendritic and synaptic development (Tissir and Goffinet 2012). Unraveling the contributions of neuronal and glial cilia biology will help understand the etiology of these disorders.

A number of questions remain regarding the roles of primary cilia in cortex and rest of the brain. How many molecules are cilia capable of sensing? Where do these molecules come from? Are cilia sensitive to concentration gradients, mechanical stimulation or both? What downstream signaling changes or transcription factors do cilia control? Is the process of neuronal/glial ciliogenesis identical between groups of cells? How is brain plasticity affected in the absence of cilia? Could these structures contribute to neurodegenerative diseases of the brain as implications arise (e.g. Amyotrophic Lateral Sclerosis, Alzheimer's, Huntington Disease (Chakravarthy et al. 2011; Keryer et al. 2011; Ma et al. 2012))? Answers to these questions will most certainly become an intense area of research over the next decade and add another level of complexity to our views of neural cell communication in the brain.

References

Adamantidis A, Thomas E, Foidart A, Tyhon A, Coumans B, Minet A, Tirelli E, Seutin V, Grisar T, Lakaye B (2005) Disrupting the melanin-concentrating hormone receptor 1 in mice leads to cognitive deficits and alterations of NMDA receptor function. Eur J Neurosci 21:2837–2844

Amador-Arjona A, Elliott J, Miller A, Ginbey A, Pazour GJ, Enikolopov G, Roberts AJ, Terskikh AV (2011) Primary cilia regulate proliferation of amplifying progenitors in adult hippocampus: implications for learning and memory. J Neurosci 31:9933–9944

Anastas SB, Mueller D, Semple-Rowland SL, Breunig JJ, Sarkisian MR (2011) Failed cytokinesis of neural progenitors in citron kinase-deficient rats leads to multiciliated neurons. Cereb Cortex 21:338–344

Arellano JI, Guadiana SM, Breunig JJ, Rakic P, Sarkisian MR (2012) Development and distribution of neuronal cilia in mouse neocortex. J Comp Neurol 590:848–873

Bachmann-Gagescu R, Ishak GE, Dempsey JC, Adkins J, O'Day D, Phelps IG, Gunay-Aygun M, Kline AD, Szczaluba K, Martorell L, Alswaid A, Alrasheed S, Pai S, Izatt L, Ronan A, Parisi MA, Mefford H, Glass I, Doherty D (2012) Genotype-phenotype correlation in CC2D2A-related Joubert syndrome reveals an association with ventriculomegaly and seizures. J Med Genet 49:126–137

Balordi F, Fishell G (2007) Hedgehog signaling in the subventricular zone is required for both the maintenance of stem cells and the migration of newborn neurons. J Neurosci 27:5936–5947

Barzi M, Berenguer J, Menendez A, Alvarez-Rodriguez R, Pons S (2010) Sonic-hedgehog-mediated proliferation requires the localization of PKA to the cilium base. J Cell Sci 123:62–69

Belgacem YH, Borodinsky LN (2011) Sonic hedgehog signaling is decoded by calcium spike activity in the developing spinal cord. Proc Natl Acad Sci USA 108:4482–4487

Bennouna-Greene V, Kremer S, Stoetzel C, Christmann D, Schuster C, Durand M, Verloes A, Sigaudy S, Holder-Espinasse M, Godet J, Brandt C, Marion V, Danion A, Dietemann JL, Dollfus H (2011) Hippocampal dysgenesis and variable neuropsychiatric phenotypes in patients with Bardet-Biedl syndrome underline complex CNS impact of primary cilia. Clin Genet 80:523–531

Berbari NF, Bishop GA, Askwith CC, Lewis JS, Mykytyn K (2007) Hippocampal neurons possess primary cilia in culture. J Neurosci Res 85:1095–1100

Berbari NF, Johnson AD, Lewis JS, Askwith CC, Mykytyn K (2008a) Identification of ciliary localization sequences within the third intracellular loop of G protein-coupled receptors. Mol Biol Cell 19:1540–1547

Berbari NF, Lewis JS, Bishop GA, Askwith CC, Mykytyn K (2008b) Bardet-Biedl syndrome proteins are required for the localization of G protein-coupled receptors to primary cilia. Proc Natl Acad Sci USA 105:4242–4246

Billingsley G, Bin J, Fieggen KJ, Duncan JL, Gerth C, Ogata K, Wodak SS, Traboulsi EI, Fishman GA, Paterson A, Chitayat D, Knueppel T, Millan JM, Mitchell GA, Deveault C, Heon E (2010) Mutations in chaperonin-like BBS genes are a major contributor to disease development in a multiethnic Bardet-Biedl syndrome patient population. J Med Genet 47:453–463

Bishop GA, Berbari NF, Lewis J, Mykytyn K (2007) Type III adenylyl cyclase localizes to primary cilia throughout the adult mouse brain. J Comp Neurol 505:562–571

Brailov I, Bancila M, Brisorgueil MJ, Miquel MC, Hamon M, Verge D (2000) Localization of 5-HT(6) receptors at the plasma membrane of neuronal cilia in the rat brain. Brain Res 872:271–275

Breunig JJ, Silbereis J, Vaccarino FM, Sestan N, Rakic P (2007) Notch regulates cell fate and dendrite morphology of newborn neurons in the postnatal dentate gyrus. Proc Natl Acad Sci USA 104:20558–20563

Breunig JJ, Sarkisian MR, Arellano JI, Morozov YM, Ayoub AE, Sojitra S, Wang B, Flavell RA, Rakic P, Town T (2008) Primary cilia regulate hippocampal neurogenesis by mediating sonic hedgehog signaling. Proc Natl Acad Sci USA 105:13127–13132

Breunig JJ, Haydar TF, Rakic P (2011) Neural stem cells: historical perspective and future prospects. Neuron 70:614–625

Cantagrel V, Silhavy JL, Bielas SL, Swistun D, Marsh SE, Bertrand JY, Audollent S, Attie-Bitach T, Holden KR, Dobyns WB, Traver D, Al-Gazali L, Ali BR, Lindner TH, Caspary T, Otto EA, Hildebrandt F, Glass IA, Logan CV, Johnson CA, Bennett C, Brancati F, Valente EM, Woods CG, Gleeson JG (2008) Mutations in the cilia gene ARL13B lead to the classical form of Joubert syndrome. Am J Hum Genet 83:170–179

Caspary T, Larkins CE, Anderson KV (2007) The graded response to sonic hedgehog depends on cilia architecture. Dev Cell 12:767–778

Chakravarthy B, Gaudet C, Menard M, Atkinson T, Chiarini A, Dal Pra I, Whitfield J (2010) The p75 neurotrophin receptor is localized to primary cilia in adult murine hippocampal dentate gyrus granule cells. Biochem Biophys Res Commun 401:458–462

Chakravarthy B, Gaudet C, Menard M, Atkinson T, Brown L, Laferla FM, Armato U, Whitfield J (2011) Amyloid-beta peptides stimulate the expression of the p75(NTR) neurotrophin receptor in SHSY5Y human neuroblastoma cells and AD transgenic mice. J Alzheimers Dis 19:915–925

Chih B, Liu P, Chinn Y, Chalouni C, Komuves LG, Hass PE, Sandoval W, Peterson AS (2012) A ciliopathy complex at the transition zone protects the cilia as a privileged membrane domain. Nat Cell Biol 14:61–72

Corbit KC, Shyer AE, Dowdle WE, Gaulden J, Singla V, Chen MH, Chuang PT, Reiter JF (2008) Kif3a constrains beta-catenin-dependent Wnt signalling through dual ciliary and non-ciliary mechanisms. Nat Cell Biol 10:70–76

D'Souza-Schorey C, Chavrier P (2006) ARF proteins: roles in membrane traffic and beyond. Nat Rev Mol Cell Biol 7:347–358

Dahl HA (1963) Fine structure of cilia in rat cerebral cortex. Z Zellforsch Mikrosk Anat 60:369–386

Davenport JR, Yoder BK (2005) An incredible decade for the primary cilium: a look at a once-forgotten organelle. Am J Physiol Renal Physiol 289:F1159–F1169

Davenport JR, Watts AJ, Roper VC, Croyle MJ, van Groen T, Wyss JM, Nagy TR, Kesterson RA, Yoder BK (2007) Disruption of intraflagellar transport in adult mice leads to obesity and slow-onset cystic kidney disease. Curr Biol 17:1586–1594

Del Cerro MP, Snider RS (1969) The Purkinje cell cilium. Anat Rec 165:127–130

Domire JS, Mykytyn K (2009) Markers for neuronal cilia. Methods Cell Biol 91:111–121

Domire JS, Green JA, Lee KG, Johnson AD, Askwith CC, Mykytyn K (2011) Dopamine receptor 1 localizes to neuronal cilia in a dynamic process that requires the Bardet-Biedl syndrome proteins. Cell Mol Life Sci 68:2951–2960

Dowdle WE, Robinson JF, Kneist A, Sirerol-Piquer MS, Frints SG, Corbit KC, Zaghloul NA, van Lijnschoten G, Mulders L, Verver DE, Zerres K, Reed RR, Attie-Bitach T, Johnson CA, Garcia-Verdugo JM, Katsanis N, Bergmann C, Reiter JF (2011) Disruption of a ciliary B9 protein complex causes Meckel syndrome. Am J Hum Genet 89:94–110

Einstein EB, Patterson CA, Hon BJ, Regan KA, Reddi J, Melnikoff DE, Mateer MJ, Schulz S, Johnson BN, Tallent MK (2010) Somatostatin signaling in neuronal cilia is critical for object recognition memory. J Neurosci 30:4306–4314

Fuchs JL, Schwark HD (2004) Neuronal primary cilia: a review. Cell Biol Int 28:111–118

Garcia-Gonzalo FR, Corbit KC, Sirerol-Piquer MS, Ramaswami G, Otto EA, Noriega TR, Seol AD, Robinson JF, Bennett CL, Josifova DJ, Garcia-Verdugo JM, Katsanis N, Hildebrandt F, Reiter JF (2011) A transition zone complex regulates mammalian ciliogenesis and ciliary membrane composition. Nat Genet 43:776–784

Gascue C, Tan PL, Cardenas-Rodriguez M, Libisch G, Fernandez-Calero T, Liu YP, Astrada S, Robello C, Naya H, Katsanis N, Badano JL (2012) Direct role of Bardet-Biedl syndrome proteins in transcriptional regulation. J Cell Sci 125:362–375

Green JA, Mykytyn K (2010) Neuronal ciliary signaling in homeostasis and disease. Cell Mol Life Sci 67:3287–3297

Green JS, Parfrey PS, Harnett JD, Farid NR, Cramer BC, Johnson G, Heath O, McManamon PJ, O'Leary E, Pryse-Phillips W (1989) The cardinal manifestations of Bardet-Biedl syndrome, a form of Laurence-Moon-Biedl syndrome. N Engl J Med 321:1002–1009

Hamon M, Doucet E, Lefevre K, Miquel MC, Lanfumey L, Insausti R, Frechilla D, Del Rio J, Verge D (1999) Antibodies and antisense oligonucleotide for probing the distribution and putative functions of central 5-HT6 receptors. Neuropsychopharmacology 21:68S–76S

Han YG, Spassky N, Romaguera-Ros M, Garcia-Verdugo JM, Aguilar A, Schneider-Maunoury S, Alvarez-Buylla A (2008) Hedgehog signaling and primary cilia are required for the formation of adult neural stem cells. Nat Neurosci 11:277–284

Handel M, Schulz S, Stanarius A, Schreff M, Erdtmann-Vourliotis M, Schmidt H, Wolf G, Hollt V (1999) Selective targeting of somatostatin receptor 3 to neuronal cilia. Neuroscience 89:909–926

Harwell CC, Parker PR, Gee SM, Okada A, McConnell SK, Kreitzer AC, Kriegstein AR (2012) Sonic hedgehog expression in corticofugal projection neurons directs cortical microcircuit formation. Neuron 73:1116–1126

Jin H, White SR, Shida T, Schulz S, Aguiar M, Gygi SP, Bazan JF, Nachury MV (2010) The conserved Bardet-Biedl syndrome proteins assemble a coat that traffics membrane proteins to cilia. Cell 141:1208–1219

Karlsson U (1966) Three-dimensional studies of neurons in the lateral geniculate nucleus of the rat. I. Organelle organization in the perikaryon and its proximal branches. J Ultrastruct Res 16:429–481

Kempermann G, Jessberger S, Steiner B, Kronenberg G (2004) Milestones of neuronal development in the adult hippocampus. Trends Neurosci 27:447–452

Keryer G, Pineda JR, Liot G, Kim J, Dietrich P, Benstaali C, Smith K, Cordelieres FP, Spassky N, Ferrante RJ, Dragatsis I, Saudou F (2011) Ciliogenesis is regulated by a huntingtin-HAP1-PCM1 pathway and is altered in Huntington disease. J Clin Invest 121:4372–4382

Kim J, Lee JE, Heynen-Genel S, Suyama E, Ono K, Lee K, Ideker T, Aza-Blanc P, Gleeson JG (2010) Functional genomic screen for modulators of ciliogenesis and cilium length. Nature 464:1048–1051

Kumamoto N, Gu Y, Wang J, Janoschka S, Takemaru K, Levine J, Ge S (2012) A role for primary cilia in glutamatergic synaptic integration of adult-born neurons. Nat Neurosci 15:399–405

Lee JE, Gleeson JG (2011) Cilia in the nervous system: linking cilia function and neurodevelopmental disorders. Curr Opin Neurol 24:98–105

Leroux MR (2010) Tubulin acetyltransferase discovered: ciliary role in the ancestral eukaryote expanded to neurons in metazoans. Proc Natl Acad Sci USA 107:21238–21239

Li A, Saito M, Chuang JZ, Tseng YY, Dedesma C, Tomizawa K, Kaitsuka T, Sung CH (2011) Ciliary transition zone activation of phosphorylated Tctex-1 controls ciliary resorption, S-phase entry and fate of neural progenitors. Nat Cell Biol 13:402–411

Louvi A, Grove EA (2011) Cilia in the CNS: the quiet organelle claims center stage. Neuron 69:1046–1060

Lugert S, Basak O, Knuckles P, Haussler U, Fabel K, Gotz M, Haas CA, Kempermann G, Taylor V, Giachino C (2010) Quiescent and active hippocampal neural stem cells with distinct morphologies respond selectively to physiological and pathological stimuli and aging. Cell Stem Cell 6:445–456

Ma X, Peterson R, Turnbull J (2012) Adenylyl cyclase type 3, a marker of primary cilia, is reduced in primary cell culture and in lumbar spinal cord in situ in G93A SOD1 mice. BMC Neurosci 12:71

Mandl L, Megele R (1989) Primary cilia in normal human neocortical neurons. Z Mikrosk Anat Forsch 103:425–430

Marley A, von Zastrow M (2010) DISC1 regulates primary cilia that display specific dopamine receptors. PLoS One 5:e10902

Marshall WF, Rosenbaum JL (2001) Intraflagellar transport balances continuous turnover of outer doublet microtubules: implications for flagellar length control. J Cell Biol 155:405–414

Massinen S, Hokkanen ME, Matsson H, Tammimies K, Tapia-Paez I, Dahlstrom-Heuser V, Kuja-Panula J, Burghoorn J, Jeppsson KE, Swoboda P, Peyrard-Janvid M, Toftgard R, Castren E, Kere J (2011) Increased expression of the dyslexia candidate gene DCDC2 affects length and signaling of primary cilia in neurons. PLoS One 6:e20580

Michaud EJ, Yoder BK (2006) The primary cilium in cell signaling and cancer. Cancer Res 66:6463–6467

Ming GL, Song H (2005) Adult neurogenesis in the mammalian central nervous system. Annu Rev Neurosci 28:223–250

Mirzadeh Z, Merkle FT, Soriano-Navarro M, Garcia-Verdugo JM, Alvarez-Buylla A (2008) Neural stem cells confer unique pinwheel architecture to the ventricular surface in neurogenic regions of the adult brain. Cell Stem Cell 3:265–278

Miyoshi K, Onishi K, Asanuma M, Miyazaki I, Diaz-Corrales FJ, Ogawa N (2006) Embryonic expression of pericentrin suggests universal roles in ciliogenesis. Dev Genes Evol 216:537–542

Miyoshi K, Kasahara K, Miyazaki I, Asanuma M (2009a) Lithium treatment elongates primary cilia in the mouse brain and in cultured cells. Biochem Biophys Res Commun 388:757–762

Miyoshi K, Kasahara K, Miyazaki I, Shimizu S, Taniguchi M, Matsuzaki S, Tohyama M, Asanuma M (2009b) Pericentrin, a centrosomal protein related to microcephalic primordial dwarfism, is required for olfactory cilia assembly in mice. FASEB J 23:3289–3297

Moser JJ, Fritzler MJ, Rattner JB (2009) Primary ciliogenesis defects are associated with human astrocytoma/glioblastoma cells. BMC Cancer 9:448

Mukhopadhyay S, Wen X, Chih B, Nelson CD, Lane WS, Scales SJ, Jackson PK (2010) TULP3 Bridges the IFT-A complex and membrane phosphoinositides to promote trafficking of G protein-coupled receptors into primary cilia. Genes Dev 24:2180–2193

Nimmerjahn A, Kirchhoff F, Helmchen F (2005) Resting microglial cells are highly dynamic surveillants of brain parenchyma in vivo. Science 308:1314–1318
Nishiyama A, Komitova M, Suzuki R, Zhu X (2009) Polydendrocytes (NG2 cells): multifunctional cells with lineage plasticity. Nat Rev Neurosci 10:9–22
Ou G, Blacque OE, Snow JJ, Leroux MR, Scholey JM (2005) Functional coordination of intraflagellar transport motors. Nature 436:583–587
Ou Y, Ruan Y, Cheng M, Moser JJ, Rattner JB, van der Hoorn FA (2009) Adenylate cyclase regulates elongation of mammalian primary cilia. Exp Cell Res 315:2802–2817
Palma V, Lim DA, Dahmane N, Sanchez P, Brionne TC, Herzberg CD, Gitton Y, Carleton A, Alvarez-Buylla A, Ruiz i Altaba A (2005) Sonic hedgehog controls stem cell behavior in the postnatal and adult brain. Development 132:335–344
Piperno G, Fuller MT (1985) Monoclonal antibodies specific for an acetylated form of alpha-tubulin recognize the antigen in cilia and flagella from a variety of organisms. J Cell Biol 101:2085–2094
Riobo NA, Lu K, Ai X, Haines GM, Emerson CP Jr (2006) Phosphoinositide 3-kinase and Akt are essential for Sonic Hedgehog signaling. Proc Natl Acad Sci USA 103:4505–4510
Rooryck C, Pelras S, Chateil JF, Cances C, Arveiler B, Verloes A, Lacombe D, Goizet C (2007) Bardet-Biedl syndrome and brain abnormalities. Neuropediatrics 38:5–9
Sanai N, Nguyen T, Ihrie RA, Mirzadeh Z, Tsai HH, Wong M, Gupta N, Berger MS, Huang E, Garcia-Verdugo JM, Rowitch DH, Alvarez-Buylla A (2011) Corridors of migrating neurons in the human brain and their decline during infancy. Nature 478:382–386
Santiago Ramón y Cajal (trans: Swanson N, Swanson LW) (1995) Histology of the nervous system, vols I & II. Oxford University Press, New York, pp 805– 806
Sharma N, Kosan ZA, Stallworth JE, Berbari NF, Yoder BK (2011) Soluble levels of cytosolic tubulin regulate ciliary length control. Mol Biol Cell 22:808–816
Shida T, Cueva JG, Xu Z, Goodman MB, Nachury MV (2010) The major alpha-tubulin K40 acetyltransferase alphaTAT1 promotes rapid ciliogenesis and efficient mechanosensation. Proc Natl Acad Sci USA 107:21517–21522
Sorokin S (1962) Centrioles and the formation of rudimentary cilia by fibroblasts and smooth muscle cells. J Cell Biol 15:363–377
Sorokin SP (1968a) Centriole formation and ciliogenesis. Aspen Emphysema Conf 11:213–216
Sorokin SP (1968b) Reconstructions of centriole formation and ciliogenesis in mammalian lungs. J Cell Sci 3:207–230
Sotelo JR, Trujillo-Cenoz O (1958) Electron microscope study on the development of ciliary components of the neural epithelium of the chick embryo. Z Zellforsch Mikrosk Anat 49:1–12
Spassky N, Han YG, Aguilar A, Strehl L, Besse L, Laclef C, Ros MR, Garcia-Verdugo JM, Alvarez-Buylla A (2008) Primary cilia are required for cerebellar development and Shh-dependent expansion of progenitor pool. Dev Biol 317:246–259
Stanic D, Malmgren H, He H, Scott L, Aperia A, Hokfelt T (2009) Developmental changes in frequency of the ciliary somatostatin receptor 3 protein. Brain Res 1249:101–112
Tissir F, Goffinet AM (2012) Cilia: conductors' batons of neuronal maturation. Nat Neurosci 15:344–345
Town T, Breunig JJ, Sarkisian MR, Spilianakis C, Ayoub AE, Liu X, Ferrandino AF, Gallagher AR, Li MO, Rakic P, Flavell RA (2008) The stumpy gene is required for mammalian ciliogenesis. Proc Natl Acad Sci USA 105:2853–2858
Tucker RW, Pardee AB, Fujiwara K (1979a) Centriole ciliation is related to quiescence and DNA synthesis in 3T3 cells. Cell 17:527–535
Tucker RW, Scher CD, Stiles CD (1979b) Centriole deciliation associated with the early response of 3T3 cells to growth factors but not to SV40. Cell 18:1065–1072
Vincensini L, Blisnick T, Bastin P (2011) 1001 model organisms to study cilia and flagella. Biol Cell 103:109–130
Wang Z, Li V, Chan GC, Phan T, Nudelman AS, Xia Z, Storm DR (2009) Adult type 3 adenylyl cyclase-deficient mice are obese. PLoS One 4:e6979
Wang Z, Phan T, Storm DR (2011) The type 3 adenylyl cyclase is required for novel object learning and extinction of contextual memory: role of cAMP signaling in primary cilia. J Neurosci 31:5557–5561

Weatherbee SD, Niswander LA, Anderson KV (2009) A mouse model for Meckel syndrome reveals Mks1 is required for ciliogenesis and Hedgehog signaling. Hum Mol Genet 18:4565–4575

Wheatley DN, Wang AM, Strugnell GE (1996) Expression of primary cilia in mammalian cells. Cell Biol Int 20:73–81

Whitfield JF (2004) The neuronal primary cilium – an extrasynaptic signaling device. Cell Signal 16:763–767

Willaredt MA, Hasenpusch-Theil K, Gardner HA, Kitanovic I, Hirschfeld-Warneken VC, Gojak CP, Gorgas K, Bradford CL, Spatz J, Wolfl S, Theil T, Tucker KL (2008) A crucial role for primary cilia in cortical morphogenesis. J Neurosci 28:12887–12900

Wilson SL, Wilson JP, Wang C, Wang B, McConnell SK (2012) Primary cilia and Gli3 activity regulate cerebral cortical size. Dev Neurobiol 72:1196–1212

Yoshimura K, Takeda S (2012) Hedgehog signaling regulates myelination in the peripheral nervous system through primary cilia. Differentiation 83:S78–S85

Yoshimura K, Kawate T, Takeda S (2011) Signaling through the primary cilium affects glial cell survival under a stressed environment. Glia 59:333–344

Zhou C, Cunningham L, Marcus AI, Li Y, Kahn RA (2006) Arl2 and Arl3 regulate different microtubule-dependent processes. Mol Biol Cell 17:2476–2487

Chapter 5
Primary Cilia and Inner Ear Sensory Epithelia

Cynthia M. Grimsley-Myers and Ping Chen

Contents

C.M. Grimsley-Myers • P. Chen (✉)
Department of Cell Biology, Emory University, Atlanta, GA 30322, USA
e-mail: ping.chen@emory.edu

K.L. Tucker and T. Caspary (eds.), *Cilia and Nervous System Development and Function*, DOI 10.1007/978-94-007-5808-7_5,

Abstract Our sense of hearing and balance depends on the proper form of the mechanosensory organs in the inner ear. Mechanic signals of sound and balance are translated to the movement of fluid or biomineral membranes overlaying the sensory organs of the inner ear. In response, mechanotransduction is activated by the deflection of polar hair bundles on the apical surface of sensory hair cells. In addition to being a transit or permanent component of the hair bundles, a cellular organelle, the primary cilia, has emerged as a central figure in the formation of various inner ear structure features critical for mechanotransduction. The entire inner ear is developed from the otic placode, which is orchestrated by a network of signaling pathways to ensure the development of a proper form for its functions. Primary cilia are essential in sculpturing the hearing organ and the polar hair bundles via the planar cell polarity (PCP) signaling pathway and in controlling biomineralization in the inner ear. Furthermore, recent studies have also implicated potential roles for primary cilia in other signaling pathways required for inner ear development, including sonic hedgehog (Shh), canonical Wnt, and Notch signaling.

Keywords Inner ear • Planar cell polarity • Hair cells • Usher genes • Development • Mouse

5.1 Introduction

The mammalian inner ear is a highly complex structure responsible for hearing and balance. In the adult, it is arranged as an intricate structure, termed the membranous labyrinth, which is housed within bony channels in the temporal bone of the skull. The membranous labyrinth is divided into dorsal vestibular and ventral auditory regions. The vestibular portion of the inner ear is composed of three orthogonally-positioned semicircular canals (anterior, posterior, and lateral) and two otolith organs (the utricle and saccule) whereas the auditory portion of the mammalian ear contains the snail-shaped cochlea. Each inner ear sensory organ consists of a precisely patterned cellular mosaic of alternating sensory hair cells and non-sensory supporting cells. Each hair cell contains an asymmetrically-positioned primary cilium, the kinocilium, either permanently or transiently during development, depending on the species and the specific type of sensory organs. Kinocilia of all of the hair cells in each of the inner ear sensory organs are polarized in a coordinated manner. The mechanotransduction function of the inner ear sensory organs relies on the structure and polarity of their sensory hair cells in both the cochlea and vestibule, and the media that overlay the sensory hair cells, the fluid or the crystal biomineral membrane, or otolith.

The entire inner ear develops from a patch of ectoderm cells near the hindbrain. The formation of the complex structure involves many signaling pathways, including sonic hedgehog (Shh), β-catenin mediated canonical Wnt, bone morphogenetic protein (BMP), retinols, fibroblast growth factor (FGF), and Notch signaling. In addition, the coordinated polarization of the hair cells in each inner ear sensory organ and the extension of the cochlea are regulated by the planar cell polarity (PCP) pathway. In this chapter, we will describe the structure of the inner ear sensory epithelia and sensory cells to show that the primary cilium is an integral structural component of inner ear sensory cells. We will speculate about the potential roles of primary cilia in mechanotransduction, summarize the involvement of primary cilia in otolith morphogenesis, and focus on the role of primary cilia in PCP regulation in the inner ear. Since primary cilia have also been implicated in Shh, canonical Wnt, and Notch signaling, we will review briefly the dynamic roles of these signaling pathways in inner ear induction, regional specification, cellular proliferation, differentiation, and patterning. The potential role of primary cilia in these signaling pathways in inner ear development will also be discussed.

5.2 The Primary Cilium as an Essential Structure Component of the Inner Ear: Sensory Epithelia and Sensory Cells

5.2.1 The Sensory Epithelia

The mechanosensory function of the five vestibular organs and one hearing organ is provided by the hair cells, which reside in sensory epithelia within each of these six structures (Fig. 5.1). The hair cells in each sensory epithelium are sensitive to a particular modality and orientation of stimulation. The auditory sensory epithelium (organ of Corti) is a narrow band of cells on the floor of the cochlear duct that runs along the length of the cochlea spiral and detects sound (Fig. 5.1a). The saccule and utricle sensory epithelia are termed maculae. These are oval or flat sheets that together detect gravity and linear acceleration. These organs are therefore responsible for our control of posture. Finally, at the base of each of the three semicircular canals are cristae, which are sensory patches that detect angular acceleration or rotation of the head (one for each plane of motion of the head). Despite their functional differences, the hair cells that serve as sensory receptors in these epithelia share significant similarities in their morphology, organization and function.

All inner ear sensory epithelia are composed of both sensory hair cells and non-sensory supporting cells. These cells are arranged in a mosaic pattern such that each hair cell is surrounded by and separated from its neighbors by supporting cells (Fig. 5.1b). The supporting cells provide structural support to the epithelium and are also thought to regulate the fluid ionic composition surrounding the hair cells (Fig. 5.1c). The cell bodies of the supporting cells are positioned underneath the hair cells and rest on the extracellular matrix underneath the sensory epithelium. They project finger-like phalangeal processes up between the hair cells to the luminal surface of the epithelium, at which point they fill in the gaps between the hair cells to form a tight, ion impermeable epithelium. Although hair cells have received

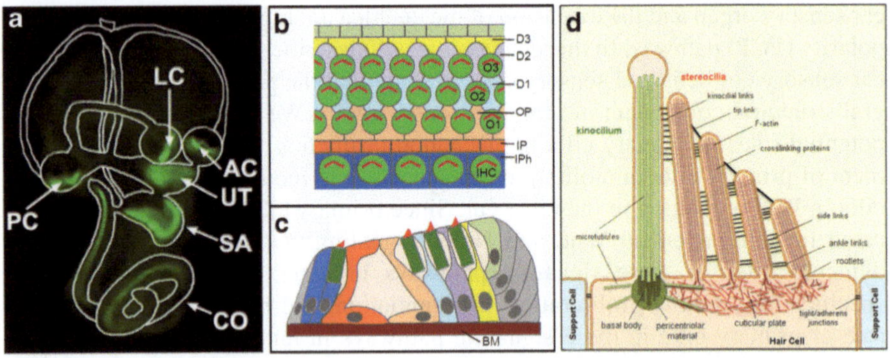

Fig. 5.1 Inner ear sensory epithelia and the hair bundle. (**a**) A tracing of an embryonic day 18 old mouse inner ear shows the five sensory organs in the vestibule and one sensory organ in the spiraled cochlea. The animal expresses green fluorescent protein (GFP) in the sensory hair cells, marking all six sensory organs of the inner ear. The white tracing outlines the fluid-filled labyrinth of the inner ear. *AC* anterior crista, *LC* lateral crista, *PC* posterior crista, *SA* saccule, *UT* utricle, CO cochlea. (**b, c**) While each of the inner ear sensory epithelia has a unique morphology, they all consist of alternating sensory hair cells and supporting cells, exemplified with a diagram of the sensory epithelium in the cochlea, known as the organ of Corti. The sensory cells are colored in green, and the various types of morphologically distinct non-sensory supporting cells are in other colors. The surface (**b**) and cross section views (**c**) of the epithelium illustrate the general cellular organization and the distinct arrangement of polar hair cells. The cross section diagram illustrates the *IHC* inner hair cell, O1–O3 the first to the third rows of outer hair cells, *IPh* inner phalangeal cells, *IP* inner pillar cells, *OP* outer pillar cells, D1-D3 the first to the third rows of Dieters' cells, *BM* basilar membrane. (**d**) The diagram highlights the structure and major cellular components of the hair bundle and associated hair cell apical domain. A prominent feature of the hair bundle is the polar positioning of the primary cilium, the kinocilium, the staircase arrangement of the stereocilia, and the extensive links between the kinocilium and the tallest stereocilia and the links among stereocilia. In addition, the stereocilia bundle is anchored in the actin-rich cuticular plate while the kinocilium is based on the basal body, which also organizes the cytoplasmic microtubule arrays (Panel *a* was adopted from a review by Jones and Chen 2008)

most of the research focus, supporting cells develop simultaneously and adjacent to the hair cells and are required for hair cell formation and survival.

At the luminal surface of the epithelium, the supporting cell processes are tightly coupled to the adjacent hair cell membrane by cell-cell junctions characteristic of both tight and adherens junctions (Nunes et al. 2006). These junctions are important for withstanding the mechanical stresses imposed on the epithelial and also for separating fluid in the lumen of the membranous labyrinth (endolymph) from the fluid in the bony channels (perilymph). These two fluids differ greatly in their ionic composition. Endolymph has a high potassium ion (K^+) concentration (~140 mM) and a low sodium ion (Na^+) concentration, which is unusual for an extracellular fluid. Perilymph, however, is closer to normal extracellular fluid and has a high Na^+/K^+ composition. This unusual endolymph composition is maintained by active ion transports in the stria vascularis of the cochlea and the dark cell regions of the vestibular system and is critical for normal mechanotransduction by the hair cells.

5.2.2 Cilia and the Mechanotransduction Apparatus – The Hair Bundle

The specialized hair cell organelle responsible for mechanotransduction is the hair or stereociliary bundle (Fig. 5.1d). Deflection of this hair bundle, resulting from either sound waves or a change in head position, leads to an increase in the rate of opening of relatively nonspecific cationic channels in the tip of the hair bundle and the flow of primarily K^+ ions from the endolymph into the hair cell, leading to depolarization and an afferent nerve signal that is transmitted to the brain. This bundle resembles fine hairs under a microscope, thus giving hair cells their name. The hair bundle is located on the apical surface of the hair cell and is composed of a single true cilium, the kinocilium, and between 50 and 200 actin-based stereocilia (Frolenkov et al. 2004; Hudspeth 1989). The kinocilium is composed of a 9+2 microtubule arrangement, similar to motile cilia. However, kinocilia lack the inner arms of the motor protein dynein required for motility are believed to be non-motile (Kikuchi et al. 1989). The stereocilia in the bundle are organized in a very precise pattern, forming graded rows arranged in a staircase-like pattern, with the kinocilium centered next to the tallest row of stereocilia (Fig. 5.1d). This bundle arrangement is not unique to mammals and is also seen in chickens, frogs and fish, indicating that its precise organization is important for mechanotransduction. Indeed, the asymmetric structure of the hair bundle renders it directionally sensitive to deflection. Deflections of the bundle towards the kinocilium and tallest stereocilia cause an opening of K^+ ion channels and depolarization whereas those in the opposite direction, towards the shortest stereocilia increase the probability of K^+ ion channel closure, resulting in hyperpolarization (Hudspeth 1989; Hudspeth and Corey 1977; Hudspeth and Jacobs 1979). The bundle is insensitive to perpendicular stimuli. Mutations in genes that disrupt this bundle morphology or orientation cause hearing impairment and vestibular malfunction (El-Amraoui and Petit 2005; Petit 2006; Yoshida and Liberman 1999).

Despite their misleading name, stereocilia are, in fact, specialized derivatives of actin-based microvilli (Fig. 5.1d). They are composed of a high density of hexagonally packed parallel actin filaments extensively crosslinked by actin-bundling proteins such as fimbrin and epsin proteins (Flock et al. 1982; Sekerkova et al. 2006; Tilney et al. 1980). The high density of actin filaments and extensive crosslinking renders rigidity to the shaft of the stereocilium. Each stereocilium also narrows or tapers immediately above its basal insertion, such that only a few central core actin fibers, termed rootlets, insert into the body of the cell (Tilney et al. 1980). When subjected to mechanical force, the stereocilium is predisposed to bend in a rod-like fashion at its tapered end rather than bending at a higher position or breaking. The stereociliary rootlets are anchored in the apical cytoplasm of the hair cell into a specialized structure called the cuticular plate, which is a large gel-like meshwork of crosslinked actin filaments and associated cytoskeletal proteins. The basal body of the kinocilium is also located directly adjacent to the cuticular plate. Together with its associated pericentriolar material, the basal body organizes the cytoplasmic microtubules that radiate along the periphery of the cell cortex and lateral hair cell membrane and anchors their minus ends (Moser et al. 2010).

There are several types of extracellular filaments between the stereocilia that help maintain the uniform integrity of the hair bundle (Fig. 5.1d). These links include top-links, side-links and ankle links, although the presence and or roles of these links vary depending upon developmental stage and hair cell type (Lefevre et al. 2008; Nayak et al. 2007; Petit 2006; Petit and Richardson 2009). In addition, a tip-link also joins the tip of a shorter stereocilium with the shaft of an adjacent longer stereocilium (Furness and Hackney 1985; Pickles et al. 1984). These tip-links are composed of homodimers of Cadherin 23 and Protocadherin 15, interacting in trans (Kazmierczak et al. 2007). They are directly connected to the mechanoelectrical transduction channels, and therefore function as gating elements that control the opening of the channels upon bundle deflection (Assad et al. 1991; Fettiplace 2006). As the bundle moves in the excitatory direction, tension of the tip-link opens the channel, allowing an influx of cations and depolarization of the hair cell. However, when the bundle moves in the opposite direction, tension is relieved and the channel closes.

In addition, there are also kinocilial links that connect the kinocilium to the adjacent stereocilia of the tallest row (Ernstson and Smith 1986; Goodyear and Richardson 2003; Hillman 1969). Although the molecular identities of these various links are not entirely clear yet, many of the links appear early in development, suggesting that they might be important for bundle formation. The prominent positioning of the kinocilium, the links between the kinocilium and stereocilia, and the organization of the microtubules by the basal body and associated pericentriole material all point to a central role for the kinocilium in hair bundle morphogenesis and integrity.

5.2.3 Cilia and the Planar Cell Polarity of Hair Cells

The gradation in height of the stereocilia and the off-center position of the kinocilium define an intrinsic polarity or orientation of the hair cell. As noted above, this intrinsic polarity of the hair cell is critical, as the hair cell response to bundle deflection is directionally sensitive. However, in addition to the intrinsic polarity of individual hair cells, hair cells are also uniformly aligned in one direction across the sensory epithelia (Fig. 5.2). In the organ of Corti, hair cells are oriented with their kinocilia closer to the lateral, or outer border of the cochlea duct (Fig. 5.2a, b). In the cristae, the hair cells are invariably oriented towards only one side of the epithelium (Fig. 5.2c, d). In the maculae, the hair cells are oriented along the mediolateral axis of the epithelium, with cells on either side of a line of polarity reversal oriented in opposite directions (Fig. 5.2e, f). This coordinated cellular orientation is a striking example of planar cell polarity (PCP), a term that refers to the polarity of cellular structures within the plane of an epithelium, perpendicular to the apical-basal axis of polarity. Although other examples exist in mammals, such as the orientation of hair follicles in the skin, the uniform orientation of hair cells in the ear, manifested with the coordinated polar positioning of the kinocilia and graded arrangement of the stereocilia relative to the kinocilium, is most often used as a model for the

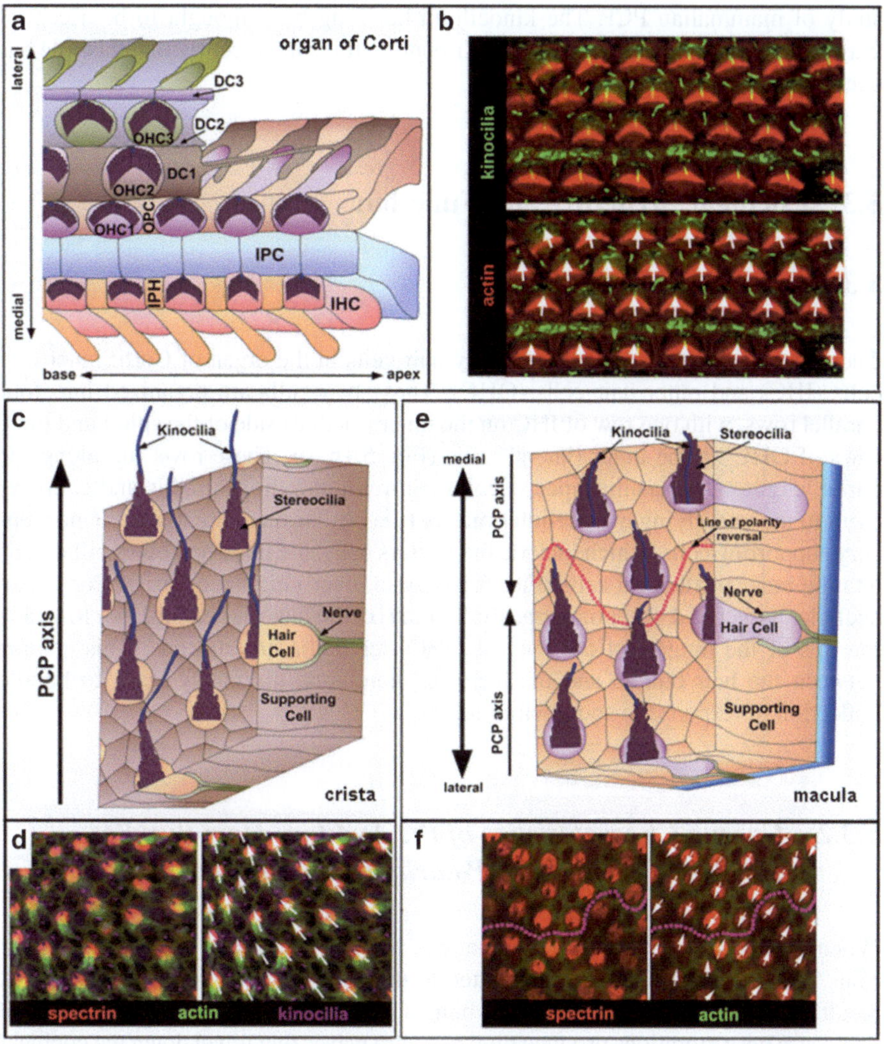

Fig. 5.2 Planar cell polarity of the inner ear sensory epithelia. (**a–f**) Schematic diagrams (**a**, **c**, and **e**) and confocal images (**b**, **d**, and **f**) of the sensory organs in the cochlea (**a**, **b**), crista (**c**, **d**), and the utricle macula (**e**, **f**). In the organ of Corti (**a**, **b**), the inner (IHCs) and outer hair cells (OHCs) are interdigitated with several distinct types of nonsensory supporting cells: the inner phalangeal cells (IPHs), inner pillar cells (IPCs), outer pillar cells (OPCs) and three rows of Deiters' cells (DC1-DC3). The kinocilium (b, *green*) is positioned near the tip of the "V"-shaped structure formed by actin-rich stereocilia with graded height. Note that all the "V"s are uniformly aligned (indicated by white arrows), showing a distinctive PCP. In the crista (c, d), actin-binding protein Spectrin (*red*) accumulates in the cuticular plates of the hair cells but is conspicuously excluded from the fonticulus within which the basal body resides and projects the kinocilium (*blue*). Stereocilia bundles were stained with actin dye phalloidin (*green*). Note that the kinocilium is positioned at the same asymmetric location on the apical surface of all the hair cells (indicated by *white arrows*). (**e**, **f**) In utricle macula, spectrin staining (*red*) permits easy visualization of the intrinsic polarity of each hair cell (*white arrows*). The hair cells on two sides of the line of polarity reversal (*pink line*) are oriented towards each other in the utricle macula (This figure is adopted from Padmashree and Chen 2009)

study of mammalian PCP. The kinocilium has long been thought to be a central player for the formation of the polar hair bundle. It is only recently that this role has been demonstrated (Jones et al. 2008).

5.3 Cochlear Structure and Function

5.3.1 Organ of Corti

There are two types of mechanosensory hair cells in the organ of Corti: inner hair cells (IHC) and outer hair cells (OHC). These hair cells are organized into four parallel rows, with one row of IHC on the inner (medial) side of the spiral and three rows of OHC on the outer (lateral) side (Fig. 5.1b, c). These rows run along the entire length of the snail-shaped cochlea. Surrounding the hair cells and coupling them to the underlying extracellular matrix (the basilar membrane) are the nonsensory supporting cells, which include the Deiters cells, the inner and outer pillar cells and the inner phalangeal cells (Fig. 5.1b). Also, overlying the organ of Corti is an acellular, structured sheet of extracellular matrix (the tectorial membrane) to which the tips of the longest of the longest OHC stereocilia are attached. This contact between the hair bundle and the tectorial membrane ultimately leads to bundle deflection in response to sound vibrations.

5.3.2 Unique Characteristics of the Auditory Hair Bundles and Their Planar Cell Polarity for Sound Detection

When a sound wave enters the external ear, it induces frequency-dependent vibrations in the eardrum that are transmitted to the bones of the middle ear (ossicles), resulting in the vibration of the stapes bone upon the oval window of the fluid-filled cochlea. These vibrations produce fluid pressure waves that travel down the cochlear duct, inducing vibrations of the basilar membrane. Movements of the basilar membrane then result in shearing displacements of the apical surface of the organ of Corti relative to the overlying tectorial membrane, which produces deflection of the stereocilia. This deflection opens mechanotranduction channels, leading to an influx of cations and depolarization of the hair cell (Hudspeth 1989).

Because of gradual changes in the features of the organ of Corti such as stereocilia height and the width and stiffness of the basilar membrane, hair cells at different locations along the cochlear spiral are selectively tuned to different stimulus frequencies. These characteristic frequencies are mapped smoothly and monotonically along the long axis of the sensory epithelium, similar to the graded wooden bars of a xylophone. Hair cells at the cochlear base response to high frequencies whereas those at the apex to low frequencies. This is how the ear discriminates differences in tone.

In the mammalian cochlea, the stereocilia in the OHC are organized into V- or crescent-shaped rows, with the kinocilium centered at the point of the V. In mature hair cells, these configurations are stabilized by the stereocilia rootlets that project into the cuticular plate at the apex of the hair cells. However, the initial patterning of the hair bundle occurs before cuticular plate formation, indicating that other factors outside of the cuticular plate guide the initial formation of this bundle pattern. Importantly, each stereociliary bundle points toward the outer (lateral) border of the cochlear duct throughout the length of the cochlea spiral (Fig. 5.2a, b). This arrangement is essential for the correct perception of sound, as disruption of this uniform bundle orientation, on even a few hair cells, has been shown to result in a significant decrease in hearing sensitivity (Yoshida and Liberman 1999). Furthermore, frequency discrimination by the cochlea requires synchronized response of the hair cells at the same location in the cochlea and the uniform orientation of the hair bundle.

It is peculiar that the kinocilium is only present during development in the organ of Corti, disappearing shortly after birth (around P10 in the mouse). In the adult, only the basal body remains in the apical cytoplasm at one side of the stereociliary bundle. This regression of the kinocilium shortly after bundle formation indicates that the kinocilium is not required for maintenance of hair bundle structure and function. However, its presence throughout the lifespan of other types of hair cells with the capacity to regenerate, such as mammalian vestibular hair cells and hair cells in chicken and frogs, suggests that it might be important to render the regeneration capacity.

5.4 Vestibular Structure and Function

5.4.1 Vestibular Sensory Hair Cells

Similar to the organ of Corti, the vestibular sensory organs display an alternating arrangement of sensory hair cells and nonsensory supporting cells, with each hair cell surrounded by a rosette of supporting cells. There are two types of hair cells in the vestibular organs, which differ in their innervation characteristics, cell shapes, hair bundle geometry and K^+ conductances (Moravec and Peterson 2004; Rusch and Eatock 1996a, b). Type II cells, which are present in all vertebrates, are cylindrically-shaped and are innervated by bouton afferent nerve terminals (Lysakowski 1996; Wersall 1956). These bouton endings make synaptic contact with only a small portion of the basolateral membrane. Type I cells, which are found in mammals, birds and reptiles, are amphora shaped and are innervated by a large cup-shaped afferent terminal called a calyx. This calyx ending frequently envelops the basolateral surface of one or more type I hair cells. Individual afferent fibers often innervate both cell types, and therefore have both bouton and calyx nerve endings.

The vestibular hair bundle is very similar to the auditory hair bundle. It consists of between 40 and 70 stereocilia and a kinocilium that lies in the middle of the stereocilia. The staircase arrangement of height-ranked stereocilia and the kinocilium placement in relation to the stereocilia is also preserved, as is the basic transduction mechanism. Mechanical bending of the stereocilia leads to the opening of ion channels and an influx of K^+ into the hair cell. In contrast to the V-shaped cochlear hair bundles, though, the vestibular hair bundles are circular in shape, with the tallest stereocilia next to the kinocilium on one side of the circle (Fig. 5.2c–f). The lengths of stereocilia and kinocilium are also much longer and the kinocilium persists throughout life. The supporting cells in the vestibular organs also have a primary cilium, but it is much shorter than the hair cell kinocilium.

5.4.2 *Crista Physiology*

There are three semicircular canals within the inner ear oriented at 90° angles to each other. Near the base of one arm of each semicircular canal is a dilation of the canal, called the ampulla. Within the ampulla and running perpendicular to the canal is the sensory epithelium, the crista ampullaris, which contains hair cells uniformly polarized in one direction. In contrast to the maculae (see below), there is no line of polarity reversal within an individual crista and all hair cells remain oriented in one direction across the epithelium. Analogous to the tectorial membrane of the auditory sensory epithelium, an accessory gelatinous structure called the cupula hangs from the roof of the ampulla above the hair cells and fills the lumen of the semicircular duct. The hair cell stereocilia are embedded in this gelatinous cupula. Rotation of the head causes the fluid in the semicircular canal to push against the cupula, which leads to the bending of the stereocilia.

The cristae of the semicircular canals work in pairs to detect angular rotations of the head. When the head moves, the fluid in one canal moves in one direction while the fluid in the corresponding canal on the opposite side of the head moves in the opposite direction. This leads to the opposite bending of the stereocilia on opposite sides. Thus, a turn of the head depolarizes hair cells in one crista and hyperpolarizes those on the other side. The uniform orientation of hair bundles in each of the cristae (Fig. 5.2c, d) is essential for a concerted readout of movement in three dimensions. The role of the kinocilium in hair bundle function, however, was uncertain as deflection of the kinocilium separately from the bundle was shown to have no effect on intracellular potential and transduction also persisted after its microsurgical ablation using a bullfrog model (Hudspeth and Jacobs 1979).

5.4.3 *Macula Physiology*

The utricle and saccule are located in the center of membranous labyrinth, between the membranous labyrinth and the semicircular ducts. The sensory epithelia of these organs are the utricular macula and the saccular macula, respectively. Within each macula, the hair cells are overlaid by the otolithic membrane, a thick gelatinous layer of glycoprotein covered by thousands of minute crystals of calcium carbonate, termed otoconia. The tips of the stereocilia and kinocilium are embedded in this otolithic membrane. The hair cells are oriented along a line of polarity reversal (Fig. 5.2e, f) such that the hair cells on either side of the line have opposite orientations.

The maculae specifically detect gravity and linear acceleration by sensing the inertial displacement of the otoconia. When the head is tilted, gravity pulls on the otoconia and the otolitic membrane, causing the stereocilia to bend. The saccule and utricle are positioned at right angles to each other within the ear. Thus, with any head position, gravity will bend the stereocilia of one patch of hair cells, leading to hair cell depolarization and an afferent nerve signal. The utricular macula is oriented horizontally when the head is upright and is most sensitive to tilt when the head is horizontal. The saccular macula is oriented vertically and is most sensitive to tilt when the head is horizontal. The saccule is also responsible for detecting linear accelerations of the head in the vertical direction (for example, when moving on an elevator). Similar to the organ of Corti, the hair bundles within the maculae are uniformly oriented along the mediolateral axis of the epithelium. However, hair cells on either side of a middle line called the striola have opposite orientations. The kinocilia are oriented toward the striola in the utricle and away from it in the saccule. As a result of this line of polarity reversal, tilting of the head in any direction will depolarize some hair cells and hyperpolarize others within the organ.

5.4.4 Primary Cilia in Controlling Biomineralization in the Vestibule

In the zebrafish, the saccule and utricle are also overlaid with biominerals known as otoliths. Otoliths in the zebrafish similarly consist of crystals of minerals, proteins and Ca^{2+}. The microvilli-derived stereocilia and microtubule-based kinocilium form a staircase pattern similar to mammal vestibular hair cells and extend from the surface of hair cells to attach to otoliths. The number, size and placement of the otoliths are precise for proper functions. Several recent studies reveal an essential role for primary cilia in controlling the morphogenesis of the otoliths (Colantonio et al. 2009; Wu et al. 2011; Yu et al. 2011).

The first link of cilia to otolith formation is revealed by a study of dynein mutants in zebrafish. Dynein regulatory complex (DRC) regulates the motility of motile cilia. Mutations in one of the regulatory subunits, Gas8, do not result in defects in ciliogenesis or placement of the cilia, but in the motility of motile cilia (Colantonio et al. 2009). In Gas8 mutants, otolith number, morphology and size are drastically changed. It was also demonstrated that cilia-driven fluid flow is essential to control otolith biomineralization. Two subsequent studies further showed that hydrodynamics influences otolith morphogenesis (Wu et al. 2011) and that otoliths can be generated in mutants devoid of cilia, but their morphology and positions are disrupted in cilia mutants (Yu et al. 2011). The role of cilia motility or hydrodynamics has not yet been reported.

5.5 Primary Cilia in Planar Cell Polarity Signaling in the Inner Ear

It is becoming increasingly clear that the kinocilium and/or basal body play a critical role in hair bundle development, both in determining the intrinsic structural polarity of hair cells as well as in coordinating uniform hair cell orientation throughout the epithelium. Initially, the kinocilium's importance was suggested by detailed descriptions of the temporal appearance and polarization of the kinocilium during bundle formation, as the kinocilium's position leads to and predicts the final polarity of the bundle (Fig. 5.3). Although the initial descriptions of hair bundle development were performed in the chicken auditory organ, the basilar papilla (Tilney et al. 1980), studies in the mouse suggest that the basic steps of bundle development have been evolutionarily conserved (Denman-Johnson and Forge 1999; Forge et al. 1997; Mbiene and Sans 1986; Zine and Romand 1996). More recently, analyses of mutant mouse strains for ciliary genes have revealed a requirement for the ciliary genes in hair bundle formation, and have also revealed a genetic link between ciliary genes and an evolutionarily conserved PCP signaling module in coordinating hair cell planar polarity during development. These studies are described in more detail below.

5.5.1 The Development of the Hair Bundle

In the mouse, formation of cochlear sensory hair cells begins around embryonic day (E) 12-E14, when precursors for the organ of Corti exit the cell cycle and commit to either a hair cell or supporting cell fate (Chen et al. 2002; Ruben 1967). Hair cell

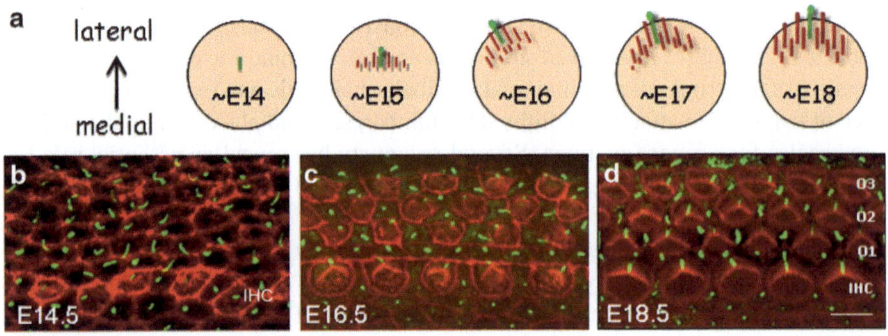

Fig. 5.3 Polarization of the kinocilium leads establishment of planar cell polarity in the cochlea. (**a**, **b**) Diagrams (**a**) and confocal images (**b**) illustrate the development of the hair bundle and its polarity. In mice, the inner hair cells (IHC) toward the center, or medial side, of the cochlear spiral, become visible first in the cochlea at embryonic day 14.5 (E14.5). At this stage, actin accumulation (*red*) in the inner hair cells is visible, and the kinocilia (*green*) are apparently randomly positioned at the apical surface of hair cells. As the cochlea develops, hair cell differentiation extends to the outer hair cells (O1–O3) from the medial-to-lateral direction. Stereocilia bundles become more distinct (*red*) and the kinocilia (*green*) are positioned to the final location at the apical surface of hair cells

differentiation then initiates in the basal region of the cochlea between E13.5 and E14.5 and over the next 3 days, hair cell differentiation proceeds in a basal to apical gradient along the entire length of the sensory epithelium, until one row of inner and three rows of outer hair cells are formed (McKenzie et al. 2004). In addition, a second gradient of differentiation moves across the medial to lateral axis of the epithelium, such that the inner hair cells appear first followed by each row of outer hair cells (Chen et al. 2002; McKenzie et al. 2004). In the vestibular organs, hair cell differentiation begins slightly earlier, at around E12.

During the early stages of hair cell differentiation, development of the hair bundle occurs (Fig. 5.3). At the start of bundle development, the hair cell apical surface is covered by very short microvilli of uniform height and a single kinocilium rests at its center. Then by around E13.5 in the vestibular organs and E15.5 in the cochlea, a few, very short precursor microvilli that will eventually give rise to stereocilia grow up around the kinocilium. Next, by around E15.5 in the vestibular organs and E16.5 in the cochlea, the kinocilium moves from the center of the luminal surface of the hair cell to the peripheral edge (Cotanche and Corwin 1991; Dabdoub and Kelley 2005). Simultaneously, the precursor microvilli immediately surrounding the developing kinocilium begin to elongate to form stereocilia, followed by the elongation of those stereocilia in rows further away from the kinocilium, initiating the staircase pattern of height ranked rows. Thus, development of the kinocilium appears to precede that of the stereocilia, and it is easy to speculate from its positioning in the bundle that it somehow helps to direct the pattern of stereocilia elongation (and the formation of the V-shaped bundle structure in the cochlea). Furthermore, the polarization of the kinocilium also precedes that of the stereocilia and parallels the PCP axis of the epithelium, suggesting that the kinocilium is a major determinant of hair cell planar orientation.

Interestingly, although the initial movement of the kinocilium to the peripheral edge of the hair cell has a significant bias in the direction of final orientation of the bundle (towards the lateral edge), the resulting position often deviates from its final position. Therefore, the kinocilium (and adjacent stereocilia) often undergo a period of reorientation in order to become fully aligned with their neighbors (Cotanche and Corwin 1991; Dabdoub and Kelley 2005). Following this reorientation period, the newly formed stereocilia continue to grow in width and also form rootlets that will anchor them into the cuticular plate. By the end of embryogenesis, most nascent hair bundles in the cochlea exhibit their characteristic 'V' shapes and have become uniformly oriented toward the outer or lateral edge of the cochlear duct. Hair bundles in the utricle and saccule have also acquired their mechanotransduction capability and appear mature in form and polarity, with the kinocilium elongated well above the longest stereocilia (Geleoc and Holt 2003). However, during the first postnatal week the hair bundles continue to undergo further row-specific differential outgrowth, enhancing the staircase pattern of height-ranked stereocilia (Denman-Johnson and Forge 1999; Forge et al. 1997; Mbiene and Sans 1986; Zine and Romand 1996). Development of mechanotransduction also occurs postnatally within the organ of Corti, in a basal to apical gradient from around P0-P6 (Lelli et al. 2009). Finally, the kinocilium retracts by around P10.

Concomitant with hair cell formation in the cochlea, elongation of the cochlea occurs by unidirectional extension of a postmitotic primordium driven by integrated radial and mediolateral intercalations (Jones and Chen 2007; McKenzie et al. 2004; Wang et al. 2005). This extension occurs in a basal-to-apical direction and from E14.5 to E18.5 the length of the cochlea increases approximately twofold. Although the molecular mechanisms that mediate this bundle development and cochlear morphogenesis are still incompletely understood, genetic studies suggest that ciliary genes function downstream of an evolutionarily conserved planar cell polarity (PCP) signaling module to regulate both polarized extension and the establishment of PCP in the organ of Corti during hair cell differentiation.

5.5.2 The Planar Cell Polarity (PCP) Signaling Pathway in the Inner Ear

Planar cell polarity (PCP) is a term that refers to the polarity of cellular structures within the plane of an epithelium, perpendicular to the apical-basal axis of polarity. A genetic pathway for PCP was first identified in *Drosophila*, where all adult cuticular tissues display PCP features (Adler 2002; Axelrod and McNeill 2002; Klein and Mlodzik 2005; Simons and Mlodzik 2008; Strutt 2008). For example, each cell in the adult wing produces a single distally pointing hair, while sensory bristles on the thorax all point in a posterior direction. In the eye, multicellular ommatidia point away from the dorsoventral midline (equator) and toward the dorsal and ventral edges (poles) of the eye. Remarkably, in all of these tissues, a common set of "core" PCP genes has been shown to be responsible for the generation of planar polarity (Adler 2002; Klein and Mlodzik 2005). This core group includes the seven-pass transmembrane receptor *frizzled* (*fz*), the atypical cadherin *flamingo* or *starry night* (*fmi* or *stan*), the four-pass transmembrane protein *van gogh* or *strabismus* (*vang* or *stbm*), as well as the intracellular proteins *disheveled* (*dsh*), *prickle* (*pk*) and *diego* (*dgo*). Activation of the core PCP genes leads to the activation of JNK (c-Jun N-terminal kinase) and Rac/Rho GTPases and involves relocalization of *Dsh* to the plasma membrane.

Recent work has since revealed that the function of at least the upstream core PCP genes has been highly evolutionarily conserved. Severe defects in planar organization and cochlear development are observed in mice with mutations in homologs of these genes. These mutants include: *Loop-tail (Lp,* mutation in *Vangl2/Stbm), spin cycle* and *crash* (independent mutations in *Celsr1/fmi*), *Dishevelled–1*$^{-/-}$; *Dishevelled-2*$^{-/-}$ double mutants and *Frizzled3*$^{-/-}$; *Frizzled6*$^{-/-}$ double mutants. In addition to misoriented hair bundles, these mutants also display a shorter and wider cochlea sensory epithelium and extra, disorganized rows of hair cells in the apical region of the cochlea (Chacon-Heszele and Chen 2009; Curtin et al. 2003; Hamblet et al. 2002; Montcouquiol et al. 2003; Wang et al. 2005, 2006a, b). This is consistent with defects in polarized extension. Interestingly, all of these mutants also display craniorachischisis, a severe defect in neural tube closure in which the entire neural tube from mid-brain to tail fails to close. This phenotype, which results in perinatal lethality, is also thought to result from failure of convergent extension during neurulation (Curtin et al. 2003; Hamblet et al. 2002; Kibar et al. 2001; Murdoch et al. 2001, 2003; Wang et al. 2006a, b). Homologs of these genes also regulate convergent extension in the frog *Xenopus laevis* and the zebrafish *Danio rerio* (Keller 2002; Mlodzik 2002; Wallingford et al. 2002), indicating that this signaling cassette has been highly conserved throughout evolution and likely plays a very fundamental role in oriented cellular movements during development.

Intriguingly, *Scrb1* (*Scribble*) and protein tyrosine kinase 7 (*PTK7*) are also both required for morphogenesis and planar cell polarity in the organ of Corti as well as neural tube closure in mice (Lu et al. 2004; Montcouquiol et al. 2003; Murdoch et al. 2003). However, orthologs of these genes do not appear to play an important role during PCP establishment in *Drosophila*. Thus, although there has been an overall conservation of the molecular mechanisms that regulate directional extension movements and epithelial organization across a variety of species and polarized cell types, differences also exist. Other novel PCP signaling mechanisms are also expected to occur in mammals.

A hallmark feature of the core PCP proteins is their asymmetric localization. In the organ of Corti, for example, Vangl2, Frizzled-3 (Fz3) and Frizzled-6 (Fz6) appear to be enriched on the medial side of the hair cell membrane, whereas Dishevelled-2 (Dvl2) is located on the lateral side (Etheridge et al. 2008; Montcouquiol et al. 2006; Wang et al. 2005, 2006a, b). Interestingly, the loss of polarized membrane localization of any of the core PCP proteins results in the loss of localization of the others, suggesting that these proteins function as a complex (Deans et al. 2007; Wang et al. 2005, 2006a, b). However, how this complex detects the potential signal(s) and orients the hair bundle is still incompletely understood. Their distinct asymmetric localizations suggest that PCP protein localization may be important for establishing the PCP axis within the sensory epithelium. Indeed, the asymmetric localization of PCP proteins has been shown to precede morphological polarization of hair cells in both the organ of Corti and the utricle (Deans et al. 2007; Jones et al. 2008). However, it is not yet clear whether the asymmetric localization of the core PCP proteins is actually required for hair cell planar polarization. One study in the mouse utricle suggests that while the localization of the core PCP proteins may correlate with the axis for PCP, their localizations may not actually dictate bundle orientation. In this study, a homolog of the Drosophila core PCP protein Pk, Pk2, was shown to preferentially localize on the medial side of hair cells across the entire epithelium and importantly, did not switch sides at the line of polarity reversal (where bundle polarity reverses) (Deans et al. 2007). Similarly, Fz6 was localized on the lateral side across the entire epithelium. The failure of these PCP proteins to switch sides supports the notion the core PCP proteins function to coordinate cell orientation at the tissue level rather than to determine the direction of cell polarity *per se*.

It is also important to note that mutations in the core PCP genes do not affect the intrinsic structural polarity of the hair bundle. In PCP mutants, the cochlear hair bundle still maintains its characteristic ‘V’ shape with the kinocilium at the vertex of the V. It is only the coordinated orientation of the bundles that are affected. While a role for PCP proteins in the development of the structural polarity of the hair bundles cannot be ruled out, due to possible gene redundancies, most likely the primary role of the core PCP genes is to establish uniform polarity between neighboring cells rather than to guide the intrinsic hair cell structural polarity.

5.5.3 The Discovery of the Link Between Cilia and PCP Signaling in the Inner Ear

The temporal appearance and polarization of the kinocilium during bundle development suggests that it has dual roles in directing both the hair cell intrinsic polarity machinery and the coordination of hair cell polarity at the tissue level. However, more direct evidence for these roles came from mice mutant for genes associated with Bardet-Biedl syndrome (BBS) (Ross et al. 2005). BBS is a pleiotropic disorder in humans characterized by retinal degeneration, obesity, polydactyly, renal and gonadal malformations and cognitive impairment (Beales 2005; Tobin and Beales 2007). At least 14 BBS genes (*Bbs1-Bbs14*) have been identified, most of whose gene products localize to the cilia and/or basal bodies. Similar to PCP mutants, null mutations in *Bbs1*, *Bbs4*, or *Bbs6* lead to stereociliary bundle orientation defects (Ross et al. 2005). These mice also displayed exencephaly, a neural tube closure defect. Moreover, both *Bbs1* and *Bbs6* were shown to genetically interact with *Vangl2* (Ross et al. 2005). *Vangl2*$^{+/Lp}$; *Bbs1*$^{+/-}$and *Vangl2*$^{+/Lp}$; *Bbs6*$^{+/-}$ double heterozygous mice had cochlear abnormalities similar to those observed in the *Bbs*$^{-/-}$ homozygous mutants, whereas no abnormalities were observed in *Bbs6*$^{+/-}$, *Bbs1*$^{+/-}$, or *Vangl2*$^{+/Lp}$ heterozygous mutants. Furthermore, in zebrafish *trilobite* mutants (*tri*, an ortholog of *vangl2*), suppression of *Bbs4* or *Bbs6* using morpholinos led to enhanced convergent extension defects compared to *tri* mutants alone (Ross et al. 2005). Together, these data indicate a genetic link between ciliary/basal body proteins and the PCP signaling pathway, perhaps as downstream effectors of the 'direction sensing' core PCP module. These finding were then further supported by a separate study in which suppression of *Bbs1*, *Bbs4*, or *Bbs6* in zebrafish led to mild convergent extension defects that were further impaired with morpholinos against *Wnt 11* or *Wnt5* (non-canonical Wnt ligands) but were partially rescued by a membrane bound form of *Dishevelled* selectively involved in PCP signaling (Axelrod et al. 1998; Gerdes et al. 2007).

A more direct role for the kinocilium in hair cell PCP was then demonstrated by genetic ablation of the kinocilium within the early developing mouse inner ear (Jones et al. 2008; Sipe and Lu 2011). The intraflagellar transport protein Ift88, or Polaris, is required for the assembly and maintenance of the primary cilia in multiple tissues. In the cochlea, Ift88 is expressed in the kinocilium axoneme and at the base of the kinocilium at E14.5 (during early hair bundle development) (Jones et al. 2008). Conditional deletion of *Ift88* in the cochlea led to the loss or underdevelopment of the kinociliary axoneme (although the basal body was still present) and

PCP phenotypes including misoriented hair bundles and a shorter and wider cochlear duct (Jones et al. 2008). *Ift88* was also shown to interact genetically with *Vangl2*, strongly suggesting that the kinocilium is a critical component of the PCP machinery in the inner ear (Jones et al. 2008). As further support for this, conditional deletion of *Kif3a*, which is a member of the kinesin-II microtubule motor complex required for anterograde IFT and ciliogenesis, was also shown to produce PCP-like phenotypes in the organ of Corti (Jones et al. 2008; Sipe and Lu 2011).

Interestingly, in both the *Ift88* and *Kif3a* mutant bundles, the core PCP proteins including Dvl2, Fz3, and Vangl2 were correctly (asymmetrically) localized with respect to the mediolateral axis of the epithelium, even in cells with misoriented bundles. Thus, PCP protein localization did not accurately reflect bundle polarity. This indicates that the *Ift88*/*Kif3a* mutants suffer from a failure to properly respond to the polarization signal coming from the core PCP proteins, and that the kinocilium probably functions downstream of the core PCP proteins. The molecular machinery that links the PCP pathway to the kinocilium is still unknown, but may involve the known PCP effector proteins Inturned and Fuzzy, which were shown to regulate apical actin assembly and the orientation of ciliary microtubules during ciliogenesis in *Xenopus laevis* (Park et al. 2006).

5.5.4 Determination of Intrinsic Cell Polarity by Ciliary Genes

In addition to PCP defects, many ciliary mutants also display defects in the intrinsic polarity of the sensory hair cells. For instance, mild defects in the orientation and V-shape of the hair bundle were recently observed in mice with mutations in the basal body protein, ALMS1 (Jagger et al. 2011). In addition, many of the stereociliary bundles in the Bbs mutant mice had an abnormally flattened shape (Ross et al. 2005). In these hair cells, the kinocilium (which was present and appeared grossly normal) was displaced and no longer closely associated with the stereocilia. These defects suggest an additional role for BBS proteins in formation of the hair bundle structure, perhaps to regulate the movement or anchoring of the kinocilium or to facilitate the release of a 'guiding' signal to the stereocilia. The BBS proteins are generally believed to mediate the trafficking of proteins and vesicles to the base of the cilium and basal body and to facilitate IFT and ciliary membrane biogenesis (Nachury et al. 2007). Therefore, they may be required for the presence of an unknown directional signal or anchoring component within the basal body. However, this role is likely independent of core PCP protein function, as no defects in bundle structure are observed in PCP mutants. Also, no bundle structural phenotype was noted in the *Lp*; *Bbs* double heterozygous mutants. However, this does not rule out a role for other downstream effectors of the core PCP proteins, such as Inturned and Fuzzy, in structural development of the hair bundle and/or ciliogenesis (see below).

The importance of the basal body in hair bundle structure is also seen in the *Ift88* mutants. In the *Ift88* mutants, most hair cells had normal bundle morphology, with organized, height-ranked rows of stereocilia (Jones et al. 2008). Very few hair cells contained bundle structural defects, i.e., flattened or 'unpatterned' bundles. However, a few bundles were circularized, with the stereocilia rows arranged in concentric rings. Others were polarized but misoriented. In both of these types of mutant bundles, the position of the basal body directly correlated with the bundle orientation. When the basal body remained in the center of the cell, the hair bundle was circularized. When the basal body migrated to the periphery, the hair bundle was polarized (but misoriented with respect to the PCP axis). This suggests that IFT88/kinocilium function is important for basal body positioning, and that the basal body, as opposed to or in the absence of the kinocilium, may play a primary role in establishing the intrinsic polarized structure of the hair bundle. This idea is reinforced by the hair bundle and basal body phenotypes in *Kif3a* mutants. In *Kif3a* mutants, the basal body was mispositioned along both the apicobasal and planar cell polarity axes. The basal body position was also uncoupled from hair bundle orientation and numerous hair bundle structural defects, including flattened or occasionally fragmented bundles were present (Sipe and Lu 2011).

5.5.5 Crosstalk Between Cilia and PCP Genes

The studies of PCP genes and ciliary genes in the inner ear reveal the distinctive roles for the core PCP gene and ciliary genes (Fig. 5.4). Essentially, core PCP genes coordinate the polarity of neighboring cells, and ciliary genes are required for the intrinsic polarity of individual cells through basal body and associated components. How the basal body communicates with the core PCP complex and the actin cytoskeleton and microtubule network in order to drive its position from the center of the cell to the periphery, and how the basal body in turn regulates the intrinsic polarity of cells, are not entirely clear. It is apparent that PCP genes can regulate polarity of cilia and ciliogenesis, and that cell-specific machinery for building the polar structure interacts with ciliary genes.

In Xenopus multiciliated cells, basal body planar polarization can be regulated by core PCP genes *Dvl2* and *Vangl2* (Mitchell et al. 2009). It is further expected that PCP effecters, such as small GTPases and motor proteins, probably play an important role downstream of core PCP genes. In accordance with this, p21-activated kinases (PAKs), cytoskeletal effectors downstream of Rac/Cdc42 small GTPases, have been shown to become asymmetrically localized to the lateral hair cell membrane in a developmental gradient that coincides with kinocilium/basal body migration (Grimsley-Myers et al. 2009). There is also an apparent feedback regulation among the players. Sipe et al. also show that *Kif3a* is required for the localized activation of PAK on the hair cell cortex, which in turn is required for basal body positioning and hair bundle morphogenesis (Sipe and Lu 2011).

In addition, several studies suggest that the PCP effector proteins Inturned and Fuzzy are required for ciliogenesis in vertebrates and thus could potentially play a role in hair bundle development in mammals. *Inturned* and *Fuzzy* are referred to as PCP effector genes because they were shown to specifically regulate wing hair and bristle orientation downstream of the core PCP genes in *Drosophila* (Adler et al.

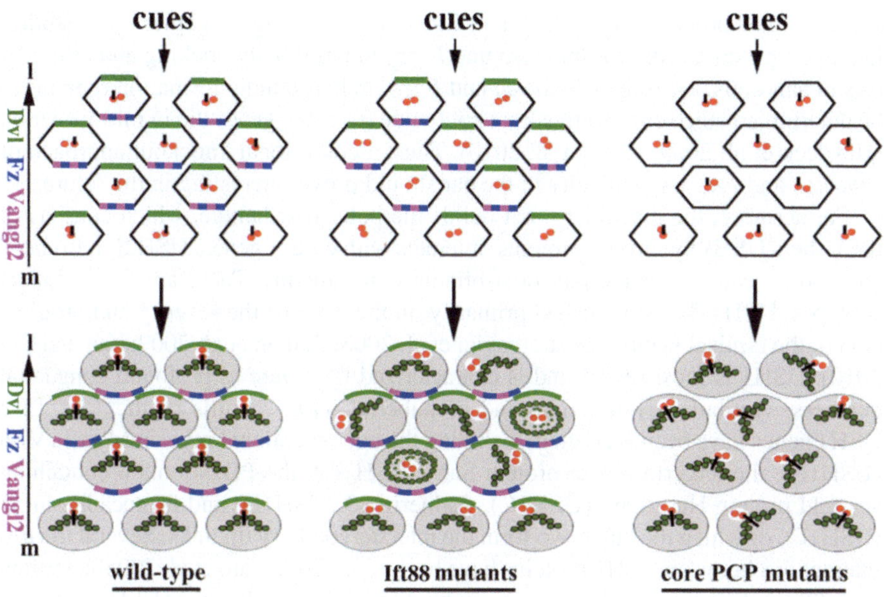

Fig. 5.4 Ciliary genes regulate the intrinsic polarity of sensory hair cells in PCP signaling PCP signaling in wild-type cells both instructs the intrinsic polarity of individual hair cells and coordinates the cellular polarity among neighboring cells. In wild-type cells, core PCP proteins receive and respond to unknown directional cues. Subsequently, core PCP proteins form membrane complexes and become asymmetrically localized in hair cells. Interactions of core PCP protein complexes across the cellular junctions formed by neighboring cells establish polar distribution of core PCP proteins across the entire organ of Corti along the planar cell polarity axis. The polar distribution of PCP protein complexes, however, is necessary but not sufficient to direct the correct positioning of the kinocilium and its basal body. In core PCP mutants, core PCP protein complexes are disrupted and the kinocilium is randomly positioned. In cilia mutants, core PCP protein complexes form and remain their asymmetric distribution. The basal body, however, is randomly positioned and appears to direct the orientation of the stereocilia bundle, or the orientation of sensory hair cells (This figure is adopted from Jones et al. 2008)

1994; Collier and Gubb 1997; Park et al. 1996). In *Xenopus*, *Inturned* and *Fuzzy* are required for convergent extension movements and for the formation of cilia in floorplate and multiciliated mucociliary epidermal cells (Park et al. 2006). More specifically, *Inturned* and *Fuzzy* were shown to mediate the assembly of an apical actin network required for the normal orientation of ciliary microtubules (Park et al. 2006). A later study also demonstrated that *Inturned* is required together with *Dvl* and the small GTPAse *RhoA* for the docking of basal bodies to the apical cell surface of epidermal cells (Park et al. 2008). In mice, targeted disruption of either *Inturned* (*Intu*) or *Fuzzy* (Fuz) leads to multiple defects, including shorter and fewer cilia, neural tube closure defects and abnormal Hh signaling (Gray et al. 2009; Heydeck et al. 2009; Zeng et al. 2010). However, neither *Inturned* nor *Fuzzy* displayed a genetic interaction with the core PCP protein Vangl2 in convergent extension, suggesting that *Inturned* and *Fuzzy* function independently of the core PCP genes in some cellular processes (Heydeck and Liu 2011). Although together these studies indicate a potential role for *Inturned* and *Fuzzy* in basal body docking and ciliogenesis in hair cells, the role of *Inturned* and *Fuzzy* in hair bundle formation remains to be determined since constitutive knockout mice for these genes die in mid-gestation (Heydeck et al. 2009; Zeng et al. 2010). Thus, a conditional knockout approach of *Inturned* and *Fuzzy* specifically in the ear should prove interesting in the future.

Furthermore, the machinery that builds the polar hair bundle, which consists of the Usher (USH) Syndrome proteins, interacts with ciliary genes. USH Syndrome is the leading cause of hereditary deaf-blindness in humans. There are three clinical subtypes, USH1-3, distinguished primarily on the basis of the severity and progression of the clinical symptoms (Leibovici et al. 2008; Saihan et al. 2009; Yan and Liu 2010). USH1 is most severe and is characterized by severe to profound congenital deafness, vestibular dysfunction, and prepubertal onset retinitis pigmentosa. Five USH1 genes have been identified, including the unconventional myosin Myosin VIIa (USH1B), the ankyrin repeat protein Sans (USH1G), the PDZ-domain containing scaffold protein Harmonin (USH1C), Cadherin 23 (USH1D) and Protocadherin 15 (USH1F), of which the latter two form the interstereociliary tip links. Several *in vitro* interactions between USH1 proteins have been reported (Adato et al. 2005b; Senften et al. 2006). In particular, Harmonin can bind to all four of the other USH1 proteins, and is thought to act as a scaffold in these protein complexes (Adato et al. 2005b; Bahloul et al. 2010; Boeda et al. 2002; Reiners et al. 2005; Siemens et al. 2002).

In mice, mutations in any of the USH1 genes results in abnormal hair bundle architecture as early as E17.5 to E18.5 (Di Palma et al. 2001; Kikkawa et al. 2008; Lefevre et al. 2008; Pawlowski et al. 2006; Senften et al. 2006). The mutant bundles are fragmented, with the stereocilia assembled into multiple clumps. Often these stereociliary clumps are misoriented with the kinocilium dissociated from the stereocilia. The kinocilium/basal body is also frequently mispositioned within the hair cell apical surface, although it is usually found near the periphery and in the lateral half of the cell surface (Chacon-Heszele et al. 2012; Kikkawa et al. 2008; Lefevre et al. 2008; Pawlowski et al. 2006; Self et al. 1998) This suggests that USH1 proteins are also important for kinocilium/basal body positioning and hair bundle formation, similar to ciliary proteins.

Although the localization patterns of USH1 proteins vary over the course of bundle development, they are generally enriched within the stereocilia, often at the tips of growing stereocilia (Ahmed et al. 2003, 2006; Bahloul et al. 2010; Boeda et al. 2002; Caberlotto et al. 2011; Lefevre et al. 2008; Michel et al. 2005; Rzadzinska et al. 2005). However, USH1 proteins have also been observed within the interstereociliary and kinociliary links and basal body during early bundle formation (Lagziel et al. 2005; Michel et al. 2005). Kikkawa et al. have also observed altered microtubule organization around the basal body in Protocadherin 15 mutant mice (Kikkawa et al. 2008). Together, these data indicate that USH1 proteins probably play multiple roles in the ear during bundle development, including stereocilia and/or kinocilium anchoring, stereociliary cohesion, trafficking of and/or anchoring of molecules within the stereocilia, signaling from the basal body, and organization of the microtubule and actin networks within the developing cuticular plate. With so many possible roles, it has been difficult to decipher the primary defect(s) in USH1 mutants and to discern the contribution of USH1 proteins to basal body/kinocilium positioning/functioning versus other functions during bundle formation.

A more direct link between USH syndrome proteins and cilia has been described in vertebrate photoreceptor cells. Similar to hair cells, photoreceptor cells are highly specialized, ciliated sensory neurons that are capable of phototransduction. These cells contain a non-motile connecting cilium that links a metabolically active inner segment to a light-sensitive outer segment. The phototransductive membranes in the outer segment are synthesized by organelles in the inner segment and then transported through the connecting cilium to the outer segment. As USH syndrome affects vision as well as hearing, the role of USH syndrome proteins in photoreceptor cell function has begun to gain scrutiny.

Intriguingly, all five of the known USH1 proteins as well several USH2 proteins have been observed at the periciliary region of mammalian photoreceptor cells (Liu et al. 2007; Maerker et al. 2008; Williams 2008; Wolfrum and Schmitt 2000). The scaffolds SANS and whirlin (USH2D) localize to the pericilliary collar of the apical inner segment and the adjacent connecting cilium and basal body complex, where they interact with the cytoplasmic domain of Usherin (USH2A) and VLGR1b (USH2C) (Liu et al. 2007; Maerker et al. 2008). The long ectodomains of Usherin and VLGR1b are thought to form fibrous links connecting the membranes of the connecting cilium and the inner segment, which are analogous in function and molecular composition to the ankle links between neighboring stereocilia in hair cells (Adato et al. 2005a; Liu et al. 2007; Maerker et al. 2008; McGee et al. 2006; Michalski et al. 2007). MyosinVIIa also localizes to the ciliary membrane and has been shown to be important for the ciliary trafficking of membrane components, including rhodopsin (Liu et al. 1997, 1999; Wolfrum and Schmitt 2000). Harmonin also localizes to the apical inner segment (Maerker et al. 2008). Together these studies support a direct role for USH proteins in cilium function and give credence to the idea that the USH proteins might also mediate primary cilium functions in other tissues, including hair cells in the ear. This notion is supported by a case study of two siblings with USH syndrome, who were found to display bronchiectasis and reduced nasal mucociliary clearance associated with immotile cilia (Bonneau et al. 1993). In addition, sperm tail structure and motility have been shown to be abnormal in Usher syndrome patients (Hunter et al. 1986). Thus, evidence is mounting that USH Syndrome may be a cilium disorder. It is therefore highly likely that the USH proteins perform critical functions at the kinocilium/basal body for normal hair bundle structure, although this remains to be formally demonstrated.

5.6 Primary Cilia and Shh, Wnt, and Notch Signaling in Inner Ear Development

The entire inner ear develops from a patch of ectodermal cells near the hindbrain, known as the otic placode. The formation of the inner ear from the otic placode is influenced by the nearby hindbrain and periotic mesenchyme tissues. In particular, diffusible morphogens, such as fibroblast growth factors (FGFs), retinoids, bone morphogenetic proteins (BMPs), Wnt, and sonic hedgehog (Shh) have been shown to play essential roles in various aspects of inner ear development, including otic induction, patterning, and cell specification. In addition, cellular interactions mediated by Notch signaling are essential for regional specification and sensory organ patterning. Among these signaling pathways, primary cilia are required for Shh signaling and implicated for β-catenin-mediated canonical Wnt and Notch signaling. The role of primary cilia in these signaling pathways during inner ear development, however, has yet to be determined.

5.6.1 Primary Cilia and Shh Signaling in the Inner Ear

Shh is a secreted morphogen that binds to the twelve-pass transmembrane receptor Patched (Ptc) receptor to release the inhibition of Patched to its associated protein, Smoothened (Smo). The derepression of Smo by Ptc results in the activation of the Ci/GLI family of transcription factors from their repressor forms, through complex interactions of Costal2 (Cos2), Fused (Fu) and Suppressor of fused [Su(fu)]. The implication for cilia in Shh signaling became apparent when a plethora of Shh defects were observed in cilia mutants. The similarities of human diseases shared by Shh or cilia mutations further supported the link between Shh signaling and cilia. Indeed, Shh signaling is defective in cilia mutants, and components of the Shh signaling pathway are localized to the cilia (Tasouri and Tucker 2011).

During ear inner ear development, Shh is one of the major signals from the ventral hindbrain to specify for the ventral inner ear fate, the cochlear fate (Riccomagno et al. 2002). It is expressed in the notochord and the floor plate during otic induction and otocyst development. In $Shh^{-/-}$ animals, the ventral cochlea is lost (Riccomagno et al. 2002), indicating a requirement for Shh in specification of the cochlea fate in the otocyst during early inner ear development. An additional role for Shh signaling in inner ear development was further revealed by examination of a mouse model in which only the repressor form of Gli3 is produced (Driver et al. 2008). The prosensory domain in the cochlea, the precursor population that gives rise to the sensory and supporting cells of the organ of Corti, is increased and ectopic sensory domains are present in this mouse model of decreased Shh signaling. Inhibition of Shh signaling results in similar consequences, further supporting a role for Shh signaling in prosensory specification.

The requirement for cilia in Shh signaling, which plays essential roles in inner ear development, predicts that Shh signaling in the inner ears from cilia mutants is likely to be defective. However, both the early and late requirement for Shh in inner ear development appears to be met in $Ift88^{CKO/CKO}$ and $Kif3a^{CKO/CKO}$ ciliary mutants (Jones et al. 2008). The cochlea is present albeit shortened and widened in cilia mutants (Jones et al. 2008). It is possible that the conditional knockout (CKO) of ciliary genes in the mice was too late to render an early cilia defect. Consequently, Shh signaling in these mutants is sufficient for cochlea specification during early inner ear development. In contrast, ciliogenesis is nearly abolished in cilia mutants during terminal differentiation of the cochlea and PCP processes are defective. It is surprising, therefore, that the late Shh phenotype, or the generation of ectopic sensory domains in the cochlea, is not observed in inducible cilia mutants. It is possible that the residual Shh signaling activity in cilia mutants is sufficient or that Shh signaling acts early for prosensory domain formation.

5.6.2 *Primary Cilia and Canonical Wnt Signaling in the Inner Ear*

While mutations in ciliary genes lead to PCP phenotypes, exactly how ciliary proteins promote PCP in the ear is unclear. Several studies have suggested that ciliary proteins may regulate a switch between PCP and canonical Wnt/β-catenin signaling. In the canonical Wnt pathway, Wnt ligand binding to the Frizzled (Fz)/low density lipoprotein receptor-related protein (LRP) complex leads to the activation of cytoplasmic Dishevelled (Dvl) and the stabilization of cytoplasmic β-catenin, which enters the nucleus and activates transcription of downstream target genes via the lymphoid enhancer-binding factor (Lef) or T cell specific transcription factor (TCF) family of proteins (Logan and Nusse 2004). Thus, the canonical Wnt pathway shares molecular components with the PCP pathway, including Fz and Dvl proteins, but produces a very different functional output. Canonical Wnt signaling is pleiotropic, with effects on multiple cellular processes including fate specification, differentiation, migration, proliferation and survival, and its precise regulation is critical for multiple developmental events during embryogenesis (Logan and Nusse 2004).

One of the first connections between cilia and decreased canonical Wnt signaling came from studies on the ankyrin-repeat protein Inversin. Inversin localizes to cilia and basal bodies, and mutations in Inversin in humans lead to Nephronophthisis type II, an infantile form of polycystic kidney disease (Morgan et al. 2002; Otto et al. 2003). In mice, recessive mutations in Inversin lead to the inversion of left-right asymmetry and kidney abnormalities, phenotypes typically associated with cilia defects (Mochizuki et al. 2002; Morgan et al. 1998). Inversin is similar to Diversin, a mammalian homologue of the *Drosophila* PCP protein Diego. Like Diego, Inversin can interact with the core PCP proteins Prickled and Strabismus (Das et al. 2004; Simons et al. 2005).

At the molecular level, Inversin inhibits canonical Wnt signaling by specifically binding and targeting cytoplasmic Dishevelled for degradation by the anaphase promoting complex. However, Inversin does not target the membrane associated form of Dishevelled involved in PCP signaling (Axelrod et al. 1998; Simons et al. 2005). This suggests that inversin may function to block canonical Wnt signaling and promote PCP signaling. Further supporting this hypothesis, overexpression of Inversin can inhibit the activation of a canonical Wnt reporter while knockdown of Inversin results in defective convergent extension movements during Xenopus gastrulation (Bergmann et al. 2008; Simons et al. 2005).

In addition to Inversin, genes required for ciliogenesis and PCP have also been linked to canonical Wnt pathway inhibition. For instance, $Kif3a^{-/-}$ mice display increased canonical Wnt reporter activity, and disruption of ciliogenesis in mice with mutations in *Kif3a*, *Ift88* or *Ofd1* leads to increased canonical pathway activation following stimulation with Wnt3a, a typical canonical Wnt ligand (Corbit et al. 2008). Loss of primary cilia in the developing mammary glands of *Ift88* mice also results in increased canonical Wnt along with branching morphogenesis defects (McDermott et al. 2010). In addition, suppression of *Bbs1*, *Bbs4* or *Bbs6* transcripts leads to an upregulation of the Wnt/β-catenin pathway in both zebrafish embryos and in mammalian cultured cells (Gerdes et al. 2007). This upregulation is thought to be due, at least in part, to perturbed proteasomal targeting of cytoplasmic β-catenin. Finally, Diversin, which is structurally related to vertebrate Inversin and *Drosophila* Diego, is required for gastrulation movements and cardiogenesis in zebrafish. Diversin was shown to mediate noncanonical Wnt signaling upstream of Rho family small GTPases and JNK through an interaction with Dvl (Moeller et al. 2006). Diversin was also shown to inhibit canonical Wnt/β-catenin signaling by promoting β-catenin phosphorylation and degradation by glycogen synthase kinase 3β (GSK3β) (Schwarz-Romond et al. 2002). Intriguingly, Diversin localizes to the centrosome/basal body and is required for basal body polarization and ciliogenesis in *Xenopus* (Itoh et al. 2009; Yasunaga et al. 2011).

Despite all of these findings, the role of ciliary genes in canonical Wnt signaling is still somewhat controversial (Wallingford and Mitchell 2011). For example, one group showed that expression of the Wnt target gene Axin2 and a quantitative Wnt reporter were completely normal in mouse embryos lacking *Kif3a*, *Ift172*, or *Ift88*. They also found no change in Wnt activity in response to Wnt3a ligand in *Ift172*, *Ift88* or *Dync2h1* mutant MEFs (Ocbina et al. 2009). Likewise, maternal-zygotic zebrafish oval (*ovl*, *Ift88*) mutants that lack all cilia display normal canonical and non-canonical Wnt signaling but show defects in Hh signaling (Huang and Schier 2009). In addition, no change in canonical Wnt signaling was seen in the kidneys of Inversin mutant mice at E16.5, P0 or P5 (Sugiyama et al. 2011), although this does not preclude a role for Inversin in canonical Wnt signaling prior to E16.5. Mice with mutations in core ciliary genes, i.e. *Ift88*, also do not generally display discernable phenotypes typical of increased canonical Wnt signaling (Chazaud and Rossant 2006; Mukhopadhyay et al. 2001; Zeng et al. 1997).

The β-catenin mediated canonical Wnt signaling pathway plays multiple roles prior to terminal differentiation of hair cells and establishment of PCP during inner ear development. Canonical Wnt signaling acts to direct an otic placodal fate against an epidermal fate during specification of the otic placode (Ohyama et al. 2006). Subsequently, Wnt signaling promotes dorsal cell identities within the otocyst (the precursor to the membranous labyrinth) (Riccomagno et al. 2005). Wnt1 and Wnt3a are expressed from the dorsal neural tube (Riccomagno et al. 2005). Wnt-responsive cells are detected in the dorsal region of the otocyst, and activate canonical Wnt signaling that antagonizes the ventral Shh signaling to specify dorsal vestibular versus ventral cochlea fate (Riccomagno et al. 2005). In addition, canonical Wnt signaling has a central role in vascular development in the ear involving Norrin and Fz4 (Xu et al. 2004). However, the role of cilia in regulating canonical Wnt signaling during inner ear development has not been carefully examined. Nor has a role for canonical Wnt signaling in PCP regulation in the inner ear been demonstrated.

In contrast to the loss of the vestibule in Wnt mutants (Riccomagno et al. 2005), the vestibule is present in *Ift88*$^{CKO/CKO}$ and *Kif3a*$^{CKO/CKO}$ mice (Jones et al. 2008). Conditional deletion of β-catenin in the otocyst or conditional expression of a non-phosphorylatable, constitutively active mutant of β-catenin in the otocyst both lead to embryonic lethality before hair bundle formation occurs (Freyer and Morrow 2010; Ohyama et al. 2006). Thus, the exact role of ciliary genes in canonical Wnt/β-catenin inhibition during hair cell orientation remains uncertain.

5.6.3 Primary Cilia and Notch Signaling in Inner Ear Development

Notch signaling is the new addition to the signaling pathways in which cilia have been implicated. The binding of membranous ligands to Notch receptors results in the cleavage of Notch intracellular domain (NICD) and its subsequent translocation to the nucleus and where NICD associates with RBP-j to activate downstream targets. Notch targets regulate cell proliferation, as well as cell differentiation through so-called lateral inhibition, or inhibit the cells with activated NICD from adopting the same fate of the neighboring cells that express Notch ligand(s). A recent study (Ezratty et al. 2011) reveals that the loss of ciliary genes compromises Notch signaling during epidermal differentiation. Furthermore, Notch3 is shown to localize to the cilia, and Presenillin2, a component of the γ-secretase that cleaves transmembrane Notch receptors to give rise to NICD and Notch signaling is enriched at the basal body.

During inner ear development, Notch signaling functions in multiple steps. It promotes the competency and restricts the boundaries of the prosensory domains in the developing inner ear (Lewis et al. 1998). It is also the major signaling pathway that prevents all of the cells in the prosensory domain from assuming a hair cell fate and directs the formation of a sensory mosaic, alternating hair and supporting cells, through lateral inhibition. In mutants defective in Notch signaling, ectopic hair cells are generated at the expense of supporting cells, proliferation is stimulated, and hair bundles are not aligned in the cochlear sensory cells. While the disorganization of hair cell polarity across the sensory organ in the cochlea may implicate Notch signaling in PCP regulation, deregulation of cell proliferation and cell fate conversion have not been observed in cilia mutants (Jones et al. 2008). The cilia mutants studied to date are conditional knockout mutants. It is possible that the inactivation of ciliogenesis machinery in these mutants is not complete and/or early enough to exhibit Notch phenotypes in these cilia mutants. It has yet to be tested whether cilia play any role in Notch signaling in the ear.

5.7 Conclusions and Perspectives

Cilia are present ubiquitously in eukaryotic cells. Their functions vary in different types of cells and they have been implicated in several signaling pathways including Shh signaling, Wnt signaling, Notch signaling, and PCP signaling. The inner ear sensory organs consist of stereotyped cellular patterns with distinct polarity features, offering an excellent opportunity to examine and dissect the underlying molecular and cellular mechanisms of the action of cilia in these signaling pathways. Indeed, the studies in the inner ear provide unequivocal evidence to support the essential roles of ciliary genes in PCP signaling and delineate the molecular events during PCP signaling. Ciliary genes are required for PCP signaling and regulate coordinated polarization of sensory hair cells and convergent extension of the cochlear duct. In particular, ciliary genes play a role in the determination of the intrinsic polarity of individual cells by directing the position of the basal body. In addition, the primary cilium of the hair cells, the kinocilium, likely has roles not only in shaping the polar mechanotransduction apparatus, the hair bundle, but also in mechanotransduction in vestibular hair cells.

The role of cilia in other signaling pathways, including Shh, canonical Wnt, and Notch signaling during inner ear development, however, has yet to be extensively tested. To date, there is no data to support the involvement of cilia in these signaling pathways during inner ear development. The limitations of the conditional knockout cilia mutant mice may account for the lack of phenotypes in the cilia mutant that are associated with these signaling pathways. Alternatively, these pathways may signal via cilia-independent mechanisms for inner ear development.

While the role for cilia genes in PCP regulation of the inner ear development is demonstrated by several studies, many key issues remain. It is not known how the cilium, basal body, or cilia proteins interact with polarized membrane-associated PCP complexes to regulate the polarity of each cell; how the cilium, basal body, or cilia proteins interact with the building machinery for the polar hair bundle; whether cilia are required for maintenance of PCP, and whether ciliary genes have cilia-independent roles in PCP signaling. Future studies addressing these issues and aiming to test and dissect the action of cilia in Shh, canonical Wnt, and Notch signaling during inner ear development will provide insightful information regarding this important cellular organelle and signaling pathways important for diverse development and disease processes.

References

Adato A, Lefevre G, Delprat B, Michel V, Michalski N, Chardenoux S, Weil D, El-Amraoui A, Petit C (2005a) Usherin, the defective protein in Usher syndrome type IIA, is likely to be a component of interstereocilia ankle links in the inner ear sensory cells. Hum Mol Genet 14:3921–3932

Adato A, Michel V, Kikkawa Y, Reiners J, Alagramam KN, Weil D, Yonekawa H, Wolfrum U, El-Amraoui A, Petit C (2005b) Interactions in the network of Usher syndrome type 1 proteins. Hum Mol Genet 14:347–356

Adler PN (2002) Planar signaling and morphogenesis in Drosophila. Dev Cell 2:525–535

Adler PN, Charlton J, Park WJ (1994) The Drosophila tissue polarity gene inturned functions prior to wing hair morphogenesis in the regulation of hair polarity and number. Genetics 137:829–836

Ahmed ZM, Riazuddin S, Ahmad J, Bernstein SL, Guo Y, Sabar MF, Sieving P, Griffith AJ, Friedman TB, Belyantseva IA, Wilcox ER (2003) PCDH15 is expressed in the neurosensory epithelium of the eye and ear and mutant alleles are responsible for both USH1F and DFNB23. Hum Mol Genet 12:3215–3223

Ahmed ZM, Goodyear R, Riazuddin S, Lagziel A, Legan PK, Behra M, Burgess SM, Lilley KS, Wilcox ER, Griffith AJ, Frolenkov GI, Belyantseva IA, Richardson GP, Friedman TB (2006) The tip-link antigen, a protein associated with the transduction complex of sensory hair cells, is protocadherin-15. J Neurosci 26:7022–7034

Assad JA, Shepherd GM, Corey DP (1991) Tip-link integrity and mechanical transduction in vertebrate hair cells. Neuron 7:985–994

Axelrod JD, McNeill H (2002) Coupling planar cell polarity signaling to morphogenesis. ScientificWorldJournal 2:434–454

Axelrod JD, Miller JR, Shulman JM, Moon RT, Perrimon N (1998) Differential recruitment of Dishevelled provides signaling specificity in the planar cell polarity and Wingless signaling pathways. Genes Dev 12:2610–2622

Bahloul A, Michel V, Hardelin JP, Nouaille S, Hoos S, Houdusse A, England P, Petit C (2010) Cadherin-23, myosin VIIa and harmonin, encoded by Usher syndrome type I genes, form a ternary complex and interact with membrane phospholipids. Hum Mol Genet 19:3557–3565

Beales PL (2005) Lifting the lid on Pandora's box: the Bardet-Biedl syndrome. Curr Opin Genet Dev 15:315–323

Bergmann C, Fliegauf M, Bruchle NO, Frank V, Olbrich H, Kirschner J, Schermer B, Schmedding I, Kispert A, Kranzlin B, Nurnberg G, Becker C, Grimm T, Girschick G, Lynch SA, Kelehan P, Senderek J, Neuhaus TJ, Stallmach T, Zentgraf H, Nurnberg P, Gretz N, Lo C, Lienkamp S, Schafer T, Walz G, Benzing T, Zerres K, Omran H (2008) Loss of nephrocystin-3 function can cause embryonic lethality, Meckel-Gruber-like syndrome, situs inversus, and renal-hepatic-pancreatic dysplasia. Am J Hum Genet 82:959–970

Boeda B, El-Amraoui A, Bahloul A, Goodyear R, Daviet L, Blanchard S, Perfettini I, Fath KR, Shorte S, Reiners J, Houdusse A, Legrain P, Wolfrum U, Richardson G, Petit C (2002) Myosin VIIa, harmonin and cadherin 23, three Usher I gene products that cooperate to shape the sensory hair cell bundle. EMBO J 21:6689–6699

Bonneau D, Raymond F, Kremer C, Klossek JM, Kaplan J, Patte F (1993) Usher syndrome type I associated with bronchiectasis and immotile nasal cilia in two brothers. J Med Genet 30:253–254

Caberlotto E, Michel V, Foucher I, Bahloul A, Goodyear RJ, Pepermans E, Michalski N, Perfettini I, Alegria-Prevot O, Chardenoux S, Do Cruzeiro M, Hardelin JP, Richardson GP, Avan P, Weil D, Petit C (2011) Usher type 1G protein sans is a critical component of the tip-link complex, a structure controlling actin polymerization in stereocilia. Proc Natl Acad Sci USA 108:5825–5830

Chacon-Heszele MF, Chen P (2009) Mouse models for dissecting vertebrate planar cell polarity signaling in the inner ear. Brain Res 1277:130–140

Chacon-Heszele MF, Ren D, Reynolds AB, Chi F, Chen P (2012) Regulation of cochlear convergent extension by the vertebrate planar cell polarity pathway is dependent on p120-catenin. Development 139:968–978

Chazaud C, Rossant J (2006) Disruption of early proximodistal patterning and AVE formation in Apc mutants. Development 133:3379–3387

Chen P, Johnson JE, Zoghbi HY, Segil N (2002) The role of Math1 in inner ear development: uncoupling the establishment of the sensory primordium from hair cell fate determination. Development 129:2495–2505

Colantonio JR, Vermot J, Wu D, Langenbacher AD, Fraser S, Chen JN, Hill KL (2009) The dynein regulatory complex is required for ciliary motility and otolith biogenesis in the inner ear. Nature 457:205–209

Collier S, Gubb D (1997) Drosophila tissue polarity requires the cell-autonomous activity of the fuzzy gene, which encodes a novel transmembrane protein. Development 124:4029–4037

Corbit KC, Shyer AE, Dowdle WE, Gaulden J, Singla V, Chen MH, Chuang PT, Reiter JF (2008) Kif3a constrains beta-catenin-dependent Wnt signalling through dual ciliary and non-ciliary mechanisms. Nat Cell Biol 10:70–76

Cotanche DA, Corwin JT (1991) Stereociliary bundles reorient during hair cell development and regeneration in the chick cochlea. Hear Res 52:379–402

Curtin JA, Quint E, Tsipouri V, Arkell RM, Cattanach B, Copp AJ, Henderson DJ, Spurr N, Stanier P, Fisher EM, Nolan PM, Steel KP, Brown SD, Gray IC, Murdoch JN (2003) Mutation of Celsr1 disrupts planar polarity of inner ear hair cells and causes severe neural tube defects in the mouse. Curr Biol 13:1129–1133

Dabdoub A, Kelley MW (2005) Planar cell polarity and a potential role for a Wnt morphogen gradient in stereociliary bundle orientation in the mammalian inner ear. J Neurobiol 64:446–457

Das G, Jenny A, Klein TJ, Eaton S, Mlodzik M (2004) Diego interacts with Prickle and Strabismus/Van Gogh to localize planar cell polarity complexes. Development 131:4467–4476

Deans MR, Antic D, Suyama K, Scott MP, Axelrod JD, Goodrich LV (2007) Asymmetric distribution of prickle-like 2 reveals an early underlying polarization of vestibular sensory epithelia in the inner ear. J Neurosci 27:3139–3147

Denman-Johnson K, Forge A (1999) Establishment of hair bundle polarity and orientation in the developing vestibular system of the mouse. J Neurocytol 28:821–835

Di Palma F, Holme RH, Bryda EC, Belyantseva IA, Pellegrino R, Kachar B, Steel KP, Noben-Trauth K (2001) Mutations in Cdh23, encoding a new type of cadherin, cause stereocilia disorganization in waltzer, the mouse model for Usher syndrome type 1D. Nat Genet 27:103–107

Driver EC, Pryor SP, Hill P, Turner J, Ruther U, Biesecker LG, Griffith AJ, Kelley MW (2008) Hedgehog signaling regulates sensory cell formation and auditory function in mice and humans. J Neurosci 28:7350–7358

El-Amraoui A, Petit C (2005) Usher I syndrome: unravelling the mechanisms that underlie the cohesion of the growing hair bundle in inner ear sensory cells. J Cell Sci 118:4593–4603

Ernstson S, Smith CA (1986) Stereo-kinociliar bonds in mammalian vestibular organs. Acta Otolaryngol 101:395–402

Etheridge SL, Ray S, Li S, Hamblet NS, Lijam N, Tsang M, Greer J, Kardos N, Wang J, Sussman DJ, Chen P, Wynshaw-Boris A (2008) Murine dishevelled 3 functions in redundant pathways with dishevelled 1 and 2 in normal cardiac outflow tract, cochlea, and neural tube development. PLoS Genet 4:e1000259

Ezratty EJ, Stokes N, Chai S, Shah AS, Williams SE, Fuchs E (2011) A role for the primary cilium in Notch signaling and epidermal differentiation during skin development. Cell 145:1129–1141

Fettiplace R (2006) Active hair bundle movements in auditory hair cells. J Physiol 576:29–36

Flock A, Bretscher A, Weber K (1982) Immunohistochemical localization of several cytoskeletal proteins in inner ear sensory and supporting cells. Hear Res 7:75–89

Forge A, Souter M, Denman-Johnson K (1997) Structural development of sensory cells in the ear. Semin Cell Dev Biol 8:225–237

Freyer L, Morrow BE (2010) Canonical Wnt signaling modulates Tbx1, Eya1, and Six1 expression, restricting neurogenesis in the otic vesicle. Dev Dyn 239:1708–1722

Frolenkov GI, Belyantseva IA, Friedman TB, Griffith AJ (2004) Genetic insights into the morphogenesis of inner ear hair cells. Nat Rev Genet 5:489–498

Furness DN, Hackney CM (1985) Cross-links between stereocilia in the guinea pig cochlea. Hear Res 18:177–188

Geleoc GS, Holt JR (2003) Developmental acquisition of sensory transduction in hair cells of the mouse inner ear. Nat Neurosci 6:1019–1020

Gerdes JM, Liu Y, Zaghloul NA, Leitch CC, Lawson SS, Kato M, Beachy PA, Beales PL, DeMartino GN, Fisher S, Badano JL, Katsanis N (2007) Disruption of the basal body compromises proteasomal function and perturbs intracellular Wnt response. Nat Genet 39:1350–1360

Goodyear RJ, Richardson GP (2003) A novel antigen sensitive to calcium chelation that is associated with the tip links and kinocilial links of sensory hair bundles. J Neurosci 23:4878–4887

Gray RS, Abitua PB, Wlodarczyk BJ, Szabo-Rogers HL, Blanchard O, Lee I, Weiss GS, Liu KJ, Marcotte EM, Wallingford JB, Finnell RH (2009) The planar cell polarity effector Fuz is essential for targeted membrane trafficking, ciliogenesis and mouse embryonic development. Nat Cell Biol 11:1225–1232

Grimsley-Myers CM, Sipe CW, Geleoc GS, Lu X (2009) The small GTPase Rac1 regulates auditory hair cell morphogenesis. J Neurosci 29:15859–15869

Hamblet NS, Lijam N, Ruiz-Lozano P, Wang J, Yang Y, Luo Z, Mei L, Chien KR, Sussman DJ, Wynshaw-Boris A (2002) Dishevelled 2 is essential for cardiac outflow tract development, somite segmentation and neural tube closure. Development 129:5827–5838

Heydeck W, Liu A (2011) PCP effector proteins inturned and fuzzy play nonredundant roles in the patterning but not convergent extension of mammalian neural tube. Dev Dyn 240:1938–1948

Heydeck W, Zeng H, Liu A (2009) Planar cell polarity effector gene Fuzzy regulates cilia formation and Hedgehog signal transduction in mouse. Dev Dyn 238:3035–3042

Hillman DE (1969) New ultrastructural findings regarding a vestibular ciliary apparatus and its possible functional significance. Brain Res 13:407–412

Huang P, Schier AF (2009) Dampened Hedgehog signaling but normal Wnt signaling in zebrafish without cilia. Development 136:3089–3098

Hudspeth AJ (1989) How the ear's works work. Nature 341:397–404

Hudspeth AJ, Corey DP (1977) Sensitivity, polarity, and conductance change in the response of vertebrate hair cells to controlled mechanical stimuli. Proc Natl Acad Sci USA 74:2407–2411

Hudspeth AJ, Jacobs R (1979) Stereocilia mediate transduction in vertebrate hair cells (auditory system/cilium/vestibular system). Proc Natl Acad Sci USA 76:1506–1509

Hunter DG, Fishman GA, Mehta RS, Kretzer FL (1986) Abnormal sperm and photoreceptor axonemes in Usher's syndrome. Arch Ophthalmol 104:385–389

Itoh K, Jenny A, Mlodzik M, Sokol SY (2009) Centrosomal localization of Diversin and its relevance to Wnt signaling. J Cell Sci 122:3791–3798

Jagger D, Collin G, Kelly J, Towers E, Nevill G, Longo-Guess C, Benson J, Halsey K, Dolan D, Marshall J, Naggert J, Forge A (2011) Alstrom Syndrome protein ALMS1 localizes to basal bodies of cochlear hair cells and regulates cilium-dependent planar cell polarity. Hum Mol Genet 20:466–481

Jones C, Chen P (2007) Planar cell polarity signaling in vertebrates. Bioessays 29:120–132

Jones C, Chen P (2008) Primary cilia in planar cell polarity regulation. Curr Top Dev Biol 85:197–224

Jones C, Roper VC, Foucher I, Qian D, Banizs B, Petit C, Yoder BK, Chen P (2008) Ciliary proteins link basal body polarization to planar cell polarity regulation. Nat Genet 40:69–77

Kazmierczak P, Sakaguchi H, Tokita J, Wilson-Kubalek EM, Milligan RA, Muller U, Kachar B (2007) Cadherin 23 and protocadherin 15 interact to form tip-link filaments in sensory hair cells. Nature 449:87–91

Keller R (2002) Shaping the vertebrate body plan by polarized embryonic cell movements. Science 298:1950–1954

Kibar Z, Vogan KJ, Groulx N, Justice MJ, Underhill DA, Gros P (2001) Ltap, a mammalian homolog of Drosophila Strabismus/Van Gogh, is altered in the mouse neural tube mutant loop-tail. Nat Genet 28:251–255

Kikkawa YS, Pawlowski KS, Wright CG, Alagramam KN (2008) Development of outer hair cells in Ames waltzer mice: mutation in protocadherin 15 affects development of cuticular plate and associated structures. Anat Rec (Hoboken) 291:224–232

Kikuchi T, Takasaka T, Tonosaki A, Watanabe H (1989) Fine structure of guinea pig vestibular kinocilium. Acta Otolaryngol 108:26–30

Klein TJ, Mlodzik M (2005) Planar cell polarization: an emerging model points in the right direction. Annu Rev Cell Dev Biol 21:155–176

Lagziel A, Ahmed ZM, Schultz JM, Morell RJ, Belyantseva IA, Friedman TB (2005) Spatiotemporal pattern and isoforms of cadherin 23 in wild type and waltzer mice during inner ear hair cell development. Dev Biol 280:295–306

Lefevre G, Michel V, Weil D, Lepelletier L, Bizard E, Wolfrum U, Hardelin JP, Petit C (2008) A core cochlear phenotype in USH1 mouse mutants implicates fibrous links of the hair bundle in its cohesion, orientation and differential growth. Development 135:1427–1437

Leibovici M, Safieddine S, Petit C (2008) Mouse models for human hereditary deafness. Curr Top Dev Biol 84:385–429

Lelli A, Asai Y, Forge A, Holt JR, Geleoc GS (2009) Tonotopic gradient in the developmental acquisition of sensory transduction in outer hair cells of the mouse cochlea. J Neurophysiol 101:2961–2973
Lewis AK, Frantz GD, Carpenter DA, de Sauvage FJ, Gao WQ (1998) Distinct expression patterns of notch family receptors and ligands during development of the mammalian inner ear. Mech Dev 78:159–163
Liu X, Vansant G, Udovichenko IP, Wolfrum U, Williams DS (1997) Myosin VIIa, the product of the Usher 1B syndrome gene, is concentrated in the connecting cilia of photoreceptor cells. Cell Motil Cytoskeleton 37:240–252
Liu X, Udovichenko IP, Brown SD, Steel KP, Williams DS (1999) Myosin VIIa participates in opsin transport through the photoreceptor cilium. J Neurosci 19:6267–6274
Liu X, Bulgakov OV, Darrow KN, Pawlyk B, Adamian M, Liberman MC, Li T (2007) Usherin is required for maintenance of retinal photoreceptors and normal development of cochlear hair cells. Proc Natl Acad Sci USA 104:4413–4418
Logan CY, Nusse R (2004) The Wnt signaling pathway in development and disease. Annu Rev Cell Dev Biol 20:781–810
Lu X, Borchers AG, Jolicoeur C, Rayburn H, Baker JC, Tessier-Lavigne M (2004) PTK7/CCK-4 is a novel regulator of planar cell polarity in vertebrates. Nature 430:93–98
Lysakowski A (1996) Synaptic organization of the crista ampullaris in vertebrates. Ann N Y Acad Sci 781:164–182
Maerker T, van Wijk E, Overlack N, Kersten FF, McGee J, Goldmann T, Sehn E, Roepman R, Walsh EJ, Kremer H, Wolfrum U (2008) A novel Usher protein network at the periciliary reloading point between molecular transport machineries in vertebrate photoreceptor cells. Hum Mol Genet 17:71–86
Mbiene JP, Sans A (1986) Differentiation and maturation of the sensory hair bundles in the fetal and postnatal vestibular receptors of the mouse: a scanning electron microscopy study. J Comp Neurol 254:271–278
McDermott KM, Liu BY, Tlsty TD, Pazour GJ (2010) Primary cilia regulate branching morphogenesis during mammary gland development. Curr Biol 20:731–737
McGee J, Goodyear RJ, McMillan DR, Stauffer EA, Holt JR, Locke KG, Birch DG, Legan PK, White PC, Walsh EJ, Richardson GP (2006) The very large G-protein-coupled receptor VLGR1: a component of the ankle link complex required for the normal development of auditory hair bundles. J Neurosci 26:6543–6553
McKenzie E, Krupin A, Kelley MW (2004) Cellular growth and rearrangement during the development of the mammalian organ of Corti. Dev Dyn 229:802–812
Michalski N, Michel V, Bahloul A, Lefevre G, Barral J, Yagi H, Chardenoux S, Weil D, Martin P, Hardelin JP, Sato M, Petit C (2007) Molecular characterization of the ankle-link complex in cochlear hair cells and its role in the hair bundle functioning. J Neurosci 27:6478–6488
Michel V, Goodyear RJ, Weil D, Marcotti W, Perfettini I, Wolfrum U, Kros CJ, Richardson GP, Petit C (2005) Cadherin 23 is a component of the transient lateral links in the developing hair bundles of cochlear sensory cells. Dev Biol 280:281–294
Mitchell B, Stubbs JL, Huisman F, Taborek P, Yu C, Kintner C (2009) The PCP pathway instructs the planar orientation of ciliated cells in the Xenopus larval skin. Curr Biol 19:924–929
Mlodzik M (2002) Planar cell polarization: do the same mechanisms regulate Drosophila tissue polarity and vertebrate gastrulation? Trends Genet 18:564–571
Mochizuki T, Tsuchiya K, Yokoyama T (2002) Molecular cloning of a gene for inversion of embryo turning (inv) with cystic kidney. Nephrol Dial Transplant 17(Suppl 9):68–70
Moeller H, Jenny A, Schaeffer HJ, Schwarz-Romond T, Mlodzik M, Hammerschmidt M, Birchmeier W (2006) Diversin regulates heart formation and gastrulation movements in development. Proc Natl Acad Sci USA 103:15900–15905
Montcouquiol M, Rachel RA, Lanford PJ, Copeland NG, Jenkins NA, Kelley MW (2003) Identification of Vangl2 and Scrb1 as planar polarity genes in mammals. Nature 423:173–177
Montcouquiol M, Sans N, Huss D, Kach J, Dickman JD, Forge A, Rachel RA, Copeland NG, Jenkins NA, Bogani D, Murdoch J, Warchol ME, Wenthold RJ, Kelley MW (2006) Asymmetric

localization of Vangl2 and Fz3 indicate novel mechanisms for planar cell polarity in mammals. J Neurosci 26:5265–5275
Moravec WJ, Peterson EH (2004) Differences between stereocilia numbers on type I and type II vestibular hair cells. J Neurophysiol 92:3153–3160
Morgan D, Turnpenny L, Goodship J, Dai W, Majumder K, Matthews L, Gardner A, Schuster G, Vien L, Harrison W, Elder FF, Penman-Splitt M, Overbeek P, Strachan T (1998) Inversin, a novel gene in the vertebrate left-right axis pathway, is partially deleted in the inv mouse. Nat Genet 20:149–156
Morgan D, Goodship J, Essner JJ, Vogan KJ, Turnpenny L, Yost HJ, Tabin CJ, Strachan T (2002) The left-right determinant inversin has highly conserved ankyrin repeat and IQ domains and interacts with calmodulin. Hum Genet 110:377–384
Moser JJ, Fritzler MJ, Ou Y, Rattner JB (2010) The PCM-basal body/primary cilium coalition. Semin Cell Dev Biol 21:148–155
Mukhopadhyay M, Shtrom S, Rodriguez-Esteban C, Chen L, Tsukui T, Gomer L, Dorward DW, Glinka A, Grinberg A, Huang SP, Niehrs C, Izpisua Belmonte JC, Westphal H (2001) Dickkopf1 is required for embryonic head induction and limb morphogenesis in the mouse. Dev Cell 1:423–434
Murdoch JN, Doudney K, Paternotte C, Copp AJ, Stanier P (2001) Severe neural tube defects in the loop-tail mouse result from mutation of Lpp 1, a novel gene involved in floor plate specification. Hum Mol Genet 10:2593–2601
Murdoch JN, Henderson DJ, Doudney K, Gaston-Massuet C, Phillips HM, Paternotte C, Arkell R, Stanier P, Copp AJ (2003) Disruption of scribble (Scrb1) causes severe neural tube defects in the circletail mouse. Hum Mol Genet 12:87–98
Nachury MV, Loktev AV, Zhang Q, Westlake CJ, Peranen J, Merdes A, Slusarski DC, Scheller RH, Bazan JF, Sheffield VC, Jackson PK (2007) A core complex of BBS proteins cooperates with the GTPase Rab8 to promote ciliary membrane biogenesis. Cell 129:1201–1213
Nayak GD, Ratnayaka HS, Goodyear RJ, Richardson GP (2007) Development of the hair bundle and mechanotransduction. Int J Dev Biol 51:597–608
Nunes FD, Lopez LN, Lin HW, Davies C, Azevedo RB, Gow A, Kachar B (2006) Distinct subdomain organization and molecular composition of a tight junction with adherens junction features. J Cell Sci 119:4819–4827
Ocbina PJ, Tuson M, Anderson KV (2009) Primary cilia are not required for normal canonical Wnt signaling in the mouse embryo. PLoS One 4:e6839
Ohyama T, Mohamed OA, Taketo MM, Dufort D, Groves AK (2006) Wnt signals mediate a fate decision between otic placode and epidermis. Development 133:865–875
Otto EA, Schermer B, Obara T, O'Toole JF, Hiller KS, Mueller AM, Ruf RG, Hoefele J, Beekmann F, Landau D, Foreman JW, Goodship JA, Strachan T, Kispert A, Wolf MT, Gagnadoux MF, Nivet H, Antignac C, Walz G, Drummond IA, Benzing T, Hildebrandt F (2003) Mutations in INVS encoding inversin cause nephronophthisis type 2, linking renal cystic disease to the function of primary cilia and left-right axis determination. Nat Genet 34:413–420
Padmashree R, Chen P (2009) Line up and listen: planar cell polarity regulation in the mammalian inner ear. Semin Cell Dev Biol 20:978–985
Park WJ, Liu J, Sharp EJ, Adler PN (1996) The Drosophila tissue polarity gene inturned acts cell autonomously and encodes a novel protein. Development 122:961–969
Park TJ, Haigo SL, Wallingford JB (2006) Ciliogenesis defects in embryos lacking inturned or fuzzy function are associated with failure of planar cell polarity and Hedgehog signaling. Nat Genet 38:303–311
Park TJ, Mitchell BJ, Abitua PB, Kintner C, Wallingford JB (2008) Dishevelled controls apical docking and planar polarization of basal bodies in ciliated epithelial cells. Nat Genet 40:871–879
Pawlowski KS, Kikkawa YS, Wright CG, Alagramam KN (2006) Progression of inner ear pathology in Ames waltzer mice and the role of protocadherin 15 in hair cell development. J Assoc Res Otolaryngol 7:83–94

Petit C (2006) From deafness genes to hearing mechanisms: harmony and counterpoint. Trends Mol Med 12:57–64
Petit C, Richardson GP (2009) Linking genes underlying deafness to hair-bundle development and function. Nat Neurosci 12:703–710
Pickles JO, Comis SD, Osborne MP (1984) Cross-links between stereocilia in the guinea pig organ of Corti, and their possible relation to sensory transduction. Hear Res 15:103–112
Reiners J, van Wijk E, Marker T, Zimmermann U, Jurgens K, te Brinke H, Overlack N, Roepman R, Knipper M, Kremer H, Wolfrum U (2005) Scaffold protein harmonin (USH1C) provides molecular links between Usher syndrome type 1 and type 2. Hum Mol Genet 14:3933–3943
Riccomagno MM, Martinu L, Mulheisen M, Wu DK, Epstein DJ (2002) Specification of the mammalian cochlea is dependent on Sonic hedgehog. Genes Dev 16:2365–2378
Riccomagno MM, Takada S, Epstein DJ (2005) Wnt-dependent regulation of inner ear morphogenesis is balanced by the opposing and supporting roles of Shh. Genes Dev 19:1612–1623
Ross AJ, May-Simera H, Eichers ER, Kai M, Hill J, Jagger DJ, Leitch CC, Chapple JP, Munro PM, Fisher S, Tan PL, Phillips HM, Leroux MR, Henderson DJ, Murdoch JN, Copp AJ, Eliot MM, Lupski JR, Kemp DT, Dollfus H, Tada M, Katsanis N, Forge A, Beales PL (2005) Disruption of Bardet-Biedl syndrome ciliary proteins perturbs planar cell polarity in vertebrates. Nat Genet 37:1135–1140
Ruben RJ (1967) Development of the inner ear of the mouse: a radioautographic study of terminal mitoses. Acta Otolaryngol Suppl 220:1–44
Rusch A, Eatock RA (1996a) A delayed rectifier conductance in type I hair cells of the mouse utricle. J Neurophysiol 76:995–1004
Rusch A, Eatock RA (1996b) Voltage responses of mouse utricular hair cells to injected currents. Ann N Y Acad Sci 781:71–84
Rzadzinska AK, Derr A, Kachar B, Noben-Trauth K (2005) Sustained cadherin 23 expression in young and adult cochlea of normal and hearing-impaired mice. Hear Res 208:114–121
Saihan Z, Webster AR, Luxon L, Bitner-Glindzicz M (2009) Update on Usher syndrome. Curr Opin Neurol 22:19–27
Schwarz-Romond T, Asbrand C, Bakkers J, Kuhl M, Schaeffer HJ, Huelsken J, Behrens J, Hammerschmidt M, Birchmeier W (2002) The ankyrin repeat protein Diversin recruits Casein kinase Iepsilon to the beta-catenin degradation complex and acts in both canonical Wnt and Wnt/JNK signaling. Genes Dev 16:2073–2084
Sekerkova G, Zheng L, Loomis PA, Mugnaini E, Bartles JR (2006) Espins and the actin cytoskeleton of hair cell stereocilia and sensory cell microvilli. Cell Mol Life Sci 63:2329–2341
Self T, Mahony M, Fleming J, Walsh J, Brown SD, Steel KP (1998) Shaker-1 mutations reveal roles for myosin VIIA in both development and function of cochlear hair cells. Development 125:557–566
Senften M, Schwander M, Kazmierczak P, Lillo C, Shin JB, Hasson T, Geleoc GS, Gillespie PG, Williams D, Holt JR, Muller U (2006) Physical and functional interaction between protocadherin 15 and myosin VIIa in mechanosensory hair cells. J Neurosci 26:2060–2071
Siemens J, Kazmierczak P, Reynolds A, Sticker M, Littlewood-Evans A, Muller U (2002) The Usher syndrome proteins cadherin 23 and harmonin form a complex by means of PDZ-domain interactions. Proc Natl Acad Sci USA 99:14946–14951
Simons M, Mlodzik M (2008) Planar cell polarity signaling: from fly development to human disease. Annu Rev Genet 42:517–540
Simons M, Gloy J, Ganner A, Bullerkotte A, Bashkurov M, Kronig C, Schermer B, Benzing T, Cabello OA, Jenny A, Mlodzik M, Polok B, Driever W, Obara T, Walz G (2005) Inversin, the gene product mutated in nephronophthisis type II, functions as a molecular switch between Wnt signaling pathways. Nat Genet 37:537–543
Sipe CW, Lu X (2011) Kif3a regulates planar polarization of auditory hair cells through both ciliary and non-ciliary mechanisms. Development 138:3441–3449
Strutt D (2008) The planar polarity pathway. Curr Biol 18:R898–R902
Sugiyama N, Tsukiyama T, Yamaguchi TP, Yokoyama T (2011) The canonical Wnt signaling pathway is not involved in renal cyst development in the kidneys of inv mutant mice. Kidney Int 79:957–965

Tasouri E, Tucker KL (2011) Primary cilia and organogenesis: is Hedgehog the only sculptor? Cell Tissue Res 345:21–40

Tilney LG, Derosier DJ, Mulroy MJ (1980) The organization of actin filaments in the stereocilia of cochlear hair cells. J Cell Biol 86:244–259

Tobin JL, Beales PL (2007) Bardet-Biedl syndrome: beyond the cilium. Pediatr Nephrol 22:926–936

Wallingford JB, Mitchell B (2011) Strange as it may seem: the many links between Wnt signaling, planar cell polarity, and cilia. Genes Dev 25:201–213

Wallingford JB, Fraser SE, Harland RM (2002) Convergent extension: the molecular control of polarized cell movement during embryonic development. Dev Cell 2:695–706

Wang J, Mark S, Zhang X, Qian D, Yoo SJ, Radde-Gallwitz K, Zhang Y, Lin X, Collazo A, Wynshaw-Boris A, Chen P (2005) Regulation of polarized extension and planar cell polarity in the cochlea by the vertebrate PCP pathway. Nat Genet 37:980–985

Wang J, Hamblet NS, Mark S, Dickinson ME, Brinkman BC, Segil N, Fraser SE, Chen P, Wallingford JB, Wynshaw-Boris A (2006a) Dishevelled genes mediate a conserved mammalian PCP pathway to regulate convergent extension during neurulation. Development 133:1767–1778

Wang Y, Guo N, Nathans J (2006b) The role of Frizzled3 and Frizzled6 in neural tube closure and in the planar polarity of inner-ear sensory hair cells. J Neurosci 26:2147–2156

Wersall J (1956) Studies on the structure and innervation of the sensory epithelium of the cristae ampulares in the guinea pig; a light and electron microscopic investigation. Acta Otolaryngol Suppl 126:1–85

Williams DS (2008) Usher syndrome: animal models, retinal function of Usher proteins, and prospects for gene therapy. Vision Res 48:433–441

Wolfrum U, Schmitt A (2000) Rhodopsin transport in the membrane of the connecting cilium of mammalian photoreceptor cells. Cell Motil Cytoskeleton 46:95–107

Wu D, Freund JB, Fraser SE, Vermot J (2011) Mechanistic basis of otolith formation during teleost inner ear development. Dev Cell 20:271–278

Xu Q, Wang Y, Dabdoub A, Smallwood PM, Williams J, Woods C, Kelley MW, Jiang L, Tasman W, Zhang K, Nathans J (2004) Vascular development in the retina and inner ear: control by Norrin and Frizzled-4, a high-affinity ligand-receptor pair. Cell 116:883–895

Yan D, Liu XZ (2010) Genetics and pathological mechanisms of Usher syndrome. J Hum Genet 55:327–335

Yasunaga T, Itoh K, Sokol SY (2011) Regulation of basal body and ciliary functions by Diversin. Mech Dev 128:376–386

Yoshida N, Liberman MC (1999) Stereociliary anomaly in the guinea pig: effects of hair bundle rotation on cochlear sensitivity. Hear Res 131:29–38

Yu X, Lau D, Ng CP, Roy S (2011) Cilia-driven fluid flow as an epigenetic cue for otolith biomineralization on sensory hair cells of the inner ear. Development 138:487–494

Zeng L, Fagotto F, Zhang T, Hsu W, Vasicek TJ, Perry WL 3rd, Lee JJ, Tilghman SM, Gumbiner BM, Costantini F (1997) The mouse Fused locus encodes Axin, an inhibitor of the Wnt signaling pathway that regulates embryonic axis formation. Cell 90:181–192

Zeng H, Hoover AN, Liu A (2010) PCP effector gene Inturned is an important regulator of cilia formation and embryonic development in mammals. Dev Biol 339:418–428

Zine A, Romand R (1996) Development of the auditory receptors of the rat: a SEM study. Brain Res 721:49–58

Chapter 6
Neuronal Cilia and Obesity

Nicolas F. Berbari, Raymond C. Pasek, and Bradley K. Yoder

Contents

Abstract Obesity is an emerging public health concern that has numerous secondary health consequences, including heart disease, high blood pressure, diabetes mellitus, osteoarthritis, and overall reduced quality of life. Historically, obesity has been viewed as increased body fat caused by overconsumption of food, combined with the sedentary lifestyle of modern society. Simply put, energy input exceeds energy output, creating an excess in fat mass. This viewpoint largely focuses on environmental and social factors in the obesity epidemic. However, it fails to take into account a growing body of evidence from several monogenetic human obesity

Authors Nicolas F. Berbari and Raymond C. Pasek contributed equally to this work.

N.F. Berbari • R. C. Pasek • B. K. Yoder, Ph.D. (✉)
Department of Cell, Developmental and Integrative Biology,
University of Alabama at Birmingham, Birmingham, AL 35294, USA
e-mail: byoder@uab.edu

K.L. Tucker and T. Caspary (eds.), *Cilia and Nervous System Development and Function*,
DOI 10.1007/978-94-007-5808-7_6, © Springer Science+Business Media Dordrecht 2013

disorders and mutant mouse and rat obesity models that indicate a profound role for genetic factors. Although most of these monogenetic human conditions are rare, it is clear that the study of their molecular and cellular etiology will offer insights into the mechanisms that regulate appetite and satiety. The objectives of this review are to discuss how mutations in genes required for the formation or function of the cilium result in obesity in human and mouse models and how the cilium may function to regulate appetite and satiation responses.

Keywords Obesity • Ciliopathy • Bardet-Biedl syndrome • Alström syndrome • Neuronal cilia • Leptin • Intraflagellar transport • Mouse models • Hedgehog • Melanin concentrating hormone • Hyperphagia • Food anticipatory behavior

6.1 Ciliopathies Associated with Obesity

The cilium is a small microtubule based cellular appendage found on most mammalian cells where it plays a crucial role as a complex sensory and signaling center (for a review on cilia signaling see Berbari et al. 2009). There is now an emerging class of genetic diseases coined ciliopathies that is associated with dysfunction of the cilium. Interestingly, the clinical features of ciliopathies range from renal cysts in humans with polycystic kidney disease (PKD1, OMIM #601313) to the cystic kidneys, skeletal anomalies, neurodevelopmental defects, heart malformations and perinatal lethality associated with Meckel-Gruber Syndrome (MKS1, OMIM # 249000) (for an in-depth review on ciliopathies see Sharma et al. 2008). This broad scope of clinical features has been attributed to both the ubiquitous nature of the cilium and the types of mutations affecting its function. PKD, for example, is associated with mutations in signaling proteins that appear to have crucial roles in renal homeostasis, while the mutations in MKS are often associated with proteins found at the base of the cilium, where they are thought to play more general roles in cilia structural integrity and the regulation of cilia protein composition. Until relatively recently much of our knowledge of cilia has come from studies in model systems focusing on the process of building and maintaining the cilium, known as Intraflagellar Transport (IFT) (for a review on IFT machinery see Pedersen and Rosenbaum 2008). While the list of ciliopathies continues to grow at an incredible pace, our current understanding of how dysfunction of the cilium leads to some of the clinically observed phenotypes remains elusive. One of the clinical features where this is most evident is the hyperphagia-associated obesity that occurs in two ciliopathies, Alström Syndrome (ALMS, OMIM #203800) and Bardet-Biedl Syndrome (BBS, OMIM #209900). The proteins affected in both of these disorders are associated with the cilium and are required for normal cilia function or formation.

Bardet-Biedl Syndrome (BBS) is a group of rare genetically heterogeneous disorders resulting in an array of seemingly unrelated symptoms and progressive degenerative defects. A triad of symptoms including obesity, retinitis pigmentosa/retinal degeneration, and polydactyly remains the hallmark for diagnosing the disease, as was the case when it was independently classified by Georges Bardet and Arthur

Biedl in the early twentieth century (Bardet 1920; Biedl 1922). Subsequent analysis has revealed that hypogonadism, renal dysfunction, and mental retardation are also highly prevalent among BBS patients. Nearly half of BBS patients are completely or partially anosmic and have deficits in thermosensation (Kulaga et al. 2004; Tan et al. 2007). Also of consequence are the secondary features of BBS, including diabetes mellitus, hypertension, and heart disease, which develop likely as a result of the obesity. Less commonly, BBS patients can have *situs inversus*, a defect often caused by dysfunctional cilia on the embryonic gastrulation stage structure called the node (Lorda-Sanchez et al. 2000; Deffert et al. 2007). Among North American and European populations, BBS is relatively uncommon with an estimated occurrence of 1 in 120,000 live births, while in Middle Eastern populations it has been reported to be as high as 1 in 13,500 (Klein and Ammann 1969; Croft et al. 1995; Beales et al. 1997). Although it is possible to detect the condition during gestation, patients are typically diagnosed when both obesity and retinal degeneration are apparent and polydactyly or mental deficits have been observed.

Confusion in diagnosing BBS exists, as a similar condition, Laurence-Moon Syndrome, is also a genetically-inherited disease in which patients present with retinitis pigmentosa and mental disability. This observation subsequently caused Solis-Cohens and Weiss to conclude that both diseases are synonymous and thus both disorders have been referred to as Laurence-Moon-Bardet-Biedl Syndrome. However, there is debate about the synonymous classification as the patients reported by Laurence and Moon displayed paraplegia of the lower extremities that is not typically associated with BBS. Furthermore, polydactyly and obesity are not generally traits of Laurence-Moon syndrome, while they are clinical hallmarks of BBS. Given that polydactyly is not fully penetrant among BBS patients, and that the obesity experienced by most BBS patients can be attenuated or eliminated with diet and exercise, the lack of these symptoms does not necessarily mean the two disorders are not allelic variations (Beales et al. 1999; Ghadami et al. 2000).

BBS has extensive genetic heterogeneity with 16 known loci (*BBS1-BBS16*, Table 6.1). Mutations in *BBS1-10* are responsible for approximately 70% of known BBS cases, and even the identification of *BBS11-16* only added fractional amounts to that number; thus it remains almost certain that other *BBS* genes have yet to be found (Chen et al. 2011; Stoetzel et al. 2006; Leitch et al. 2008; Chiang et al. 2006). Conversely, the genetic basis of Laurence-Moon is currently unknown, so it remains to be seen if the symptoms of Laurence-Moon are caused by mutations in *BBS* genes.

The mode of *BBS* inheritance is also complicated. Although originally believed to be a typical recessive Mendelian disorder, analysis by Katsanis et al. demonstrated this may not always be the case (Katsanis et al. 2001). In their study, the authors reported families where both affected and unaffected individuals could carry homozygous mutant alleles of *BBS2*. However, the affected individuals were also heterozygous for a mutation in a second BBS gene, *BBS6*. This phenomenon, which the authors named 'triallelic inheritance', demonstrates that manifestations of BBS phenotypes sometimes rely on mutations at a second locus and thus BBS may result from the overall genetic mutational load in a patient. However, analysis of the triallelic inheritance hypothesis in several other BBS cohorts did not reveal evidence of triallelism suggesting this may be an exception (Mykytyn et al. 2003; Abu-Safieh et al. 2012;

Table 6.1 Bardet-Biedl syndrome genes

BBS	Gene	Human locus	Motifs/domains	References
BBS1	BBS1	11q13.2	Beta propeller[a]	Mykytyn et al. (2002)
BBS2	BBS2	16q12.2	Beta propeller[a]	Nishimura et al. (2001)
BBS3	ARL6	3q11.2	ADP ribosylation factor-like 6	Chiang et al. (2004)
BBS4	BBS4	15q24.1	TPR repeats[a]	Mykytyn et al. (2001)
BBS5	BBS5	2q31.1	Pleckstrin homology[a]	Kulaga et al. (2004)
BBS6	MKKS	20p12.2	Chaperone-like[b]	Slavotinek et al. (2000)
BBS7	BBS7	4q27	Beta propeller[a]	Badano et al. (2003)
BBS8	TTC8	14q31.3	TPR repeats[a]	Ansley et al. (2003)
BBS9	PTHB1	7p14.3	Beta propeller[a]	Nishimura et al. (2005)
BBS10	BBS10	12q21.2	Chaperone-like[b]	Stoetzel et al. (2006)
BBS11	TRIM32	9q33.1	E3 Ubiquitin Ligase	Chiang et al. (2006)
BBS12	BBS12	4q27	Chaperone-like[b]	Stoetzel et al. (2007)
BBS13	MKS1	17q22	B9/C2	Leitch et al. (2008)
BBS14	CEP290	12q21.32	Coiled-coil	Leitch et al. (2008)
BBS15	WDPCP	2p15	Coiled-coil, WD40	Kim et al. (2010)
BBS16	SDCCAG8	1q43	Globular, coiled-coil	Schaefer et al. (2011)

Known *BBS* genes are listed with their location in the human genome as well as their known protein motifs and domains. Members of the BBSome are indicated with an[a] and BBS chaperone proteins are indicated with an[b]

[a]Indicates a member of the BBSome protein complex (Nachury et al. 2007)

[b]Indicates a member of a BBS chaperone protein complex (Seo et al. 2010)

Smaoui et al. 2006; Laurier et al. 2006). Information from these cohorts suggests that BBS is an autosomal recessive inherited disease. But these data cannot rule out possible contributions from unidentified *BBS* alleles and it remains possible that BBS can manifest through both an autosomal recessive fashion and triallelic inheritance.

The great degree of variability in both inheritance and symptoms presented by BBS patients leads to the question of whether there exists any correlation between the genes and mutations involved and the way the disease manifests itself. For example, while obesity and renal abnormalities are frequent, the degree of mental retardation or learning disabilities varies greatly, with patients from some families showing little or no mental deficits (Riise et al. 1997). Thus, the relationship between BBS mutations and the expressivity of traits has been of great interest, but largely remains inconclusive with limited genotype-phenotype correlation.

Birth weight in BBS patients is usually normal, with obesity developing during childhood continuing into adulthood (Beales et al. 1999). This observation indicates that the obese phenotype may not be a direct consequence of defects in development, but rather due to errors in energy metabolism or appetite regulation. Evidence for this hypothesis is found in the fact that when compared to BMI matched controls, BBS patients did not possess significant differences in body fat or resting metabolic rate (RMR) (Grace et al. 2003). Although this information indicates that body fat and RMR is the same, subsequent work demonstrated that circulating leptin and triglyceride levels were significantly higher in BBS patients compared to other BMI matched obese individuals, despite the fact that glucose tolerance and insulin resistance was comparable between the two groups (Feuillan et al. 2011). Leptin has a known role in suppressing appetite, and this finding supports the possibility that BBS patients have a higher degree of leptin resistance. Recent work in animal models have strongly implicated primary cilia as being necessary for regulating appetite, and that cilia regulated signaling can be disrupted in mouse models of BBS (Davenport et al. 2007; Weatherbee et al. 2009). However, the molecular mechanism causing this disease in BBS patients remains uncertain.

The other human ciliopathy associated with obesity is Alström syndrome (ALMS), which was first classified in 1959 (for a review of ALMS see Girard and Petrovsky 2011). Human ALMS patients manifest with several symptoms including obesity, retinitis pigmentosa, and hearing loss with a tendency towards shorter stature, and a disruption in the growth hormone/Insulin-like growth factor 1 signaling axis. They also exhibit phenotypes likely related to their obesity that include diabetes mellitus and elevated leptin levels when compared to unaffected individuals (Maffei et al. 2007). ALMS is a rare autosomal recessive disorder with an occurrence at less than 1 in 100,000 and is caused by mutations in the gene *ALMS1* (Collin et al. 2002; Hearn et al. 2002). To date, 81 different disease causing mutations in *ALMS1* have been reported (Joy et al. 2007; Marshall et al. 2007). The exact function of ALMS1 remains unknown, but clues to its possible cellular function were uncovered when it was found that ALMS1 is widely expressed and localizes to the centrosome and the basal body of the primary cilium in cultured human function cells (Hearn et al. 2005; Knorz et al. 2010). Interestingly, dermal fibroblasts derived from an ALMS patient had normal basal body localization and primary cilia assembly suggesting that ALMS1 might be involved in ciliary related signaling pathways, but not in establishing cilia architecture itself (Hearn et al. 2005). In contrast, knockdown of *Alms1* by siRNA in mouse inner medullary collecting duct (mIMCD3) cells caused a stunted cilia phenotype, and also impaired their mechanical stimuli sensing abilities. This discrepancy in phenotype could be due to the nature of the mutation, which may not have caused a complete lack of protein function.

Although BBS and ALMS are relatively rare diseases, understanding how these proteins normally regulate satiation responses will provide important insights into molecular pathways that could be manipulated to control satiation and obesity.

6.2 Ciliopathy Mouse Models of Obesity

To better understand the causes of human obesity, genetically manipulated mouse models are continuously being developed with the ultimate goal of elucidating the molecular mechanisms driving the phenotype. Surprisingly, over the past decade the primary cilium has emerged as a key factor regulating satiation responses. The *Bbs* and *Alms1* mouse models along with mutations affecting the Intraflagellar transport

88 (*Ift88*) gene have established a strong link between defects in cilia mediated sensory or signaling activity and obesity (Table 6.2). In this section, we will review data derived from several of the obesity mouse models that have been associated with ciliary dysfunction and highlight the proposed function of the affected proteins.

6.2.1 *Bbs* Mutant Mouse Models

Seminal work in the BBS field by Nachury et al. has shown that BBS proteins 1, 2, 4, 5, 7, 8, and 9 form a ~450 kDa complex called the BBSome (Nachury et al. 2007). The BBSome is thought to function in transport of membrane along with specific transmembrane proteins to the cilium (Jin et al. 2010). Of these BBSome genes, mouse models of *Bbs1*, *2*, *4*, and *8* have been generated and characterized. The validity of utilizing mice to model human BBS was demonstrated when a knock-in allele of the *Bbs1* M390R mutation, one of the most common single human BBS disease alleles, was created and replicated many of the human symptoms of BBS, including retinal degeneration, male infertility, and obesity (Davis et al. 2007). Strikingly, these hallmark features of BBS are shared phenotypes among other mouse models of BBS such as *Bbs2* and *Bbs4* mutants (Mykytyn et al. 2004; Nishimura et al. 2004) (Fig. 6.1a). Neurological defects were also observed. For example, disruption of *Bbs1* or *Bbs4* caused cilia loss on the olfactory epithelium, and the same report demonstrated partial or total anosmia in a cohort of human BBS patients. These same studies found a common social dominance defect among *Bbs2* and *Bbs4* mutant mice, in which the mutants were more submissive to control mice. Although the olfactory and behavior phenotypes may not directly be related to the obesity seen in these mice, it does reflect the importance of the BBSome genes in the regulation of behavior.

As in BBS patients, obesity is not present in young *Bbs* mutant mice. In fact, most *Bbs* mutant mice are initially runted, and it has been proposed that this is possibly due to olfactory defects that make it difficult for pups to accomplish nipple searching and suckling (Eichers et al. 2006). However, mutants eventually developed hyperphagia and became obese. The obesity phenotype also correlated with hyperleptinemia in *Bbs1* M390R knock-in mice. More recently, *Bbs8*-null mice have been reported that also have defects in olfactory function, as has been shown in *Bbs1* and *4* mutant mice. When these *Bbs8* mutant mice were crossed to an olfactory receptor reporter line ($M72^{TL}$), severe defects in the targeting of olfactory sensory neurons became apparent, and individual axonal fibers seemed to wander, instead of terminating at a single glomerulus as in the control mice (Tadenev et al. 2011). The axonal targeting defects reported in the *Bbs8* mutant mice further confirm the importance of the BBSome in proper neuronal development and signaling.

Other genes involved in human BBS that do not encode direct BBSome components have been identified. These proteins share homology to chaperones, and there are indications that these too are necessary for normal activity of satiation pathways. These genes include *BBS6*, *10*, and *12*, and they encode proteins that may be

Table 6.2 Cilia associated mouse models of obesity

Gene	Allele (MGI)	Type of allele	Mouse phenotypes	References
AC3	$Adcy3^{tm1Drs}$	Knock-out	Obesity, anosmia	Wang et al. (2009) and Wong et al. (2000)
Alms1	$Alms1^{Gt(XH152)Byg}$	Genetrap	Obesity, retinopathy, male infertility, late onset hearing loss	Collin et al. (2005)
Alms1	$Alms1^{foz}$	Spontaneous	Obesity, male infertility, late onset hearing loss	Arsov et al. (2006)
Bbs1	$Bbs1^{tm1Vcs}$	Knock-in	Obesity, retinopathy, male infertility, ventriculomegaly	Davis et al. (2007)
Bbs2	$Bbs2^{tm1Vcs}$	Knock-out	Obesity, retinopathy, renal cysts, male infertility, anosmia, social submissiveness, ventriculomegaly	Nishimura et al. (2004)
[a]*Bbs3/Arl6*	$Arl6^{tm2Vcs}$	Knock-out	Increased fat mass, retinopathy, male infertility, hydrocephalus, elevated blood pressure	Zhang et al. (1994)
[b]*Bbs4*	$Bbs4^{tm1Vcs}$	Knock-out	Obesity, retinopathy, male infertility, social submissiveness, ventriculomegaly	Mykytyn et al. (2004)
[b]*Bbs4*	$Bbs4^{Gt1Nk}$	Genetrap	Obesity (sex dependent penetrance and severity), retinopathy, social submissiveness, increased anxiety	Kulaga et al. (2004) and Eichers et al. (2006)
Bbs6/Mkks	$Mkks^{tm1Vcs}$	Knock-out	Obesity, retinopathy, male infertility, anosmia, elevated blood pressure, social submissiveness, ventriculomegaly	Fath et al. (2005)
[a]*Bbs11/Trim32*	$Trim32^{Gt(BGA355)Byg}$	Genetrap	Increased body weight, muscular myopathy	Kudryashova et al. (2011)
Ift88	$Ift88^{tm1.1Bky}$	Conditional knock-out	Obesity, renal cysts, hepatic cysts	Davenport et al. (2007)
Kif3a	$Kif3a^{tm1Gsn}$	Conditional knock-out	Obesity, renal cysts, hepatic cysts	Davenport et al. (2007)
Tub	Tub^{tub}	Spontaneous	Obesity, retinopathy, late onset hearing loss	Coleman and Eicher (1990)
Tub	Tub^{tm1Rok}	Knock-out	Obesity, retinopathy, late onset hearing loss	Stubdal et al. (2000)

Mutant mice with defects in known or potential cilia genes are shown alphabetically, along with the allele listing according to Mouse Genome Informatics (MGI http://www.informatics.jax.org/). Please note that the mouse phenotypes column includes some of the more prominent features, but for the sake of brevity, is not comprehensive

[a]*Bbs3* mutant mice do not become obese, but do display increased fat mass. Likewise, *Bbs11* mutants have not been reported to be obese, but do display a significant and persistent increase in body weight starting at 2 months of age

[b]Two different $Bbs4^{-/-}$ knockout mouse lines have been independently generated and different penetrance and severity of obesity have been reported for each

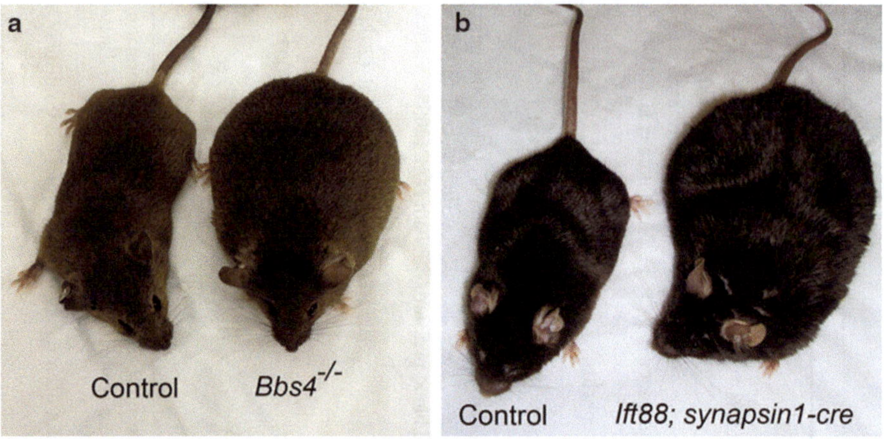

Fig. 6.1 Obese cilia mutant mouse models. (**a**) An obese *Bbs4*$^{-/-}$ mutant (*right*) next to his wild-type littermate (*left*). (**b**) An obese conditional *Ift88* mutant that has lost cilia throughout the central nervous system as the result of synapsin1-cre expression (*right*) next to a wild-type littermate (*left*)

necessary to stabilize the BBSome (Seo et al. 2010). Of these three, only a mouse mutant of *Bbs6* (previously referred to as Mkks for its involvement in McKusick-Kaufman syndrome) has been reported. As seen in the other BBSome mutant mice, *Bbs6* mutants display age dependent retinal degeneration and exhibit hyperphagic behavior leading to obesity with elevated leptin levels (Fath et al. 2005). Furthermore, male infertility was reported due to a failure in the formation of spermatozoa flagella, similar to the findings of *Bbs1, 2*, and *4* mutant mice.

Not all known *BBS* genes fall into the category of being a BBSome complex member or having chaperone-like properties. This includes the *BBS3* gene that encodes the small GTPase ARL6. Mouse *Bbs3* mutants exhibit both retinal degeneration and male infertility due to loss of sperm flagella (Zhang et al. 2011). In addition, severe hydrocephalus accompanied by altered beating of ependymal cilia was found, but no loss or obvious defects in primary cilia morphology were evident. Most strikingly however, was the apparent lack of an overt obesity phenotype in the *Bbs3* mutants. Likewise, leptin levels in the mutants were not statistically different than controls (Zhang et al. 2011). Although the *Bbs3* mutants do not have a significant increase in body weight, they do have an increase in the amount of gonadal and retroperitoneal fat. The reason that *Bbs3* mutant mice do not display an obesity phenotype is unknown. It was proposed this may be due to the early onset hydrocephalus; however, it should be noted that obesity along with hydrocephalus is seen in some of the other Bbs models. The lack of an obesity phenotype could also be related to different functions of the Bbs proteins and differential effects they have on protein trafficking. Melanin concentrating hormone receptor1 (Mchr1) is a ciliary localized G protein coupled receptor (GPCR) known to have orexigenic effects. Intriguingly, in obese models such as *Bbs2* and *Bbs4* mutants, Mchr1 is not present in the cilium while it does localize to in the cilia of cultured neurons from *Bbs3* mutants (Zhang et al. 2011). It is also important to note that Bbs3 is neither a BBSome complex member nor a BBS chaperone protein, raising the possibility that Bbs3 has functions independent of the other BBS proteins.

In some cases, previously identified genes are now being recognized as belonging to the BBS family. For example, mutations in the E3 ubiquitin ligase TRIM32 was identified in BBS patient (hence called *BBS11*) through the use of homozygosity mapping with SNP arrays (Chiang et al. 2006). *Trim32/Bbs11* mutant mice display muscular dystrophy and a decreased concentration of neurofilaments, as well as a reduction in myelinated motoraxon diameters (Kudryashova et al. 2009, 2011). A small increase in body weight was found in the *Bbs11* mice when compared to controls, but this was only a 10% difference at 8 weeks of age (Kudryashova et al. 2009). Much like Bbs3, Bbs11 is neither a BBSome complex member nor a BBS chaperon protein, and thus may also have independent or unique functions from the rest of the BBSome. However, it remains possible that the muscular dystrophy and reduction in motor axon myelination are precluding the emergence of an obesity phenotype. In addition, mutations in BBS11 can cause two distinct clinical disorders; BBS and limb-girdle muscular dystrophy type 2H (LGMD2H). BBS phenotypes were associated with N-terminal mutations (P130S) while LGMD2H appears to be caused by mutations in the C-terminal region (R394H, D487N, D588del, or T520TfsX13) that do not disrupt its ability to function in ubiquitination.

Studies of more recently identified BBS genes include *MKS1/BBS13* and *CEP290/BBS14*. However, current reports utilize *Bbs13* and *Bbs14* mutant mice that are either embryonically lethal and/or not true genetic nulls, thus making the potential role of *Bbs13* and 14 in obesity and appetite regulation ambiguous (Tadenev et al. 2011; Weatherbee et al. 2009; Lancaster et al. 2011). Regardless, the fact that not all *Bbs* mutant mice have the same phenotypes indicates complexity and diversity in the functions of the different *BBS* genes along with differential effects of the specific mutations on gene function.

6.2.2 Mouse Model of Alström Syndrome

The other ciliopathy with obesity as a symptom is Alström syndrome. In contrast to BBS, Alström syndrome appears to be caused by mutations in a single gene *ALMS1*. The ALMS1 protein localizes to the basal body of ciliated cells, but the function of the protein is not certain (Hearn et al. 2005; Collin et al. 2005). Mouse models corresponding to Alström syndrome have also been reported and include a gene-trapped allele (*Alms1*$^{-/-}$) and a spontaneous mutant (*fat aussie, foz*) (Arsov et al. 2006; Collin et al. 2005). Mice lacking functional Alms1 are born at a normal weight much like their human counterparts. However, hyperphagic behavior and obesity ensue that is accompanied by hyperinsulinemia and type II diabetes. The *Alms1*$^{-/-}$ mutant mice also have enlarged livers with the accumulation of lipid deposits, and the pancreas is hyperplastic. In addition to obesity, mice lacking Alms1 display male infertility, as well as retinal and cochlear defects, all of which are reminiscent of cilia associated defects in human patients.

6.2.3 Obesity in the Intraflagellar Transport Mutants

Cilia formation and maintenance, and possibly its signaling activity, is dependent on the intraflagellar transport (IFT) system to mediate bidirectional transport of proteins between the base and tip of the cilium. Null alleles of the *Ift88* gene (originally referred to as *Tg737* in mouse) caused early embryonically lethality and even hypomorphic alleles caused death prior to adulthood with systemic effects (Moyer et al. 1994; Murcia et al. 2000; Lehman et al. 2008). The necessity of cilia for normal mammalian development has made analyzing possible roles of the cilium in satiation and obesity difficult. This problem was circumvented with the creation of conditional alleles of *Ift88* and the IFT motor *Kif3a* (Marszalek et al. 1999; Haycraft et al. 2007). Using a tamoxifen-inducible cre recombinase expressed from the actin promoter (CAGG-creER^{TM}) (Davenport et al. 2007), cilia loss could be induced systemically in adult mice. This was found to cause hyperphagia within 3 weeks of inducing cilia loss and subsequently caused obesity. Furthermore, the obesity phenotype was prevented by maintaining adult conditional cilia mutant mice on a restricted diet, wherein they were provided the same daily amount of food as normal controls consumed. This observation indicates that the obesity phenotype is caused by the lack of a satiation response that leads to the overconsumption and not a general alteration in metabolic or locomotor activity.

The change in feeding behavior observed in the *Ift88* and *Kif3a* conditional mutant mice led to the possibility that cilia on neurons may be responsible for the obesity phenotype. To test this hypothesis, conditional *Ift88* and *Kif3a* mutant mice were crossed to synapsin1-cre mice to cause loss of cilia exclusively in neurons (Zhu et al. 2001). As with the systemically induced cilia mutants, neuronal specific cilia mutant mice became morbidly obese and strongly implicated a previously unappreciated role for neuronal cilia in regulating appetite (Fig. 6.1b).

The hypothalamus is a critically important signaling center of the brain known to regulate appetite. This action is done in large part by neurons that express either pro-opiomelanocortin (POMC) or agouti-related protein (AgRP) that release signaling factors ultimately suppressing or enhancing appetite, respectively (for a review see Mountjoy 2010). Importantly, hypothalamic neurons each possess a single primary cilium, although the function of the cilium on these neurons is largely unexplored. To address the role of neuronal cilia and appetite, *Ift88* and *Kif3a* conditional mutants were crossed to POMC-cre or AgRP-cre expressing mice, to conditionally ablate cilia on POMC or AgRP expressing neurons, respectively (Xu et al. 2005b). By 6 weeks of age, both male and female POMC cilia mutant mice weighed significantly more than control mice, and continued to become morbidly obese into adulthood. This was not evident in the mice lacking cilia on AgRP neurons (Berbari and Yoder, unpublished data). Another observation that was reported in the POMC cilia mutant mice was an increase in the levels of leptin, fasting serum glucose, and insulin (Davenport et al. 2007). This was observed only in the obese state and not in mice kept lean by pair-feeding, indicating that these elevated levels were a secondary consequence of the obesity. This report was significant for providing some of the initial evidence indicating the importance of neuronal cilia in regulating obesity.

6.2.4 Other Obesity Mouse Models Associated with Ciliary Proteins

In addition to the *Bbs*, *Alms1*, and *Ift88* mouse models there are several other mutant mouse strains supporting a connection between cilia and obesity. One prime example is a mutation in the type III adenylyl cyclase (ACIII). ACIII localizes to the primary cilia throughout the adult rodent brain (Bishop et al. 2007) and loss of ACIII causes anosmia and obesity by 3 months of age. Interestingly, even when ACIII mutants do not weigh significantly more than wildtype siblings, they have an increased level of serum leptin (Wang et al. 2009). It is interesting to note that a recent Genome Wide Association Study revealed there is a SNP near ACIII that is associated with obesity in humans (Hebebrand et al. 2010; Nordman et al. 2008).

Another example may be the *tubby* mouse that was first identified at the Jackson Laboratory as a spontaneous mutant causing a maturity-onset obesity phenotype (Coleman and Eicher 1990) and subsequently regenerated by gene targeting (Ashrafi et al. 2003). The *tubby* mutants have progressive loss of hearing and vision (Ohlemiller et al. 1995), similarities that are also shared with the *Bbs* and *Alms1* mutant mice. The spontaneous *tubby* mouse possesses a single base pair mutation within a splice site of the gene *Tub* (named after the mutant mouse) resulting in the expression of an aberrant transcript. The functions of the Tub protein remain ambiguous, but it is found to be highly expressed in portions in the brain, including the arcuate nucleus of the hypothalamus (Kleyn et al. 1996). The Tub protein is dispensable in assembly of the cilia, and no defects in the cilia assembly process of intraflagellar transport (IFT) have been reported. Intriguingly, despite the fact that Tub has yet to be reported in mammalian primary cilia, a physical association between Tub and the IFT complex has been reported in an immortalized human cell line (Mukhopadhyay et al. 2010). Further evidence for a ciliary role of Tub can be found with the *C. elegans* homolog *tub-1* that undergoes transport along the ciliary axoneme (Mukhopadhyay et al. 2005). Like the *tubby* mouse, *C. elegans* with a deletion of *tub-1* show an increase in fat content, suggesting an evolutionarily conserved role of the gene in regulating fat storage. Other proteins in the tubby family of proteins have also been implicated as having ciliary roles. For example, tubby-like protein 3 (*Tulp3*) localizes to the primary cilia during mouse development, and is necessary for proper Shh signaling but its association with obesity has not yet been determined (Norman et al. 2009).

6.3 Potential Molecular Mechanisms of Ciliopathy Associated Obesity

The hyperphagia induced obesity is one of the more intriguing phenotypes of ciliopathies that remains to be fully explained. There are indeed several possibilities described in the literature as to how loss of the cilia could alter states of satiety and appetite. Here we review a few of the candidate molecular pathways that may play roles in obesity associated with cilia dysfunction. These possibilities include primary deficits in leptin signaling, altered G-protein coupled receptor (GPCR) signaling, and abnormal regulation of mTor and hedgehog signaling pathways (Table 6.3).

6.3.1 Cilia and the Leptin Signaling Axis

While the conditional allele of *Ift88* implicated a role for neuronal cilia in satiety and more specifically a role for POMC neuronal cilia, this study did not specify a molecular framework for the underlying hyperphagia phenotype (Davenport et al. 2007). More recent data has aimed to accomplish this by showing that Bbs1, a component of the BBSome, directly binds to the leptin receptor and that BBS proteins may have a role in leptin receptor trafficking (Fig. 6.2a) (Seo et al. 2009). The initial identification of the leptin gene encoding a small protein hormone in the spontaneous obese mouse mutant *ob/ob* was the source of much excitement (Zhang et al. 1994). Importantly, leptin suppresses feeding activity and it is secreted into serum at levels proportionate to the amount of adipose tissue, the hormone's primary source (Considine et al. 1996). Interestingly, these recent studies in *Bbs2*, *Bbs4*, and *Bbs6* mutant mice also revealed that they are hyper-leptinemic and importantly, they fail to reduce food intake in response to IP or ICV injection of leptin (Rahmouni et al. 2008). Thus, defects in leptin signaling were thought to contribute directly to the obesity phenotype in BBS.

The excitement surrounding leptin's initial discovery was attenuated when it was determined that nearly all obese mice and humans have markedly elevated levels of circulating leptin, yet do not have normal leptin-mediated repression of appetite (Considine et al. 1996; Maffei et al. 1995). This barrier phenomenon is known as *leptin resistance*, the mechanism of which remains an active area of research. Thus, in obesity research, one challenge is determining whether leptin resistance is a primary cause or a consequence of the obesity. One approach used to overcome this situation is to decrease the amount of adipose tissue, and consequently the levels of circulating leptin, through caloric restriction. Interestingly, when this was performed on the BBS mutant mouse models to maintain body weight and leptin levels as seen in controls, they were still resistant to leptin, suggesting leptin signaling defects are a primary cause of the phenotype (Seo et al. 2009). However, the study did not take into account a phenomenon called *food anticipatory behavior* wherein the mice alter their meal structure in response to the calorie restriction such that they consume

Table 6.3 Neuronal cilia signaling proteins that may contribute to obesity

Candidate molecule	Known receptors	Localization	Potential mechanism	References
Leptin	Lep-R	Cell membrane, 'peri'-ciliary membrane	Changes in leptin receptor localization	Seo et al. (2009)
Melanin concentrating hormone	Mchr1, [a]Mchr2	Ciliary membrane	Changes in Mchr1 localization	Berbari et al. (2008)
Dopamine	D_1	Ciliary	Changes in Drd1 cilia localization	Domire et al. (2011)
Serotonin	5-HT_6	Ciliary membrane in certain brain regions	Changes in ciliary receptor 5HT6 signaling	Brailov et al. (2000)
Adeylate cyclase III	Olfactory GPCRs	Cilia throughout the brain	Altered coupling downstream of ciliary GPCRs	Bishop et al. (2007)
Hedgehog	Patched	Ciliary membrane in regions of adult neurogenesis	Altered adult neurogenesis Altered hypothalamic development Non-canonical hedgehog signaling	N/A

Potential pathways behind the obesity phenotype are listed with the components that localize to the cilium or at the base of the cilium, as well as a current hypothesized mechanism.
[a]Mchr2 is common to primates, canines, and other carnivores, but is absent among rodents

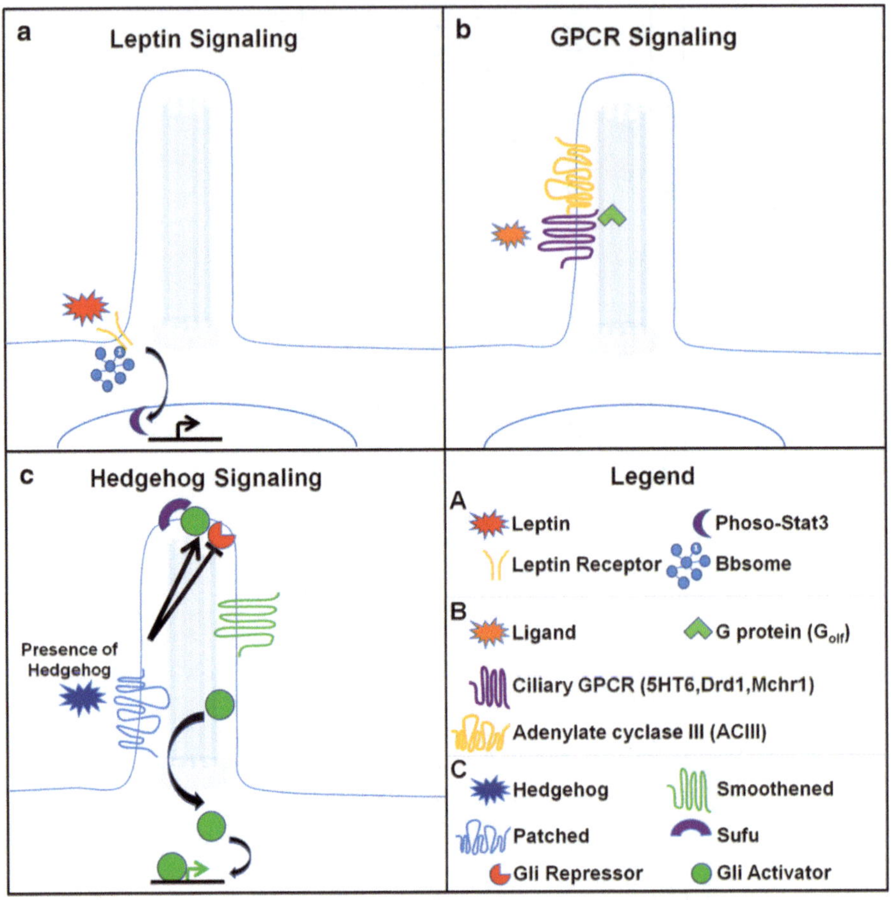

Fig. 6.2 Potential ciliary signaling pathways necessary for appetite regulation. (**a**) Depicts the leptin receptor interacting with Bbs1 of the BBSome near the base of the cilia where it has been proposed to be available for leptin activation and subsequent phospho-Stat3 induction and translocation to the nucleus. (**b**) A depiction of cilia specific GPCRs such as Mchr1, Sstr3, Drd1, and 5HT6, and their potential effector ACIII and G proteins such as $G_{s/q/i,o}$. (**c**) The Hedgehog signaling pathway, with patched repressing smoothened translocation into the cilium until ligand stimulation, upon which Gli transcription factors are processed from the inhibitor to the activator forms followed by subsequent translocation to the nucleus

nearly all of their calories within the first few hours of food access (for a review on anticipatory feeding behavior and methods see Mistlberger 2009). Interestingly, during this entrained period the mice consume nearly the same amount of food as they were given during calorie restriction, even when they are given *ad libitum* amounts of food. This entrained behavior persists for over a week and during this period the mice do not reduce food intake in response to leptin and thus appear leptin resistant (Berbari and Yoder unpublished). This feeding behavior resulting from the anticipation of food is in large part thought to be the result of a feeding clock, somewhat analogous to but distinct from the circadian clock (Mistlberger 2009). It remains to be seen whether either *Bbs* or *Ift* conditional mice would respond to leptin when both body weight and anticipatory feeding behavior are experimentally controlled. This would require testing animals for leptin sensitivity after deterioration of the food anticipatory behavior.

The BBS studies utilized congenital mutants and reported a loss of approximately 20% of the leptin responsive pro-opiomelanocortin (POMC) neurons in the arcuate nucleus of the hypothalamus (Seo et al. 2009). Thus, the authors propose that improper leptin receptor trafficking in POMC neurons of the arcuate nucleus leads to their inability to sense leptin and thus mediate its anorectic effects. While several possibilities exist for the loss of 20% of POMC neurons in *Bbs* mice, it is of note that alteration in the Foxo1/Insulin signaling pathway have resulted in similar changes in adult POMC neuronal populations (Plum et al. 2012). Loss of POMC neurons in *Bbs* models could be the result of neurodevelopmental changes or possibly maintenance of POMC neuron population through altered adult neurogenesis that lead to the hyperphagic phenotype. Similarly, the obesity phenotype observed in the hyperphagic *Ift88* conditional POMC-cre model could be a hypothalamic developmental phenotype. However, the rapid onset of obesity upon ubiquitous loss of cilia induced by the actin promoter driving cre line (CaGG-CreER) in adult mice suggests that cilia play a direct role in appetite and satiety. What is needed is an investigation using inducible cilia mutants in POMC neurons or other neuronal populations implicated in feeding behavior in order to elucidate the role of primary cilia signaling in appetite and satiety.

6.3.2 Cilia and Melanin-Concentrating Hormone Pathway

While both genetic models and biochemical approaches have informed our current understanding of BBS the precise mechanism behind BBS-associated obesity is unknown. Many of the BBS proteins form a protein complex (BBSome) that is involved in proper cilia membrane formation (Nachury et al. 2007). Indeed there is evidence that the BBSome may be a membrane vesicle coat complex that is critical for establishing and maintaining the cilia membrane's signaling capabilities by directing specific receptors to this compartment (Jin et al. 2010). Furthermore, *Bbs* mutant mouse models appear to improperly localize several cilia-specific GPCRs, and most relevant with regard to obesity is the melanin-concentrating hormone receptor 1 (Mchr1) (Berbari et al. 2008). Mchr1 couples through Gαi to reduce cAMP and decreases the frequency of spontaneous action potentials in hypothalamus (Gao and van den Pol 2001, 2002). Mch injections induce feeding behavior while Mchr1 mutant mice are resistant to diet induced obesity (Gomori et al. 2003; Chen et al. 2002). Intriguingly, Mchr1 fails to localize normally in neuronal cilia of *Bbs2*, *Bbs3*, and *Bbs4* mutant mice (Berbari et al. 2008; Zhang et al. 2011). Thus in both *Ift88* and *Bbs* obese mutants, Mchr1 fails to be properly localized creating a defect in Mchr1 signaling, possibly leading to the hyperphagic behavior in these models. Both pharmacological and genetic agonism of the Mchr1 pathway are associated with hyperphagia while antagonism is associated with anorectic behavior, as such antagonism of this receptor has been of interest to the pharmaceutical industry (Qu et al. 1996; Borowsky et al. 2002; Ludwig et al. 2001; Shimada et al. 1998; Chen et al. 2002). However, assuming that the obese phenotype behind both *Bbs* and *Ift* models is driven by a similar molecular pathway, then one would have to propose that in the absence of the cilium or the ability to reach the cilium that the Mchr1 pathway is ectopically activated or not efficiently desensitized after activation.

While there is circumstantial evidence for Mchr1 cilia mis-localization driving hyperphagia in cilia mutant mouse models, it is interesting to note that there is an emerging list of GPCRs that preferentially localize to neuronal cilia in different regions of the brain (Fig. 6.2b). Some of these neuronal cilia specific GPCRs include somatostatin receptor 3 (Sstr3), serotonin receptor 6 (5HT6), and dopamine receptor 1(Drd1) (Handel et al. 1999; Schulz et al. 2000; Hamon et al. 1999; Brailov et al. 2000; Marley and von Zastrow 2010; Domire et al. 2011). While the significance of localizing these receptors within the ciliary compartment remain unknown, it is enticing to speculate that perhaps they may play a role in appetite and satiety, especially when one considers that the somatostatin, serotonin, and dopaminergic systems have all been implicated in either reward or feeding behaviors directly (Vijayan and McCann 1977; Aponte et al. 1984; Pollock and Rowland 1981; Salamone et al. 1990).

6.3.3 Cilia and the Mammalian Target of Rapamycin (mTOR) Pathway

Another pathway that may be involved in neuronal cilia regulation of satiation is the mammalian target of rapamycin (mTOR) pathway. mTOR signaling is complex and involves many factors (for an in depth review of mTOR and disease see Dazert and Hall 2011). mTOR is a serine/threonine protein kinase which as its name implies can be inhibited by the antifungal rapamycin. It has been established as a regulator/coordinator of cellular metabolic activity that responds to both the energy and stress levels experienced by the cell. It carries out these regulatory roles by participating in two protein complexes, the rapamycin-sensitive mTOR Complex 1 and the rapamycin-insensitive mTORC2. In general mTORC1 regulates translational control and mTORC2 is involved in cytoskeleton organization. While the functions of mTOR and its interactors have been determined in considerable detail at the genetic and cellular levels, the effects of mTOR signaling on the organismal level continue to emerge. Interestingly, there are several reports associating the cilium or its signaling proteins with overactivation of mTOR activity or in changes in the cytoskeleton and cell size (Sharma et al. 2011; Bell et al. 2011; Boehlke et al. 2010). In addition, rapamycin is able to partially rescue renal cystic disease in mouse models of PKD, further supporting a connection between cilia and mTOR (Shillingford et al. 2006, 2010). While the *in vivo* relevance of cilia and mTOR signaling with regards to the obesity phenotype in cilia mutants remains to be determined, it is of note that mTOR signaling within the hypothalamus has been associated with obesity in other animal models (Cota et al. 2006). It will be interesting to determine if mTOR signaling activity within adult neurons is regulated through neuronal cilia and influences feeding behavior.

6.3.4 Hedgehog Signaling and the Cilium

The final pathway that we will discuss with regard to neuronal cilia and obesity is the hedgehog (Hh) pathway. Hh and its role in cilia and neuronal development is reviewed in Chap. 2 of this work by Mariani and Caspary. Several groups have demonstrated that canonical hedgehog signaling in mammalian cells utilizes the ciliary compartment. This is best demonstrated by the transient localization of several of the pathway components to the cilium and altered pathway activity when the cilium has been disrupted (reviewed in detail by Goetz and Anderson 2010). With regard to hedgehog signaling and neuronal cilia, there is a consensus emerging that primary cilia within the adult central nervous system sense hedgehog ligand and mediate the process of adult neurogenesis (Breunig et al. 2008; Han et al. 2008). Furthermore when neuronal cilia-mediated hedgehog signaling is altered in a gain of function fashion it can result in medulloblastoma and when it is disrupted in a loss of function fashion in the developing brain it is associated with a range of neurodevelopmental phenotypes (Chizhikov et al. 2007; Han et al. 2009).

If altered hedgehog signaling and cilia mutations have such profound effects on the adult and developing nervous system, how may they account for the hyperphagia associated obesity in adults? In both *Ift* conditional and *Bbs* models the possibility that altered hedgehog signaling in the developing hypothalamus can lead to obesity in adulthood has yet to be thoroughly investigated. For example, POMC-cre conditional *Ift88* mutant models appear normal other than the onset of hyperphagia and obesity, but the potential for a developmental phenotype remains. This becomes important when the expression pattern of POMC is taken into account. POMC is known to be expressed in places outside of the arcuate nucleus, of the hypothalamus such as the nucleus tract solitarius of the hindbrain, and the anterior and intermediate lobe of the pituitary in neonatal animals (Xu et al. 2005a). This becomes particularly important when the crucial role of hedgehog not only in the developing neural tube but also in the developing hypothalamus is taken into account (Szabo et al. 2009; Alvarez-Bolado et al. 2012). To address these potentials both *Ift* and *Bbs* conditional models need to be tested with inducible POMC-cre deletion. These experiments would also be useful in assessing whether different molecular mechanisms may be involved in driving obesity in *Bbs* and *Ift* mouse models.

The possibility remains that hedgehog signaling, which is required for adult neurogenesis is disrupted, thus contributing to hyperphagia. There are also reports of adult neurogenesis within the hypothalamus (Kokoeva et al. 2005; Xu et al.

2005c; Pierce and Xu 2010; Lee et al. 2012). Perhaps this process is compromised in CAGG-creER;*Ift88* conditional mutants. However, hyperphagic behavior is observed within 3 weeks of cilia loss in these mice and thus it may not have sufficient time to be a result of altered adult neurogenesis.

Finally a third potential for a non-canonical form of hedgehog signaling exists in the adult hypothalamus that requires neuronal cilia. The relevance of a non-canonical Hh pathway emerges when the expression pattern of certain pathway components in the adult brain is analyzed. For example, the Hh receptor, Patched, is expressed in regions of the brain that do not co-express the Hh effector Smoothened (Traiffort et al. 1998, 1999). This incongruence in pathway component expression pattern is especially true with regard to the hypothalamus (for a review of hedgehog in the adult brain see Traiffort et al. 2010). Furthermore it has been shown that hedgehog can directly alter neuronal activity (Bezard et al. 2003; Pascual et al. 2005). In developing spinal neurons, Hh stimulation causes a transient increase in Ca^{2+} activity that was dependent on Smo and Gαi (Belgacem and Borodinsky 2011). Since hedgehog pathway components such as Patched are expressed in the adult hypothalamus, it is feasible that cilia may alter satiation responses through regulation of neuronal firing activity (Traiffort et al. 1998, 1999). Exploring whether loss of cilia alters this increase in Ca^{2+} in response to Hh in POMC neurons could prove to be a very fruitful avenue of investigation to connect cilia dysfunction to abnormal satiation.

6.4 Non-Mammalian Ciliopathy Models of BBS

Although more distantly related to human beings than mice, non-mammalian models have proven to be invaluable to the study of the role of human cilia and their relation to disease. This is particularly evident in the study of the assembly and maintenance of the cilium through IFT, and how disruption of this event can lead to certain phenotypes. This process, referred to as intraflagellar transport (IFT), was largely characterized biochemically using the small green algae, *Chlamydomonas*, and genetically using *C. elegans* (for an in depth review see Pedersen and Rosenbaum 2008). In this section we focus solely on the genes and proteins known to be associated with the obesity phenotype observed in ciliopathies, as such it will largely focus on the functional roles of the BBS genes in both *Chlamydomonas reinhardtii* and *Caenorhabditis elegans*, two of the most well studied non-mammalian organisms in regards to cilia/flagella biology

While *Chlamydomonas* has served as good model for biochemical purification of flagellar and IFT components it has also proven useful for comparative genomics studies in discovering new ciliopathy genes, such as *BBS5* (Kulaga et al. 2004). Through the use of this simple model, elegant studies have begun to shed new light on how the BBS proteins may function as modulators of ciliary signaling and even serve as structural components of the transition zone (Lechtreck et al. 2009; Craige et al. 2010).

Much of what we know about the molecular motors that mediate cilia formation and maintenance has come from studies visualizing IFT movement in the cilia of *C. elegans*. In *C. elegans* cilia of the sensory neurons it has been demonstrated that both BBS7 and BBS8 serve as adaptors to the IFT complexes and their cargoes (Blacque et al. 2004), however, whether they play similar roles in mammalian systems has not been determined. Interestingly, it has been shown that *C. elegans* ciliary morphology can change dependent on cilia-mediated signaling and that the phenotypes of *bbs* mutant worms can be ameliorated by altering the downstream second messengers (Tan et al. 2007; Mukhopadhyay et al. 2008; Mok et al. 2011). Recent work has also suggested that altered neuroendocrine signaling and exocytosis of factors such as insulin drives many of the phenotypes observed in *bbs* mutant worms (Lee et al. 2011). Interestingly, another study points to more general roles for bbs proteins in cilia membrane homeostasis (Kaplan et al. 2012). Although the invertebrates lack many of the organ systems present in mammals, it is clear that both *C. elegans* and *Chlamydomonas* models offer advantages in both cost, time and in some instances genetic tractability. These models will continue to provide insights into the fundamental processes that are mediated by the ciliopathy proteins and the cilium and thus further serve to inform our understanding of complex phenotypes such as feeding behavior and the regulation of appetite and satiety.

6.5 Conclusion

In summary, remarkable progress has been made in the past 20 years demonstrating the clinic importance of the cilium in may tissues and developmental processes. Despite this progress, there remain several key questions that must be addressed before we can understand the molecular and cellular mechanisms responsible. Hopefully, as research on the rare ciliopathies advances we will gain an understanding of fundamental processes such as satiety and appetite that we can then apply to direct therapeutic strategies for an exceedingly common clinical feature such as obesity.

Acknowledgements We thank Yoder lab members for suggestions and critical review of the chapter. This work was supported in part by NIH RO1 DK075996 to BKY. NFB was supported by NRSA F32 DK088404, RCP was supported by T32 GM008111 to (Dr. Yoder).

References

Abu-Safieh L, Al-Anazi S, Al-Abdi L, Hashem M, Alkuraya H, Alamr M, Sirelkhatim MO, Al-Hassnan Z, Alkuraya B, Mohamed JY, Al-Salem A, Alrashed M, Faqeih E, Softah A, Al-Hashem A, Wali S, Rahbeeni Z, Alsayed M, Khan AO, Al-Gazali L, Taschner PE, Al-Hazzaa S, Alkuraya FS (2012) In search of triallelism in Bardet-Biedl syndrome. Eur J Hum Genet 20:420–427

Alvarez-Bolado G, Paul FA, Blaess S (2012) Sonic hedgehog lineage in the mouse hypothalamus: from progenitor domains to hypothalamic regions. Neural Dev 7:4

Ansley SJ, Badano JL, Blacque OE, Hill J, Hoskins BE, Leitch CC, Kim JC, Ross AJ, Eichers ER, Teslovich TM, Mah AK, Johnsen RC, Cavender JC, Lewis RA, Leroux MR, Beales PL, Katsanis N (2003) Basal body dysfunction is a likely cause of pleiotropic Bardet-Biedl syndrome. Nature 425:628–633

Aponte G, Leung P, Gross D, Yamada T (1984) Effects of somatostatin on food intake in rats. Life Sci 35:741–746

Arsov T, Silva DG, O'Bryan MK, Sainsbury A, Lee NJ, Kennedy C, Manji SS, Nelms K, Liu C, Vinuesa CG, de Kretser DM, Goodnow CC, Petrovsky N (2006) Fat aussie – a new Alstrom syndrome mouse showing a critical role for ALMS1 in obesity, diabetes, and spermatogenesis. Mol Endocrinol 20:1610–1622

Ashrafi K, Chang FY, Watts JL, Fraser AG, Kamath RS, Ahringer J, Ruvkun G (2003) Genome-wide RNAi analysis of Caenorhabditis elegans fat regulatory genes. Nature 421:268–272

Badano JL, Ansley SJ, Leitch CC, Lewis RA, Lupski JR, Katsanis N (2003) Identification of a novel Bardet-Biedl syndrome protein, BBS7, that shares structural features with BBS1 and BBS2. Am J Hum Genet 72:650–658

Bardet G (1920) Sur un syndrome d'obesite' congenitale avec polydactylie et retinite pigmentaire (contribution a l'etude des formes cliniques de l'obesite hypophysaire). University of Paris, Paris

Beales PL, Warner AM, Hitman GA, Thakker R, Flinter FA (1997) Bardet-Biedl syndrome: a molecular and phenotypic study of 18 families. J Med Genet 34:92–98

Beales PL, Elcioglu N, Woolf AS, Parker D, Flinter FA (1999) New criteria for improved diagnosis of Bardet-Biedl syndrome: results of a population survey. J Med Genet 36:437–446

Belgacem YH, Borodinsky LN (2011) Sonic hedgehog signaling is decoded by calcium spike activity in the developing spinal cord. Proc Natl Acad Sci USA 108:4482–4487

Bell PD, Fitzgibbon W, Sas K, Stenbit AE, Amria M, Houston A, Reichert R, Gilley S, Siegal GP, Bissler J, Bilgen M, Chou PC, Guay-Woodford L, Yoder B, Haycraft CJ, Siroky B (2011) Loss of primary cilia upregulates renal hypertrophic signaling and promotes cystogenesis. J Am Soc Nephrol 22:839–848

Berbari NF, Lewis JS, Bishop GA, Askwith CC, Mykytyn K (2008) Bardet-Biedl syndrome proteins are required for the localization of G protein-coupled receptors to primary cilia. Proc Natl Acad Sci USA 105:4242–4246

Berbari NF, O'Connor AK, Haycraft CJ, Yoder BK (2009) The primary cilium as a complex signaling center. Curr Biol 19:R526–R535

Bezard E, Baufreton J, Owens G, Crossman AR, Dudek H, Taupignon A, Brotchie JM (2003) Sonic hedgehog is a neuromodulator in the adult subthalamic nucleus. FASEB J 17:2337–2338

Biedl A (1922) Geschwisterpaar mit adipose-genitaler Dystrophie. Dtsch Med Wochenschr 48:1630

Bishop GA, Berbari NF, Lewis J, Mykytyn K (2007) Type III adenylyl cyclase localizes to primary cilia throughout the adult mouse brain. J Comp Neurol 505:562–571

Blacque OE, Reardon MJ, Li C, McCarthy J, Mahjoub MR, Ansley SJ, Badano JL, Mah AK, Beales PL, Davidson WS, Johnsen RC, Audeh M, Plasterk RH, Baillie DL, Katsanis N, Quarmby LM, Wicks SR, Leroux MR (2004) Loss of C. elegans BBS-7 and BBS-8 protein function results in cilia defects and compromised intraflagellar transport. Genes Dev 18:1630–1642

Boehlke C, Kotsis F, Patel V, Braeg S, Voelker H, Bredt S, Beyer T, Janusch H, Hamann C, Godel M, Muller K, Herbst M, Hornung M, Doerken M, Kottgen M, Nitschke R, Igarashi P, Walz G, Kuehn EW (2010) Primary cilia regulate mTORC1 activity and cell size through Lkb1. Nat Cell Biol 12:1115–1122

Borowsky B, Durkin MM, Ogozalek K, Marzabadi MR, DeLeon J, Lagu B, Heurich R, Lichtblau H, Shaposhnik Z, Daniewska I, Blackburn TP, Branchek TA, Gerald C, Vaysse PJ, Forray C (2002) Antidepressant, anxiolytic and anorectic effects of a melanin-concentrating hormone-1 receptor antagonist. Nat Med 8:825–830

Brailov I, Bancila M, Brisorgueil MJ, Miquel MC, Hamon M, Verge D (2000) Localization of 5-HT(6) receptors at the plasma membrane of neuronal cilia in the rat brain. Brain Res 872:271–275

Breunig JJ, Sarkisian MR, Arellano JI, Morozov YM, Ayoub AE, Sojitra S, Wang B, Flavell RA, Rakic P, Town T (2008) Primary cilia regulate hippocampal neurogenesis by mediating sonic hedgehog signaling. Proc Natl Acad Sci USA 105:13127–13132

Chen Y, Hu C, Hsu CK, Zhang Q, Bi C, Asnicar M, Hsiung HM, Fox N, Slieker LJ, Yang DD, Heiman ML, Shi Y (2002) Targeted disruption of the melanin-concentrating hormone receptor-1 results in hyperphagia and resistance to diet-induced obesity. Endocrinology 143:2469–2477

Chen J, Smaoui N, Hammer MB, Jiao X, Riazuddin SA, Harper S, Katsanis N, Riazuddin S, Chaabouni H, Berson EL, Hejtmancik JF (2011) Molecular analysis of Bardet-Biedl syndrome families: report of 21 novel mutations in 10 genes. Invest Ophthalmol Vis Sci 52:5317–5324

Chiang AP, Nishimura D, Searby C, Elbedour K, Carmi R, Ferguson AL, Secrist J, Braun T, Casavant T, Stone EM, Sheffield VC (2004) Comparative genomic analysis identifies an ADP-ribosylation factor-like gene as the cause of Bardet-Biedl syndrome (BBS3). Am J Hum Genet 75:475–484

Chiang AP, Beck JS, Yen HJ, Tayeh MK, Scheetz TE, Swiderski RE, Nishimura DY, Braun TA, Kim KY, Huang J, Elbedour K, Carmi R, Slusarski DC, Casavant TL, Stone EM, Sheffield VC (2006) Homozygosity mapping with SNP arrays identifies TRIM32, an E3 ubiquitin ligase, as a Bardet-Biedl syndrome gene (BBS11). Proc Natl Acad Sci USA 103:6287–6292

Chizhikov VV, Davenport J, Zhang Q, Shih EK, Cabello OA, Fuchs JL, Yoder BK, Millen KJ (2007) Cilia proteins control cerebellar morphogenesis by promoting expansion of the granule progenitor pool. J Neurosci 27:9780–9789

Coleman DL, Eicher EM (1990) Fat (fat) and tubby (tub): two autosomal recessive mutations causing obesity syndromes in the mouse. J Hered 81:424–427

Collin GB, Marshall JD, Ikeda A, So WV, Russell-Eggitt I, Maffei P, Beck S, Boerkoel CF, Sicolo N, Martin M, Nishina PM, Naggert JK (2002) Mutations in ALMS1 cause obesity, type 2 diabetes and neurosensory degeneration in Alstrom syndrome. Nat Genet 31:74–78

Collin GB, Cyr E, Bronson R, Marshall JD, Gifford EJ, Hicks W, Murray SA, Zheng QY, Smith RS, Nishina PM, Naggert JK (2005) Alms1-disrupted mice recapitulate human Alstrom syndrome. Hum Mol Genet 14:2323–2333

Considine RV, Sinha MK, Heiman ML, Kriauciunas A, Stephens TW, Nyce MR, Ohannesian JP, Marco CC, McKee LJ, Bauer TL et al (1996) Serum immunoreactive-leptin concentrations in normal-weight and obese humans. N Engl J Med 334:292–295

Cota D, Proulx K, Smith KA, Kozma SC, Thomas G, Woods SC, Seeley RJ (2006) Hypothalamic mTOR signaling regulates food intake. Science 312:927–930

Craige B, Tsao CC, Diener DR, Hou Y, Lechtreck KF, Rosenbaum JL, Witman GB (2010) CEP290 tethers flagellar transition zone microtubules to the membrane and regulates flagellar protein content. J Cell Biol 190:927–940

Croft JB, Morrell D, Chase CL, Swift M (1995) Obesity in heterozygous carriers of the gene for the Bardet-Biedl syndrome. Am J Med Genet 55:12–15

Davenport JR, Watts AJ, Roper VC, Croyle MJ, van Groen T, Wyss JM, Nagy TR, Kesterson RA, Yoder BK (2007) Disruption of intraflagellar transport in adult mice leads to obesity and slow-onset cystic kidney disease. Curr Biol 17:1586–1594

Davis RE, Swiderski RE, Rahmouni K, Nishimura DY, Mullins RF, Agassandian K, Philp AR, Searby CC, Andrews MP, Thompson S, Berry CJ, Thedens DR, Yang B, Weiss RM, Cassell MD, Stone EM, Sheffield VC (2007) A knockin mouse model of the Bardet-Biedl syndrome 1 M390R mutation has cilia defects, ventriculomegaly, retinopathy, and obesity. Proc Natl Acad Sci USA 104:19422–19427

Dazert E, Hall MN (2011) mTOR signaling in disease. Curr Opin Cell Biol 23:744–755

Deffert C, Niel F, Mochel F, Barrey C, Romana C, Souied E, Stoetzel C, Goossens M, Dollfus H, Verloes A, Girodon E, Gerard-Blanluet M (2007) Recurrent insertional polydactyly and situs inversus in a Bardet-Biedl syndrome family. Am J Med Genet A 143:208–213

Domire JS, Green JA, Lee KG, Johnson AD, Askwith CC, Mykytyn K (2011) Dopamine receptor 1 localizes to neuronal cilia in a dynamic process that requires the Bardet-Biedl syndrome proteins. Cell Mol Life Sci 68:2951–2960

Eichers ER, Abd-El-Barr MM, Paylor R, Lewis RA, Bi W, Lin X, Meehan TP, Stockton DW, Wu SM, Lindsay E, Justice MJ, Beales PL, Katsanis N, Lupski JR (2006) Phenotypic characterization of Bbs4 null mice reveals age-dependent penetrance and variable expressivity. Hum Genet 120:211–226

Fath MA, Mullins RF, Searby C, Nishimura DY, Wei J, Rahmouni K, Davis RE, Tayeh MK, Andrews M, Yang B, Sigmund CD, Stone EM, Sheffield VC (2005) Mkks-null mice have a phenotype resembling Bardet-Biedl syndrome. Hum Mol Genet 14:1109–1118

Feuillan PP, Ng D, Han JC, Sapp JC, Wetsch K, Spaulding E, Zheng YC, Caruso RC, Brooks BP, Johnston JJ, Yanovski JA, Biesecker LG (2011) Patients with Bardet-Biedl syndrome have hyperleptinemia suggestive of leptin resistance. J Clin Endocrinol Metab 96:E528–E535

Gao XB, van den Pol AN (2001) Melanin concentrating hormone depresses synaptic activity of glutamate and GABA neurons from rat lateral hypothalamus. J Physiol 533:237–252

Gao XB, van den Pol AN (2002) Melanin-concentrating hormone depresses L-, N-, and P/Q-type voltage-dependent calcium channels in rat lateral hypothalamic neurons. J Physiol 542:273–286

Ghadami M, Tomita HA, Najafi MT, Damavandi E, Farahvash MS, Yamada K, Majidzadeh AK, Niikawa N (2000) Bardet-Biedl syndrome type 3 in an Iranian family: clinical study and confirmation of disease localization. Am J Med Genet 94:433–437

Girard D, Petrovsky N (2011) Alstrom syndrome: insights into the pathogenesis of metabolic disorders. Nat Rev Endocrinol 7:77–88

Goetz SC, Anderson KV (2010) The primary cilium: a signalling centre during vertebrate development. Nat Rev Genet 11:331–344

Gomori A, Ishihara A, Ito M, Mashiko S, Matsushita H, Yumoto M, Ito M, Tanaka T, Tokita S, Moriya M, Iwaasa H, Kanatani A (2003) Chronic intracerebroventricular infusion of MCH causes obesity in mice. Melanin-concentrating hormone. Am j Physiol Endocrinol Metab 284:E583–E588

Grace C, Beales P, Summerbell C, Jebb SA, Wright A, Parker D, Kopelman P (2003) Energy metabolism in Bardet-Biedl syndrome. Int J Obes Relat Metab Disord 27:1319–1324

Hamon M, Doucet E, Lefevre K, Miquel MC, Lanfumey L, Insausti R, Frechilla D, Del Rio J, Verge D (1999) Antibodies and antisense oligonucleotide for probing the distribution and putative functions of central 5-HT6 receptors. Neuropsychopharmacology 21:68S–76S

Han YG, Spassky N, Romaguera-Ros M, Garcia-Verdugo JM, Aguilar A, Schneider-Maunoury S, Alvarez-Buylla A (2008) Hedgehog signaling and primary cilia are required for the formation of adult neural stem cells. Nat Neurosci 11:277–284

Han YG, Kim HJ, Dlugosz AA, Ellison DW, Gilbertson RJ, Alvarez-Buylla A (2009) Dual and opposing roles of primary cilia in medulloblastoma development. Nat Med 15:1062–1065

Handel M, Schulz S, Stanarius A, Schreff M, Erdtmann-Vourliotis M, Schmidt H, Wolf G, Hollt V (1999) Selective targeting of somatostatin receptor 3 to neuronal cilia. Neuroscience 89:909–926

Haycraft CJ, Zhang Q, Song B, Jackson WS, Detloff PJ, Serra R, Yoder BK (2007) Intraflagellar transport is essential for endochondral bone formation. Development 134:307–316

Hearn T, Renforth GL, Spalluto C, Hanley NA, Piper K, Brickwood S, White C, Connolly V, Taylor JF, Russell-Eggitt I, Bonneau D, Walker M, Wilson DI (2002) Mutation of ALMS1, a large gene with a tandem repeat encoding 47 amino acids, causes Alstrom syndrome. Nat Genet 31:79–83

Hearn T, Spalluto C, Phillips VJ, Renforth GL, Copin N, Hanley NA, Wilson DI (2005) Subcellular localization of ALMS1 supports involvement of centrosome and basal body dysfunction in the pathogenesis of obesity, insulin resistance, and type 2 diabetes. Diabetes 54:1581–1587

Hebebrand J, Volckmar AL, Knoll N, Hinney A (2010) Chipping away the 'missing heritability': GIANT steps forward in the molecular elucidation of obesity – but still lots to go. Obes Facts 3:294–303

Jin H, White SR, Shida T, Schulz S, Aguiar M, Gygi SP, Bazan JF, Nachury MV (2010) The conserved Bardet-Biedl syndrome proteins assemble a coat that traffics membrane proteins to cilia. Cell 141:1208–1219

Joy T, Cao H, Black G, Malik R, Charlton-Menys V, Hegele RA, Durrington PN (2007) Alstrom syndrome (OMIM 203800): a case report and literature review. Orphanet J Rare Dis 2:49

Kaplan OI, Doroquez DB, Cevik S, Bowie RV, Clarke L, Sanders AA, Kida K, Rappoport JZ, Sengupta P, Blacque OE (2012) Endocytosis genes facilitate protein and membrane transport in C. elegans sensory cilia. Curr Biol 22:451–460

Katsanis N, Ansley SJ, Badano JL, Eichers ER, Lewis RA, Hoskins BE, Scambler PJ, Davidson WS, Beales PL, Lupski JR (2001) Triallelic inheritance in Bardet-Biedl syndrome, a Mendelian recessive disorder. Science 293:2256–2259

Kim SK, Shindo A, Park TJ, Oh EC, Ghosh S, Gray RS, Lewis RA, Johnson CA, Attie-Bittach T, Katsanis N, Wallingford JB (2010) Planar cell polarity acts through septins to control collective cell movement and ciliogenesis. Science 329:1337–1340

Klein D, Ammann F (1969) The syndrome of Laurence-Moon-Bardet-Biedl and allied diseases in Switzerland. Clinical, genetic and epidemiological studies. J Neurol Sci 9:479–513

Kleyn PW, Fan W, Kovats SG, Lee JJ, Pulido JC, Wu Y, Berkemeier LR, Misumi DJ, Holmgren L, Charlat O, Woolf EA, Tayber O, Brody T, Shu P, Hawkins F, Kennedy B, Baldini L, Ebeling C, Alperin GD, Deeds J, Lakey ND, Culpepper J, Chen H, Glucksmann-Kuis MA, Carlson GA, Duyk GM, Moore KJ (1996) Identification and characterization of the mouse obesity gene tubby: a member of a novel gene family. Cell 85:281–290

Knorz VJ, Spalluto C, Lessard M, Purvis TL, Adigun FF, Collin GB, Hanley NA, Wilson DI, Hearn T (2010) Centriolar association of ALMS1 and likely centrosomal functions of the ALMS motif-containing proteins C10orf90 and KIAA1731. Mol Biol Cell 21:3617–3629

Kokoeva MV, Yin H, Flier JS (2005) Neurogenesis in the hypothalamus of adult mice: potential role in energy balance. Science 310:679–683

Kudryashova E, Wu J, Havton LA, Spencer MJ (2009) Deficiency of the E3 ubiquitin ligase TRIM32 in mice leads to a myopathy with a neurogenic component. Hum Mol Genet 18:1353–1367

Kudryashova E, Struyk A, Mokhonova E, Cannon SC, Spencer MJ (2011) The common missense mutation D489N in TRIM32 causing limb girdle muscular dystrophy 2H leads to loss of the mutated protein in knock-in mice resulting in a Trim32-null phenotype. Hum Mol Genet 20:3925–3932

Kulaga HM, Leitch CC, Eichers ER, Badano JL, Lesemann A, Hoskins BE, Lupski JR, Beales PL, Reed RR, Katsanis N (2004) Loss of BBS proteins causes anosmia in humans and defects in olfactory cilia structure and function in the mouse. Nat Genet 36:994–998

Lancaster MA, Gopal DJ, Kim J, Saleem SN, Silhavy JL, Louie CM, Thacker BE, Williams Y, Zaki MS, Gleeson JG (2011) Defective Wnt-dependent cerebellar midline fusion in a mouse model of Joubert syndrome. Nat Med 17:726–731

Laurier V, Stoetzel C, Muller J, Thibault C, Corbani S, Jalkh N, Salem N, Chouery E, Poch O, Licaire S, Danse JM, Amati-Bonneau P, Bonneau D, Megarbane A, Mandel JL, Dollfus H (2006) Pitfalls of homozygosity mapping: an extended consanguineous Bardet-Biedl syndrome family with two mutant genes (BBS2, BBS10), three mutations, but no triallelism. Eur J Hum Genet 14:1195–1203

Lechtreck KF, Johnson EC, Sakai T, Cochran D, Ballif BA, Rush J, Pazour GJ, Ikebe M, Witman GB (2009) The Chlamydomonas reinhardtii BBSome is an IFT cargo required for export of specific signaling proteins from flagella. J Cell Biol 187:1117–1132

Lee BH, Liu J, Wong D, Srinivasan S, Ashrafi K (2011) Hyperactive neuroendocrine secretion causes size, feeding, and metabolic defects of C. elegans Bardet-Biedl syndrome mutants. PLoS Biol 9:e1001219

Lee DA, Bedont JL, Pak T, Wang H, Song J, Miranda-Angulo A, Takiar V, Charubhumi V, Balordi F, Takebayashi H, Aja S, Ford E, Fishell G, Blackshaw S (2012) Tanycytes of the hypothalamic median eminence form a diet-responsive neurogenic niche. Nat Neurosci 15(5):700–702

Lehman JM, Michaud EJ, Schoeb TR, Aydin-Son Y, Miller M, Yoder BK (2008) The Oak Ridge Polycystic Kidney mouse: modeling ciliopathies of mice and men. Dev Dyn 237:1960–1971

Leitch CC, Zaghloul NA, Davis EE, Stoetzel C, Diaz-Font A, Rix S, Alfadhel M, Lewis RA, Eyaid W, Banin E, Dollfus H, Beales PL, Badano JL, Katsanis N (2008) Hypomorphic mutations in syndromic encephalocele genes are associated with Bardet-Biedl syndrome. Nat Genet 40:443–448

Lorda-Sanchez I, Ayuso C, Ibanez A (2000) Situs inversus and hirschsprung disease: two uncommon manifestations in Bardet-Biedl syndrome. Am J Med Genet 90:80–81

Ludwig DS, Tritos NA, Mastaitis JW, Kulkarni R, Kokkotou E, Elmquist J, Lowell B, Flier JS, Maratos-Flier E (2001) Melanin-concentrating hormone overexpression in transgenic mice leads to obesity and insulin resistance. J Clin Invest 107:379–386

Maffei M, Halaas J, Ravussin E, Pratley RE, Lee GH, Zhang Y, Fei H, Kim S, Lallone R, Ranganathan S et al (1995) Leptin levels in human and rodent: measurement of plasma leptin and ob RNA in obese and weight-reduced subjects. Nat Med 1:1155–1161

Maffei P, Boschetti M, Marshall JD, Paisey RB, Beck S, Resmini E, Collin GB, Naggert JK, Milan G, Vettor R, Minuto F, Sicolo N, Barreca A (2007) Characterization of the IGF system in 15 patients with Alstrom syndrome. Clin Endocrinol (Oxf) 66:269–275

Marley A, von Zastrow M (2010) DISC1 regulates primary cilia that display specific dopamine receptors. PLoS One 5:e10902

Marshall JD, Hinman EG, Collin GB, Beck S, Cerqueira R, Maffei P, Milan G, Zhang W, Wilson DI, Hearn T, Tavares P, Vettor R, Veronese C, Martin M, So WV, Nishina PM, Naggert JK (2007) Spectrum of ALMS1 variants and evaluation of genotype-phenotype correlations in Alstrom syndrome. Hum Mutat 28:1114–1123

Marszalek JR, Ruiz-Lozano P, Roberts E, Chien KR, Goldstein LS (1999) Situs inversus and embryonic ciliary morphogenesis defects in mouse mutants lacking the KIF3A subunit of kinesin-II. Proc Natl Acad Sci USA 96:5043–5048

Mistlberger RE (2009) Food-anticipatory circadian rhythms: concepts and methods. Eur J Neurosci 30:1718–1729

Mok CA, Healey MP, Shekhar T, Leroux MR, Heon E, Zhen M (2011) Mutations in a guanylate cyclase GCY-35/GCY-36 modify Bardet-Biedl syndrome-associated phenotypes in Caenorhabditis elegans. PLoS Genet 7:e1002335

Mountjoy KG (2010) Functions for pro-opiomelanocortin-derived peptides in obesity and diabetes. Biochem J 428:305–324

Moyer JH, Lee-Tischler MJ, Kwon HY, Schrick JJ, Avner ED, Sweeney WE, Godfrey VL, Cacheiro NL, Wilkinson JE, Woychik RP (1994) Candidate gene associated with a mutation causing recessive polycystic kidney disease in mice. Science 264:1329–1333

Mukhopadhyay A, Deplancke B, Walhout AJ, Tissenbaum HA (2005) C. elegans tubby regulates life span and fat storage by two independent mechanisms. Cell Metab 2:35–42

Mukhopadhyay S, Lu Y, Shaham S, Sengupta P (2008) Sensory signaling-dependent remodeling of olfactory cilia architecture in C. elegans. Dev Cell 14:762–774

Mukhopadhyay S, Wen X, Chih B, Nelson CD, Lane WS, Scales SJ, Jackson PK (2010) TULP3 bridges the IFT-A complex and membrane phosphoinositides to promote trafficking of G protein-coupled receptors into primary cilia. Genes Dev 24:2180–2193

Murcia NS, Richards WG, Yoder BK, Mucenski ML, Dunlap JR, Woychik RP (2000) The Oak Ridge Polycystic Kidney (orpk) disease gene is required for left-right axis determination. Development 127:2347–2355

Mykytyn K, Braun T, Carmi R, Haider NB, Searby CC, Shastri M, Beck G, Wright AF, Iannaccone A, Elbedour K, Riise R, Baldi A, Raas-Rothschild A, Gorman SW, Duhl DM, Jacobson SG, Casavant T, Stone EM, Sheffield VC (2001) Identification of the gene that, when mutated, causes the human obesity syndrome BBS4. Nat Genet 28:188–191

Mykytyn K, Nishimura DY, Searby CC, Shastri M, Yen HJ, Beck JS, Braun T, Streb LM, Cornier AS, Cox GF, Fulton AB, Carmi R, Luleci G, Chandrasekharappa SC, Collins FS, Jacobson SG, Heckenlively JR, Weleber RG, Stone EM, Sheffield VC (2002) Identification of the gene (BBS1) most commonly involved in Bardet-Biedl syndrome, a complex human obesity syndrome. Nat Genet 31:435–438

Mykytyn K, Nishimura DY, Searby CC, Beck G, Bugge K, Haines HL, Cornier AS, Cox GF, Fulton AB, Carmi R, Iannaccone A, Jacobson SG, Weleber RG, Wright AF, Riise R, Hennekam RC, Luleci G, Berker-Karauzum S, Biesecker LG, Stone EM, Sheffield VC (2003) Evaluation of complex inheritance involving the most common Bardet-Biedl syndrome locus (BBS1). Am J Hum Genet 72:429–437

Mykytyn K, Mullins RF, Andrews M, Chiang AP, Swiderski RE, Yang B, Braun T, Casavant T, Stone EM, Sheffield VC (2004) Bardet-Biedl syndrome type 4 (BBS4)-null mice implicate

Bbs4 in flagella formation but not global cilia assembly. Proc Natl Acad Sci USA 101:8664–8669

Nachury MV, Loktev AV, Zhang Q, Westlake CJ, Peranen J, Merdes A, Slusarski DC, Scheller RH, Bazan JF, Sheffield VC, Jackson PK (2007) A core complex of BBS proteins cooperates with the GTPase Rab8 to promote ciliary membrane biogenesis. Cell 129:1201–1213

Nishimura DY, Searby CC, Carmi R, Elbedour K, Van Maldergem L, Fulton AB, Lam BL, Powell BR, Swiderski RE, Bugge KE, Haider NB, Kwitek-Black AE, Ying L, Duhl DM, Gorman SW, Heon E, Iannaccone A, Bonneau D, Biesecker LG, Jacobson SG, Stone EM, Sheffield VC (2001) Positional cloning of a novel gene on chromosome 16q causing Bardet-Biedl syndrome (BBS2). Hum Mol Genet 10:865–874

Nishimura DY, Fath M, Mullins RF, Searby C, Andrews M, Davis R, Andorf JL, Mykytyn K, Swiderski RE, Yang B, Carmi R, Stone EM, Sheffield VC (2004) Bbs2-null mice have neurosensory deficits, a defect in social dominance, and retinopathy associated with mislocalization of rhodopsin. Proc Natl Acad Sci USA 101:16588–16593

Nishimura DY, Swiderski RE, Searby CC, Berg EM, Ferguson AL, Hennekam R, Merin S, Weleber RG, Biesecker LG, Stone EM, Sheffield VC (2005) Comparative genomics and gene expression analysis identifies BBS9, a new Bardet-Biedl syndrome gene. Am J Hum Genet 77:1021–1033

Nordman S, Abulaiti A, Hilding A, Langberg EC, Humphreys K, Ostenson CG, Efendic S, Gu HF (2008) Genetic variation of the adenylyl cyclase 3 (AC3) locus and its influence on type 2 diabetes and obesity susceptibility in Swedish men. Int J Obes (Lond) 32:407–412

Norman RX, Ko HW, Huang V, Eun CM, Abler LL, Zhang Z, Sun X, Eggenschwiler JT (2009) Tubby-like protein 3 (TULP3) regulates patterning in the mouse embryo through inhibition of Hedgehog signaling. Hum Mol Genet 18:1740–1754

Ohlemiller KK, Hughes RM, Mosinger-Ogilvie J, Speck JD, Grosof DH, Silverman MS (1995) Cochlear and retinal degeneration in the tubby mouse. Neuroreport 6:845–849

Pascual O, Traiffort E, Baker DP, Galdes A, Ruat M, Champagnat J (2005) Sonic hedgehog signalling in neurons of adult ventrolateral nucleus tractus solitarius. Eur J Neurosci 22:389–396

Pedersen LB, Rosenbaum JL (2008) Chapter two intraflagellar transport (IFT) role in ciliary assembly, resorption and signalling. Curr Top Dev Biol 85:23–61

Pierce AA, Xu AW (2010) De novo neurogenesis in adult hypothalamus as a compensatory mechanism to regulate energy balance. J Neurosci 30:723–730

Plum L, Lin HV, Aizawa KS, Liu Y, Accili D (2012) InsR/FoxO1 signaling curtails hypothalamic POMC neuron number. PLoS One 7:e31487

Pollock JD, Rowland N (1981) Peripherally administered serotonin decreases food intake in rats. Pharmacol Biochem Behav 15:179–183

Qu D, Ludwig DS, Gammeltoft S, Piper M, Pelleymounter MA, Cullen MJ, Mathes WF, Przypek R, Kanarek R, Maratos-Flier E (1996) A role for melanin-concentrating hormone in the central regulation of feeding behaviour. Nature 380:243–247

Rahmouni K, Fath MA, Seo S, Thedens DR, Berry CJ, Weiss R, Nishimura DY, Sheffield VC (2008) Leptin resistance contributes to obesity and hypertension in mouse models of Bardet-Biedl syndrome. J Clin Invest 118:1458–1467

Riise R, Andreasson S, Borgastrom MK, Wright AF, Tommerup N, Rosenberg T, Tornqvist K (1997) Intrafamilial variation of the phenotype in Bardet-Biedl syndrome. Br J Ophthalmol 81:378–385

Salamone JD, Zigmond MJ, Stricker EM (1990) Characterization of the impaired feeding behavior in rats given haloperidol or dopamine-depleting brain lesions. Neuroscience 39:17–24

Schaefer E, Zaloszyc A, Lauer J, Durand M, Stutzmann F, Perdomo-Trujillo Y, Redin C, Bennouna Greene V, Toutain A, Perrin L, Gerard M, Caillard S, Bei X, Lewis RA, Christmann D, Letsch J, Kribs M, Mutter C, Muller J, Stoetzel C, Fischbach M, Marion V, Katsanis N, Dollfus H (2011) Mutations in SDCCAG8/NPHP10 cause Bardet-Biedl syndrome and are associated with penetrant renal disease and absent polydactyly. Mol Syndromol 1:273–281

Schulz S, Handel M, Schreff M, Schmidt H, Hollt V (2000) Localization of five somatostatin receptors in the rat central nervous system using subtype-specific antibodies. J Physiol Paris 94:259–264

Seo S, Guo DF, Bugge K, Morgan DA, Rahmouni K, Sheffield VC (2009) Requirement of Bardet-Biedl syndrome proteins for leptin receptor signaling. Hum Mol Genet 18(7): 1323–1331

Seo S, Baye LM, Schulz NP, Beck JS, Zhang Q, Slusarski DC, Sheffield VC (2010) BBS6, BBS10, and BBS12 form a complex with CCT/TRiC family chaperonins and mediate BBSome assembly. Proc Natl Acad Sci USA 107:1488–1493

Sharma N, Berbari NF, Yoder BK (2008) Ciliary dysfunction in developmental abnormalities and diseases. Curr Top Dev Biol 85:371–427

Sharma N, Kosan ZA, Stallworth JE, Berbari NF, Yoder BK (2011) Soluble levels of cytosolic tubulin regulate ciliary length control. Mol Biol Cell 22:806–816

Shillingford JM, Murcia NS, Larson CH, Low SH, Hedgepeth R, Brown N, Flask CA, Novick AC, Goldfarb DA, Kramer-Zucker A, Walz G, Piontek KB, Germino GG, Weimbs T (2006) The mTOR pathway is regulated by polycystin-1, and its inhibition reverses renal cystogenesis in polycystic kidney disease. Proc Natl Acad Sci USA 103:5466–5471

Shillingford JM, Piontek KB, Germino GG, Weimbs T (2010) Rapamycin ameliorates PKD resulting from conditional inactivation of Pkd1. J Am Soc Nephrol 21:489–497

Shimada M, Tritos NA, Lowell BB, Flier JS, Maratos-Flier E (1998) Mice lacking melanin-concentrating hormone are hypophagic and lean. Nature 396:670–674

Slavotinek AM, Stone EM, Mykytyn K, Heckenlively JR, Green JS, Heon E, Musarella MA, Parfrey PS, Sheffield VC, Biesecker LG (2000) Mutations in MKKS cause Bardet-Biedl syndrome. Nat Genet 26:15–16

Smaoui N, Chaabouni M, Sergeev YV, Kallel H, Li S, Mahfoudh N, Maazoul F, Kammoun H, Gandoura N, Bouaziz A, Nouiri E, M'Rad R, Chaabouni H, Hejtmancik JF (2006) Screening of the eight BBS genes in Tunisian families: no evidence of triallelism. Invest Ophthalmol Vis Sci 47:3487–3495

Stoetzel C, Laurier V, Davis EE, Muller J, Rix S, Badano JL, Leitch CC, Salem N, Chouery E, Corbani S, Jalk N, Vicaire S, Sarda P, Hamel C, Lacombe D, Holder M, Odent S, Holder S, Brooks AS, Elcioglu NH, Da Silva E, Rossillion B, Sigaudy S, de Ravel TJ, Alan Lewis R, Leheup B, Verloes A, Amati-Bonneau P, Megarbane A, Poch O, Bonneau D, Beales PL, Mandel JL, Katsanis N, Dollfus H (2006) BBS10 encodes a vertebrate-specific chaperonin-like protein and is a major BBS locus. Nat Genet 38(5):521–524

Stoetzel C, Muller J, Laurier V, Davis EE, Zaghloul NA, Vicaire S, Jacquelin C, Plewniak F, Leitch CC, Sarda P, Hamel C, de Ravel TJ, Lewis RA, Friederich E, Thibault C, Danse JM, Verloes A, Bonneau D, Katsanis N, Poch O, Mandel JL, Dollfus H (2007) Identification of a novel BBS gene (BBS12) highlights the major role of a vertebrate-specific branch of chaperonin-related proteins in Bardet-Biedl syndrome. Am J Hum Genet 80:1–11

Stubdal H, Lynch CA, Moriarty A, Fang Q, Chickering T, Deeds JD, Fairchild-Huntress V, Charlat O, Dunmore JH, Kleyn P, Huszar D, Kapeller R (2000) Targeted deletion of the tub mouse obesity gene reveals that tubby is a loss-of-function mutation. Mol Cell Biol 20:878–882

Szabo NE, Zhao T, Cankaya M, Theil T, Zhou X, Alvarez-Bolado G (2009) Role of neuroepithelial Sonic hedgehog in hypothalamic patterning. J Neurosci 29:6989–7002

Tadenev AL, Kulaga HM, May-Simera HL, Kelley MW, Katsanis N, Reed RR (2011) Loss of Bardet-Biedl syndrome protein-8 (BBS8) perturbs olfactory function, protein localization, and axon targeting. Proc Natl Acad Sci USA 108:10320–10325

Tan PL, Barr T, Inglis PN, Mitsuma N, Huang SM, Garcia-Gonzalez MA, Bradley BA, Coforio S, Albrecht PJ, Watnick T, Germino GG, Beales PL, Caterina MJ, Leroux MR, Rice FL, Katsanis N (2007) From the cover: loss of Bardet-Biedl syndrome proteins causes defects in peripheral sensory innervation and function. Proc Natl Acad Sci USA 104:17524–17529

Traiffort E, Charytoniuk DA, Faure H, Ruat M (1998) Regional distribution of Sonic Hedgehog, patched, and smoothened mRNA in the adult rat brain. J Neurochem 70:1327–1330

Traiffort E, Charytoniuk D, Watroba L, Faure H, Sales N, Ruat M (1999) Discrete localizations of hedgehog signalling components in the developing and adult rat nervous system. Eur J Neurosci 11:3199–3214

Traiffort E, Angot E, Ruat M (2010) Sonic Hedgehog signaling in the mammalian brain. J Neurochem 113:576–590

Vijayan E, McCann SM (1977) Suppression of feeding and drinking activity in rats following intraventricular injection of thyrotropin releasing hormone (TRH). Endocrinology 100: 1727–1730

Wang Z, Li V, Chan GC, Phan T, Nudelman AS, Xia Z, Storm DR (2009) Adult type 3 adenylyl cyclase-deficient mice are obese. PLoS One 4:e6979

Weatherbee SD, Niswander LA, Anderson KV (2009) A mouse model for Meckel syndrome reveals Mks1 is required for ciliogenesis and Hedgehog signaling. Hum Mol Genet 18:4565–4575

Wong ST, Trinh K, Hacker B, Chan GC, Lowe G, Gaggar A, Xia Z, Gold GH, Storm DR (2000) Disruption of the type III adenylyl cyclase gene leads to peripheral and behavioral anosmia in transgenic mice. Neuron 27:487–497

Xu AW, Kaelin CB, Morton GJ, Ogimoto K, Stanhope K, Graham J, Baskin DG, Havel P, Schwartz MW, Barsh GS (2005a) Effects of hypothalamic neurodegeneration on energy balance. PLoS Biol 3:e415

Xu AW, Kaelin CB, Takeda K, Akira S, Schwartz MW, Barsh GS (2005b) PI3K integrates the action of insulin and leptin on hypothalamic neurons. J Clin Invest 115:951–958

Xu Y, Tamamaki N, Noda T, Kimura K, Itokazu Y, Matsumoto N, Dezawa M, Ide C (2005c) Neurogenesis in the ependymal layer of the adult rat 3rd ventricle. Exp Neurol 192:251–264

Zhang Y, Proenca R, Maffei M, Barone M, Leopold L, Friedman JM (1994) Positional cloning of the mouse obese gene and its human homologue. Nature 372:425–432

Zhang Q, Nishimura D, Seo S, Vogel T, Morgan DA, Searby C, Bugge K, Stone EM, Rahmouni K, Sheffield VC (2011) Bardet-Biedl syndrome 3 (Bbs3) knockout mouse model reveals common BBS-associated phenotypes and Bbs3 unique phenotypes. Proc Natl Acad Sci USA 108:20678-20683

Zhu Y, Romero MI, Ghosh P, Ye Z, Charnay P, Rushing EJ, Marth JD, Parada LF (2001) Ablation of NF1 function in neurons induces abnormal development of cerebral cortex and reactive gliosis in the brain. Genes Dev 15:859–876

Chapter 7
Motile Cilia and Brain Function: Ependymal Motile Cilia Development, Organization, Function and Their Associated Pathologies

Nathalie Spassky

Contents

Abstract Unlike immotile cilia, which protrude from most cells in our body, motile cilia are restricted to sperm cells and epithelial cells lining the airways, the oviduct, the paranasal sinuses, and the brain ventricles. The best-known function of these cilia is that their coordinated beating generates extracellular flow that clears mucus from the airways, moves ova from the oviducts toward the uterus, and propels cerebrospinal fluid (CSF) through the cerebral ventricles. The vertebrate brain forms around a ventricular cavity in which the CSF, secreted by the choroid plexus in each ventricle, flows continuously. The CSF, which contains many growth factors and morphogens, is present from the first stages of brain development and plays crucial roles throughout life. Ependymal

N. Spassky (✉)
INSERM U1024, CNRS UMR 8197, Institut de Biologie de l'Ecole
Normale Supérieure, 46 rue d'Ulm, 75005 Paris, France
e-mail: nathalie.spassky@ens.fr

K.L. Tucker and T. Caspary (eds.), *Cilia and Nervous System Development and Function*,
DOI 10.1007/978-94-007-5808-7_7, © Springer Science+Business Media Dordrecht 2013

cells are specialized glial cells that extend multiple motile cilia into the cerebral ventricles. These cells appear at early postnatal stages and line all cerebral ventricles in adult mammals. In this chapter, we will review current knowledge on ependymal motile cilia development, organization, functions, and their associated pathologies.

Keywords Ependyma • Motile cilia • Cerebral ventricles • 9+2 • Cerebrospinal fluid • Multiciliated cells

7.1 Introduction

Motile cilia are microtubule-based structures protruding from the apical surface of epithelial cells in different organs such as the airways, the female reproductive tract and the brain. Each cilium is arranged in the classic 9+2 configuration (9 pairs of microtubules in the periphery and 2 pairs in the center) and coordinately beat to produce large-scale fluid flows crucial for the development and physiology of these organs. Indeed, genetic diseases in which ciliary beating is affected often leads to loss of ciliary flow linked to bronchiestasis and chronic sinusitis, ectopic pregnancies, hydrocephalus and defects in neurogenesis. In the brain, multiciliated ependymal cells line all ventricular walls in which CSF flows continuously thanks to the coordinated beating of their motile cilia. We discuss the main current knowledge on development and functions of brain motile cilia and ependyma cells.

7.2 The Brain Ventricular System

7.2.1 The Brain Ventricles

The vertebrate brain forms around a ventricular system composed of four interconnected cavities filled with cerebrospinal fluid (CSF): The two lateral ventricles located in the cerebrum are linked to the third ventricle located in the diencephalon via the interventricular foramen of Monro. The cerebral aqueduct links the third and fourth ventricle, which lies between the cerebellum and pons. The fourth ventricle joins the spinal cord canal and the subarachnoid space that envelops the brain via the left lateral aperture (foramen of Lushka) and the median aperture (foramen of Magendie and the spinal cord canal) (Fig. 7.1)

7.2.2 The Cerebrospinal Fluid (CSF)

The CSF is constantly produced by the choroid plexus in each ventricle, by filtering blood through choroidal capillaries and epithelial cells. The bulk CSF flows unidirectionally, from the lateral to the third and fourth ventricles and then out into the

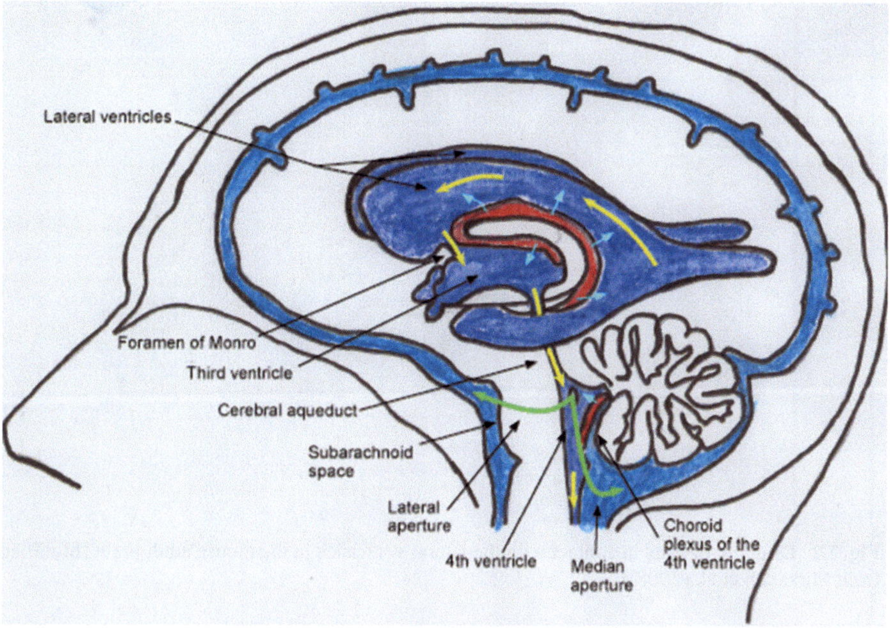

Fig. 7.1 Drawing of a sagittal view of mammalian cerebral ventricular system. CSF is secreted (*blue arrows*) by the choroid plexuses (*red*) in each ventricle. The bulk CSF flows unidirectionally (*yellow arrows*) and into the suarachnoid space via the apertures (*green arrows*)

subarachnoid space surrounding the brain where it is absorbed into the hematopoietic circulatory and lymphatic systems. The driving force of the CSF bulk flow is actively supported by arterial pulsations and pressure gradients produced by secretion and absorption (Bering 1955; Wagshul et al. 2006). Inside each ventricle and depending on their size and shape (i.e. in large mammals like humans or sheep compared to rodents), the CSF may flow locally against the bulk flow, creating local turbulence. Along the ventricular walls, the laminar flow of CSF is mainly directed by the coordinated beating of ependymal cilia in the direction of the bulk flow (Nguyen et al. 2001; Fig. 7.1). The total volume of the CSF in humans is around 140 ml and constitutes 18% of the volume of an adult human brain: about 30 ml is inside the ventricles, the remainder surrounds the brain (Segal 2000). The CP produces around 500 ml of CSF per day in a young adult, suggesting that human CSF is completely replaced about three times every 24 h. In the young adult rat, which has about 200 μl CSF, it takes about 2 h to replace the CSF (Veening and Barendregt 2010). In a healthy normal individual, the CSF contains more than 2,600 different peptides and proteins (Schutzer et al. 2010). The CSF is vital to the health of the brain; it protects it from physical damage, facilitates the removal of metabolites and provides nutrients (Segal 2001). It also contributes to brain development, homeostasis, adult neurogenesis and repair by providing physical and chemical cues (Miyan et al. 2003).

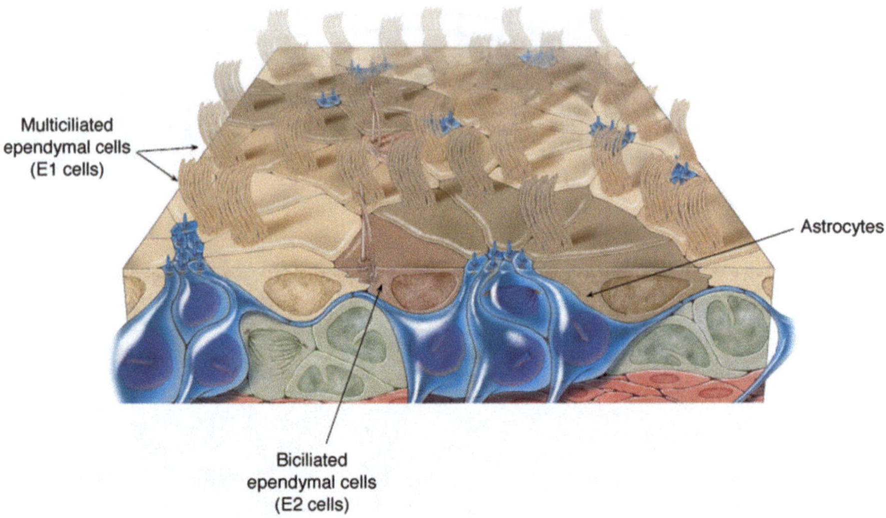

Fig. 7.2 Drawing of cells in contact with the lateral ventricles in the mouse adult brain (Modified from Mirzadeh et al. (2008))

7.2.3 Cells in Contact with the Cerebral Ventricles

7.2.3.1 Cells Extending Motile Cilia

- Multiciliated ependymal cells (E1 cells) are the most abundant epithelial cells lining the lateral, third and fourth ventricles, and are $CD24^+$, $S100\beta^+$, $FoxJ1^+$, $Sox2^+$ and $CD133^+$ but $nestin^-$ and $GFAP^-$. These cells are cuboidal with a light cytoplasm and spherical uninvaginated nuclei containing dispersed chromatin. Abundant mitochondria are localized near the basal bodies (Doetsch et al. 1997; Mirzadeh et al. 2008). The lateral processes of adjacent ependymal cells are heavily interdigitated and contain apical adherens and tight junctional complexes and gap junctions. Ependymal cilia are motile and measure 8–15 μm in length (average size: 11.5 μm), protruding in the cerebral ventricles (Mirzadeh et al. 2008, 2010; Hirota et al. 2010; Fig. 7.2). The surface exposed to the ventricular cavity is variable in size (from 100 to 600 μm^2, average size: 265 μm^2) and contains a patch of 32–73 (mean: 49) basal bodies separated by 250 nm. In contrast to other epithelia containing motile cilia, in which basal bodies cover the entire apical surface, the basal body patch occupies only 4–35% of the total apical surface of the ependymal cells and is positioned downstream with respect to the direction of ciliary beating. In the adult, multiciliated ependymal cells do not divide (Capela and Temple 2002; Spassky et al. 2005) and the best known function of ependymal cilia is to create a constant laminar flow of CSF through the cerebral ventricles. For efficient fluid flow, the ependymal cilia must beat in the same direction and in

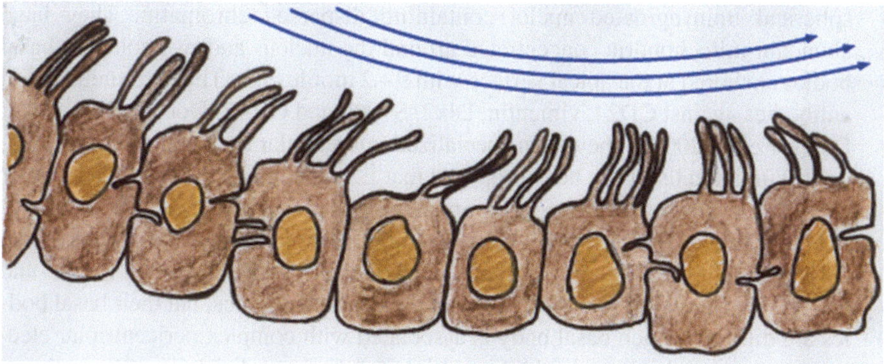

Fig. 7.3 Drawing of ependymal ciliary beating coordination leading to metachronal waves propagating over the surface of the cells and efficient fluid flow in the cerebral ventricles

a coordinated fashion. This coordinated or "metachronic" beating of adjacent cilia, seen as waves propagating over the surface of ciliated cells, is possibly mediated by tight coupling of the cells through gap junctions (Scott et al. 1974; Fig. 7.3). The close spatial relationships among cilia and the hydrodynamic interactions generated by their beating are believed, however, to be the most important factors for ciliary coordination on ependymal surfaces (Sanderson and Sleigh 1981; Salathe 2007; Guirao and Joanny 2007; Guirao et al. 2010; Fig. 7.3). The beating of ependymal cilia is highly asymmetrical, similar to that of motile cilia in other organs, such as airway epithelia and the oviduct. The beat begins with the formation of a large curve at the base of the cilium that sweeps the cilium forward in an effective stroke. The cilium then returns to its initial position through the propagation of the curve during the recovery stroke (Lechtreck et al. 2008). The waveform of the ependymal ciliary beat is nearly planar, creating a constant laminar flow of fluid: the effective stroke significantly displaces the fluid in the direction of the beat, whereas the recovery stroke has a minimal effect on fluid displacement. The frequency of ciliary beats is around 30 Hz under basal conditions, but may be modulated by external signals; for example, it may increase to 50 Hz after application of serotonin (5-HT, which causes a prolonged increase of intracellular Ca^{2+} due to the opening of Ca^{2+} release-activated Ca^{2+} channels on the plasma membrane; Nguyen et al. 2001) or adenosine receptor agonists (due to A_{2B} receptor activation, Genzen et al. 2009). It can be decreased by application of ATP through intracellular Ca^{2+}-independent, purinergic receptors and an intracellular cAMP-mediated pathway (Nguyen et al. 2001).

– Bi-ciliated ependymal cells: are sparsely distributed throughout the lateral wall of the lateral ventricles (5% of cells contacting the lateral ventricle; called E2 cells; Fig. 7.2) and are the most common cells observed in the adult mouse spinal cord (Ecc cells). At the ultrastructural level, E2 and Ecc have deeply interdigitated cell membranes with long lateral extensions, light cytoplasm, small dictyosomes (abundant cisternae of rough endoplasmic reticulum and polyribosomes) and spherical uninvaginated nuclei containing dispersed chromatin. They have abundant mitochondria concentrated around the nucleus and two complex basal bodies anchored at the apical surface with 9+2 motile cilia. They are stained with antibodies against CD24, vimentin, Dlx2, S100β and GFAP (Doetsch et al. 1997; Danilov et al. 2009). They form specialized intercellular tight and adherens junctions with E1 cells. It has been proposed that E2 cells may serve as mechanical or chemical sensors of CSF flow or composition (Bruni 1998; Mirzadeh et al. 2008). It is not known whether these cells are mitotically active. In the mouse spinal cord, Ecc cells are $Vimentin^+$, $CD24^+$, $FoxJ1^+$, $Sox2^+$ and $CD133^+$ but $nestin^-$ and $GFAP^-$. These cells resemble E2 cells in the lateral ventricles, but their basal bodies are different: Each basal body is associated with complex pericentriolar electron-dense particles organized into multiple (up to six) radial spikes. Ecc cells are mitotically active, but the timing of their greatest proliferation together with the distribution of pairs of BrdU-labeled cells at different time points after injection suggest that proliferation of these cells contributes to the extension of the central canal during spinal cord growth (Alfaro-Cervello et al. 2012).

7.2.3.2 Other Cells

- Tanycytes are bipolar cells with microvilli and primary cilia (9+0 structure) that mostly contact the third ventricle, although a few have been observed in the lateral walls of the lateral ventricles (Doetsch et al. 1997) and in discrete locations around the central canal of the spinal cord. Tanycytes share some features with radial glia and astrocytes: they stain strongly with antibodies against GFAP, nestin, Glast, vimentin Sox2, S100β and the functional receptors for the neurotransmitters γ-aminobutyrate (GABA) and glutamate (Rodriguez et al. 2005; Hamilton et al. 2009). In the third ventricle, tanycytes connects the CSF to the hypothalamus portal system and several studies provide molecular features suggesting that they link the CSF to neuroendocrine events (Rodriguez et al. 2005).
- Astrocytes are multipolar cells localized in the subventricular zone of the lateral ventricles. These cells represent one-third of all lateral ventricle-contacting cells and were demonstrated to be the adult neural stem cells (Doetsch et al. 1999). These cells were also observed along the central canal of the adult mouse spinal cord (Martens et al. 2002), the most dorsal of which were identified as neural stem cells (Sabourin et al. 2009). In the lateral ventricle, their apical process contains a primary cilium and form rosette-like clusters surrounded by multiciliated ependymal cells, although the functional significance of this architecture is still unknown (Mirzadeh et al. 2008; Fig. 7.2).
- An extensive plexus of axons originating mainly from serotoninergic neurons in the dorsal raphe nucleus in the reticular formation have been found on the ventricular surface of multiciliated ependymal cells, in direct contact with the cerebral ventricles. These neurons have been shown to use serotonin and glutamate as co-transmitters. Interestingly, their axons often terminate in expansions

near the base of ependymal cilia and tufts of cilia are often encircled by these axons. It has been proposed that glutamate provided by supra-ependymal axons may supplement metabolic pathways in multiciliated ependymal cells to fuel the high energy demands of ciliary beating (Harandi et al. 1986; Robinson et al. 1996). However, the roles of these supra-ependymal axons and their possible interactions with ependymal cells and CSF still need to be understood.

7.3 Development of Multiciliated Ependymal Cells

7.3.1 Description of the Main Stages of Ependymal Cells Development

During embryonic development, bipolar radial glial cells express the radial glial markers RC2 and GLAST and extend both a pial and ventricular process with a 9 + 0 cilium that project into the ventricular lumen (Tramontin et al. 2003). A subpopulation of radial glial cells is specified for the ependymal lineage around E15 in mice (Rakic and Sidman 1968 ; Spassky et al. 2005). Around birth, the nuclei of ependymal progenitors invaginate, and electron-dense aggregates and multiple deuterosomes appear in the cytoplasm. Deuterosomes are spherical structures of unknown composition, which act as nucleation centers for ciliary basal bodies in multiciliated cells (Dirksen 1971). These cells express both the radial glial cell marker GLAST and the ependymal cell markers S100β and CD24. Future basal bodies then detach from deuterosomes and migrate toward the apical surface where they dock to the cell membrane and start to extend 9 + 2 cilia. Basal bodies are first broadly distributed in the apical membrane and gradually become asymmetrically localized within the apical membrane, accumulating at the anterior region of each ependymal cell (Hirota et al. 2010). This aspect of ependymal cells planar polarity is called "translational polarity" and is unique to multiciliated cells in the brain; cells with motile cilia in the trachea and oviduct contain basal bodies covering their entire apical surface. Concomitant to the establishment of translational polarity, cilia increase in length and start beating, producing fluid flow that orients the basal bodies in the same direction. This is referred to "rotational polarity". The orientation of each basal body is determined by the positioning of its accessory structure called the basal foot, which points in the direction of the effective stroke (Mitchell et al. 2007). The cilia first beat in independent directions and progressively orient in a common and fixed direction (Guirao et al. 2010). At this mature stage, the features characteristic of immature cells (deuterosomes and aggregates) have disappeared and the planar polarized beating of the cilia directs the flow of CSF through the cerebral ventricles, which is crucial for brain development and function (Ibanez-Tallon et al. 2004; Sawamoto et al. 2006).

7.3.2 *Transcription Factors Involved in Ependymal Cell Development*

Some transcription factors required for ependymal cells development (Vax1, Sox2, Six3) have been identified, although the molecular mechanisms by which radial glial cells become ependymal cells still remain to be elucidated (Soria et al. 2004; Ferri et al. 2004; Lavado and Oliver 2011), the transcription factors FoxJ1 (also known as hepatocyte nuclear factor-3 and forkhead homolog 4) and RFX (regulatory factor X homologs) appear to be implicated. In FoxJ1 mutant mice, axonemal dyneins and kinesins are downregulated, causing defective apical migration of basal bodies and defects in the genesis of motile cilia (Brody et al. 2000; Stubbs et al. 2008; Jacquet et al. 2009; Yu et al. 2008). It was recently proposed that FoxJ1 contributes to ependymal cell differentiation by controlling ankyrin G (Ank3) expression in ependymal cells. Ank3 is a large adaptor molecule that binds to E- and N-cadherin at cell membranes (Paez-Gonzalez et al. 2011). RFX3 belongs to the regulatory factor X (RFX) family of transcription factors, which are involved in the assembly and function of cilia in nematodes, drosophila and mice (Swoboda et al. 2000; Dubruille et al. 2002; Bonnafe et al. 2004). In mice, an RFX3 deficiency leads to hydrocephalus, partly due to defects in the differentiation of the subcommissural organ and choroid plexuses (Baas et al. 2006). Interestingly, RFX3 binds to FoxJ1 promoter and to the promoters of the genes encoding two axonemal dyneins involved in ciliary motility, suggesting that RFX3 regulates both ciliary assembly and motility (Zein El et al. 2009).

7.3.3 *Mechanisms Regulating Ependymal Cell Polarity*

- Translational polarity: Although the functional significance of the ependymal translational polarity in the adult is still unknown, the mechanisms involved in its regulation are starting to be deciphered. Specific ablation of radial glial cells primary cilia using mouse genetics leads to random distribution of the patch of basal bodies at the apical surface of ependymal cells, suggesting that the primary cilium provides important positional information for correct basal body accumulation in ependymal cells (Mirzadeh et al. 2010). It was also shown that during mouse brain development, non-muscle myosin II (NMII) is involved in the rostral migration of basal bodies, independently of the planar cell polarity pathway (Hirota et al. 2010; Fig. 7.4).
- Rotational polarity: The oriented and coordinated ependymal ciliary beating is crucial for CSF flow and adult neurogenesis, and depends on the rotational polarity of each basal body in the cell. During development, ependymal cilia grow and start beating in random orientations and then progressively align in each cell and along the tissue axis. Using mutant mice, siRNA, or dominant-negative constructs, it was recently shown that the core planar cell polarity genes Dishevelled2,

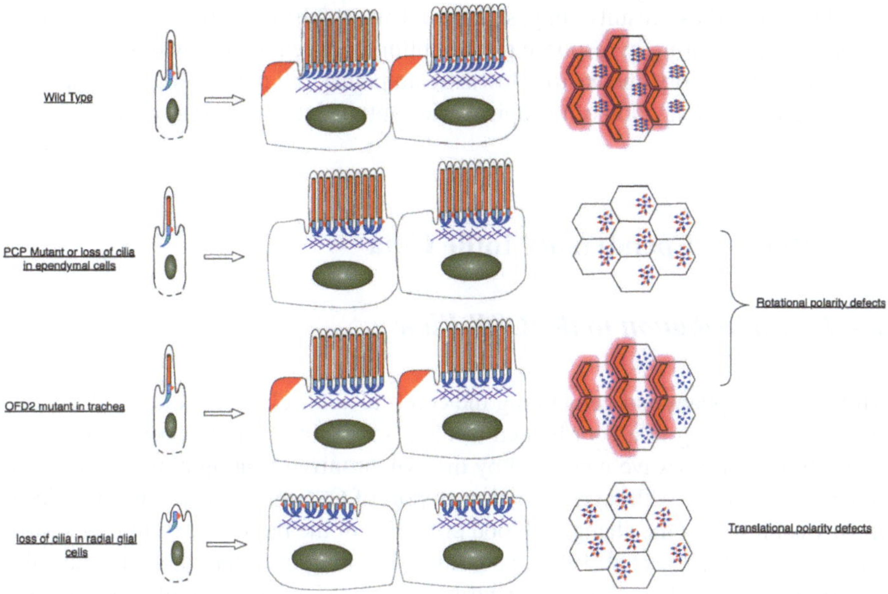

Fig. 7.4 Development of ependymal cells in wild-type and mutant animals. Radial glial cells extending a primary cilium transform into multiciliated ependymal cells. Motile cilia orient their beating in the direction of the flow: patch of basal bodies migrate to the anterior side of the cell and basal feet (*red*) point in the direction of the flow. Pink indicate the microtubule lattices and orange indicate the localisation of Vangl proteins at the apical and posterior side of the cells. In PCP mutants or in cilia defective ependymal cells or in OFD2 mutant trachea cells, motile cilia beat in random directions and leads to rotational polarity defects. In cilia defective radial glial cells, patch of basal bodies are randomly distributed along the apical surface of ependymal cells and leads to translational polarity defects

Van Gogh-like2 (Vangl2), cadherin, EGF, laminin, seven-pass receptors2 and 3 (Celsr2 and Celsr3) are required for alignment of ependymal basal feet during development (Guirao et al. 2010; Hirota et al. 2010; Tissir et al. 2010; Fig. 7.4). Most interestingly, it was shown that hydrodynamic forces among beating cilia help determine the direction of ependymal ciliary beating during maturation. Since Vangl2 is localized along cilia at these stages, it was hypothesized that Vangl2 may be part of a mechanosensor complex able to convert hydrodynamic forces into intracellular mechanisms for global basal foot alignment (Guirao et al. 2010). How PCP proteins interact is unknown, but perturbation of the cellular distribution of Vangl2 in cells in which Celsr2 and 3 are invalidated suggests that these genes may cooperate in the regulation of rotational polarity (Tissir et al. 2010). Most interestingly, it was recently shown that targeted elimination of exons 6 and 7 of Odf2, encoding a basal body and centrosome-associated protein Odf2/cenexin, leads to basal bodies lacking basal feet, disruption of the polarized organization of the apical microtubule lattice and uncoordinated beating of motile cilia in the trachea without affecting planar cell polarity.

Altogether, these results suggest a model in which maturing motile cilia may orient their beating in response to a coupling between hydrodynamic forces and planar cell polarity, thanks to a downstream planar cell polarity-based basal foot-dependent organization of the apical microtubule lattice (Kunimoto et al. 2012).

7.4 Roles of Ependymal Motile Cilia

7.4.1 Contribution to the CSF Flow

Multiciliated ependymal cells lining the cerebral ventricles extend around 50 long motile cilia beating in a coordinated fashion, indicating that cilia beat as part of a metachronal wave (wave generated by the synchronized beating along the longitudinal ciliary rows). Although the bulk transport of CSF along the brain ventricles is mostly achieved by the changing blood pressures of the brain vessels during systole and diastole (Bradley et al. 1986), coordinated beating of ependymal cilia may also contribute to CSF flow along the ventricular walls (Sawamoto et al. 2006; Guirao et al. 2010). Depending on the size and shape of each brain ventricle (especially in large mammals), and given that ependymal cilia are around 12 μm long, ependymal ciliary beating may contribute to CSF flow at least in the narrowest regions of the ventricular system, i.e. the cerebral aqueduct. As proof of principle, it is noteworthy that in mutant mouse models in which cilia formation or cilia motility is affected (e.g. Hydin, stumpy protein, $Tg737^{orpk}$, Mdnah5 mutant mice), mice develop hydrocephalus in postnatal stages due to aqueduct closure and massive dilatation of the third and lateral ventricles (Ibanez-Tallon et al. 2004; Banizs et al. 2005; Lechtreck et al. 2008; Town et al. 2008). It is thus clear that ependymal ciliary beating is crucial for CSF laminar flow along all brain ventricular walls, especially in the narrowest regions of the ventricular system (the cerebral aqueduct) where it prevents aqueduct stenosis and ventricle enlargement (Ibanez-Tallon et al. 2003, 2004).

7.4.2 Contribution to CSF-Brain Exchanges

In human airway epithelia, motile cilia beat to propel harmful inhaled material out of the lung. It was recently demonstrated that motile cilia in the airway are able to sense noxious substances entering the airways and rapidly increase ciliary beat frequency (CBF) as a defense mechanism (Shah et al. 2009). Similarly, an increase in ependymal ciliary beating in brain ventricles might minimize the layer of unstirred CSF covering the ependyma and optimize the dispersion or spreading of neural messengers in the CSF to other regions of the CNS or move debris in the direction of CSF bulk flow. Conversely, reducing CBF may allow ciliated ependymal cells to retain or localize certain messengers in the CSF, generating short-range effects on the exchange of neural messengers at the CSF-brain interface (Cathcart and Worthington 1964; Roth et al. 1985). These possibilities suggest that motile cilia in the brain, as in the airways, might have sensory roles leading to their active participation in brain-CSF exchanges. It was observed, for example, that ependymal cells express morphogens and growth factors (Noggin, Fibroblast Growth Factor 2, Vascular Endothelial Growth Factor, Hepatocyte Growth Factor, Insulin Growth Factor Binding Protein, etc....) (Lim et al. 2000; Hayamizu et al. 2001; Arai et al. 1998; Calvo et al. 2011), suggesting that ependymal cells may contribute to autocrine or paracrine trophic support, perhaps through exocytosis from motile cilia. Interestingly, PDGF-α and EGF receptors are localized along ependymal cilia although these cells are post-mitotic; their roles in ependymal cilia are still unknown (del Bigio 2010; Danilov et al. 2009). It was recently shown that ependymal cells along the central canal of the spinal cord extend two motile cilia (Alfaro-Cervello et al. 2012), although it is unknown whether ciliary beating in the spinal cord is coordinated and whether this might contribute to exchanges between CSF and neural cells.

7.4.3 Contribution to Neuronal Migration in the Adult

In the adult mammalian brain, new neurons are generated in the subventricular zone (SVZ) of the lateral ventricles (see Chap. 4 on adult neurogenesis). These progenitors migrate long distances toward the olfactory bulb where they differentiate into interneurons. The choroid plexus is a source of chemorepulsive factors, including members of the Slit family, which influence SVZ cell migration (Hu 1999; Nguyen-Ba-Charvet et al. 2004). Infusion of a recombinant Slit2-alkaline phosphatase fusion protein into the adult lateral ventricles revealed that the alkaline phosphatase signal is mainly found on the surface of the ependymal layer and in a gradient along the dorsal SVZ, where neuroblasts form longitudinal arrays of chains and migrate predominantly in the rostral direction. Most importantly, this gradient does not form when Slit2-AP is injected in $Tg737^{orpk}$ adult mutant mice, in which cilia and CSF are abnormal, suggesting that ependymal ciliary beating is crucial for CSF flow in adult mouse brain and for the formation of the Slit2 gradient in the SVZ in vivo (Sawamoto et al. 2006).

7.5 Associated Pathologies

In humans, the main pathology associated with defective ependymal cilia is hydrocephalus. Hydrocephalus is a progressive pathological condition characterized by the excessive accumulation of CSF, which can be caused by impaired CSF flow, excess CSF production or a lack of CSF reabsorption (Bruni et al. 1985). Current treatments involve surgical insertion of a ventricular shunt to facilitate drainage of excess CSF. Among ciliopathies, primary ciliary dyskinesia is the main pathology involving hydrocephalus. It arises from ultrastructural defects that perturb the motility of all motile cilia and cause a variety of symptoms that include fertility problems (ectopic pregnancy, sperm immotility) sinusitis, otitis, bronchiectasis, hydrocephalus and situs inversus (Zariwala et al. 2007).

In mice, all mutant in which cilia formation or motility is affected develop hydrocephalus (Ibanez-Tallon et al. 2004; Banizs et al. 2005; Lechtreck et al. 2008; Town et al. 2008). The lack of ependymal flow causes a secondary closure of the aqueduct and subsequent formation of triventricular hydrocephalus during early postnatal brain development. In humans, ependymal ciliary dysmotility is not sufficient to cause hydrocephalus but increases the risk of aqueduct closure; there is a 1:40 incidence of hydrocephalus caused by aqueduct stenosis in primary ciliary dyskinesia patients compared to 3:10,000 in the general population (Ibanez-Tallon et al. 2003). Defects in ependymal motile cilia are responsible for only some forms of hydrocephalus. For example, excess CSF production by the choroid plexus also leads to hydrocephalus but is primarily due to chemosensory defects of primary cilia in the choroid plexus cells (Banizs et al. 2005; Narita et al. 2010).

References

Alfaro-Cervello C, Soriano-Navarro M, Mirzadeh Z et al (2012) Biciliated ependymal cell proliferation contributes to spinal cord growth. J Comp Neurol. doi:10.1002/cne.23104 (ahead of publication)

Arai Y, Deguchi K, Takashima S (1998) Vascular endothelial growth factor in brains with periventricular leukomalacia. Pediatr Neurol 19:45–49

Baas D, Meiniel A, Benadiba C et al (2006) A deficiency in RFX3 causes hydrocephalus associated with abnormal differentiation of ependymal cells. Eur J Neurosci 24:1020–1030. doi:10.1111/j.1460-9568.2006.05002.x

Banizs B, Pike MM, Millican CL et al (2005) Dysfunctional cilia lead to altered ependyma and choroid plexus function, and result in the formation of hydrocephalus. Development 132:5329–5339

Bering EA Jr (1955) Choroid plexus and arterial pulsation of cerebrospinal fluid; demonstration of the choroid plexuses as a cerebrospinal fluid pump. AMA Arch Neurol Psychiatry 73(2):165–172

Bonnafe E, Touka M, AitLounis A et al (2004) The transcription factor RFX3 directs nodal cilium development and left-right asymmetry specification. Mol Cell Biol 24:4417–4427. doi:10.1128/MCB.24.10.4417-4427.2004

Bradley WJ, Kortman K, Burgoyne B (1986) Flowing cerebrospinal fluid in normal and hydrocephalic states: appearance on MR images. Radiology 159(3):611–616

Brody SL, Yan XH, Wuerffel MK et al (2000) Ciliogenesis and left-right axis defects in forkhead factor HFH-4-null mice. Am J Respir Cell Mol Biol 23:45–51

Bruni JE (1998) Ependymal development, proliferation, and functions: a review. Microsc Res Tech 41:2–13. doi:10.1002/(SICI)1097-0029(19980401)41:1<2::AID-JEMT2>3.0.CO;2-Z

Bruni JE, Del Bigio MR, Clattenburg R (1985) Ependyma: normal and pathological. A review of the literature. Brain Res 356(1):1–19

Calvo CF, Fontaine RH, Soueid J et al (2011) Vascular endothelial growth factor receptor 3 directly regulates murine neurogenesis. Genes Dev 25:831–844. doi:10.1101/gad.615311

Capela A, Temple S (2002) LeX/ssea-1 is expressed by adult mouse CNS stem cells, identifying them as nonependymal. Neuron 35:865–875

Cathcart RS 3rd, Worthington WC Jr (1964) Ciliary movement in the rat cerebral ventricles: clearing action and directions of currents. J Neuropathol Exp Neurol 23:609–618

Danilov AI, Gomes-Leal W, Ahlenius H et al (2009) Ultrastructural and antigenic properties of neural stem cells and their progeny in adult rat subventricular zone. Glia 57:136–152. doi:10.1002/glia.20741

Del Bigio MR (2010) Ependymal cells: biology and pathology. Acta Neuropathol 119:55–73. doi:10.1007/s00401-009-0624-y

Dirksen ER (1971) Centriole morphogenesis in developing ciliated epithelium of the mouse oviduct. J Cell Biol 51:286–302

Doetsch F, Garcia-Verdugo JM, Alvarez-Buylla A (1997) Cellular composition and three-dimensional organization of the subventricular germinal zone in the adult mammalian brain. J Neurosci 17:5046–5061

Doetsch F, Caille I, Lim DA et al (1999) Subventricular zone astrocytes are neural stem cells in the adult mammalian brain. Cell 97:703–716

Dubruille R, Laurençon A, Vandaele C et al (2002) Drosophila regulatory factor X is necessary for ciliated sensory neuron differentiation. Development 129:5487–5498. doi:10.1242/dev.00148

Ferri AL, Cavallaro M, Braida D et al (2004) Sox2 deficiency causes neurodegeneration and impaired neurogenesis in the adult mouse brain. Development 131:3805–3819. doi:10.1242/dev.01204

Genzen JR, Yang D, Ravid K, Bordey A (2009) Activation of adenosine A2B receptors enhances ciliary beat frequency in mouse lateral ventricle ependymal cells. Cerebrospinal Fluid Res 6:15. doi:10.1186/1743-8454-6-15

Guirao B, Joanny JF (2007) Spontaneous creation of macroscopic flow and metachronal waves in an array of cilia. Biophys J 92:1900–1917. doi:10.1529/biophysj.106.084897

Guirao B, Meunier A, Mortaud S et al (2010) Coupling between hydrodynamic forces and planar cell polarity orients mammalian motile cilia. Nat Cell Biol 12:341–350. doi:10.1038/ncb2040

Hamilton LK, Truong MKV, Bednarczyk MR et al (2009) Cellular organization of the central canal ependymal zone, a niche of latent neural stem cells in the adult mammalian spinal cord. Neuroscience 164:1044–1056. doi:10.1016/j.neuroscience.2009.09.006

Harandi M, Didier M, Aguera M et al (1986) GABA and serotonin (5-HT) pattern in the supraependymal fibers of the rat epithalamus: combined radioautographic and immunocytochemical studies. Effect of 5-HT content on [3] GABA accumulation. Brain Res 370(2):241–249

Hayamizu TF, Chan PT, Johanson CE (2001) FGF-2 immunoreactivity in adult rat ependyma and choroid plexus: responses to global forebrain ischemia and intraventricular FGF-2. Neurol Res 23:353–358

Hirota Y, Meunier A, Huang S et al (2010) Planar polarity of multiciliated ependymal cells involves the anterior migration of basal bodies regulated by non-muscle myosin II. Development 137:3037–3046. doi:10.1242/dev.050120

Hu H (1999) Chemorepulsion of neuronal migration by Slit2 in the developing mammalian forebrain. Neuron 23:703–711

Ibanez-Tallon I, Heintz N, Omran H (2003) To beat or not to beat: roles of cilia in development and disease. Hum Mol Genet 12(Spec No 1):R27–R35

Ibanez-Tallon I, Pagenstecher A, Fliegauf M et al (2004) Dysfunction of axonemal dynein heavy chain Mdnah5 inhibits ependymal flow and reveals a novel mechanism for hydrocephalus formation. Hum Mol Genet 13:2133–2141

Jacquet BV, Salinas-Mondragon R, Liang H et al (2009) FoxJ1-dependent gene expression is required for differentiation of radial glia into ependymal cells and a subset of astrocytes in the postnatal brain. Development 136:4021–4031. doi:10.1242/dev.041129

Kunimoto K, Yamazaki Y, Nishida T et al (2012) Coordinated ciliary beating requires Odf2-mediated polarization of basal bodies via basal feet. Cell 148:189–200. doi:10.1016/j.cell.2011.10.052

Lavado A, Oliver G (2011) Six3 is required for ependymal cell maturation. Development 138:5291–5300. doi:10.1242/dev.067470

Lechtreck KF, Delmotte P, Robinson ML et al (2008) Mutations in Hydin impair ciliary motility in mice. J Cell Biol 180:633–643. doi:10.1083/jcb.200710162

Lim DA, Tramontin AD, Trevejo JM et al (2000) Noggin antagonizes BMP signaling to create a niche for adult neurogenesis. Neuron 28:713–726

Martens DJ, Seaberg RM, van der Kooy D (2002) In vivo infusions of exogenous growth factors into the fourth ventricle of the adult mouse brain increase the proliferation of neural progenitors around the fourth ventricle and the central canal of the spinal cord. Eur J Neurosci 16:1045–1057

Mirzadeh Z, Merkle FT, Soriano-Navarro M et al (2008) Neural stem cells confer unique pinwheel architecture to the ventricular surface in neurogenic regions of the adult brain. Cell Stem Cell 3:265–278. doi:10.1016/j.stem.2008.07.004

Mirzadeh Z, Han YG, Soriano-Navarro M et al (2010) Cilia organize ependymal planar polarity. J Neurosci 30:2600–2610. doi:10.1523/JNEUROSCI.3744-09.2010

Mitchell B, Jacobs R, Li J et al (2007) A positive feedback mechanism governs the polarity and motion of motile cilia. Nature 447:97–101

Miyan JA, Nabiyouni M, Zendah M (2003) Development of the brain: a vital role for cerebrospinal fluid. Can J Physiol Pharmacol 81:317–328. doi:10.1139/y03-027

Narita K, Kawate T, Kakinuma N, Takeda S (2010) Multiple primary cilia modulate the fluid transcytosis in choroid plexus epithelium. Traffic 11:287–301. doi:10.1111/j.1600-0854.2009.01016.x

Nguyen T, Chin WC, O'Brien JA et al (2001) Intracellular pathways regulating ciliary beating of rat brain ependymal cells. J Physiol 531:131–140

Nguyen-Ba-Charvet KT (2004) Multiple roles for slits in the control of cell migration in the rostral migratory stream. J Neurosci 24:1497–1506. doi:10.1523/JNEUROSCI.4729-03.2004

Paez-Gonzalez P et al (2011) Ank3-dependent SVZ niche assembly is required for the continued production of new neurons. Neuron 71:61–75. doi:10.1016/j.neuron.2011.05.029

Rakic P, Sidman RL (1968) Subcommissural organ and adjacent ependyma: autoradiographic study of their origin in the mouse brain. Am J Anat 122:317–335. doi:10.1002/aja.1001220210

Robinson SR, Noone DF, O'Dowd BS (1996) Ependymocytes and supra-ependymal axons in rat brain contain glutamate. Glia 17:345–348. doi:10.1002/(SICI)1098-1136(199608)17:4<345::AID-GLIA9>3.0.CO;2-Y

Rodríguez EM, Blázquez JL, Pastor FE et al (2005) Hypothalamic tanycytes: a key component of brain-endocrine interaction. Int Rev Cytol 247:89–164. doi:10.1016/S0074-7696(05)47003-5

Roth Y, Kimhi Y, Edery H, Aharonson E (1985) Ciliary motility in brain ventricular system and trachea of hamsters. Brain Res 330(2):291–297

Sabourin J-C, Ackema KB, Ohayon D et al (2009) A mesenchymal-like ZEB1+niche harbors dorsal radial glial fibrillary acidic protein-positive stem cells in the spinal cord. Stem Cells 27:2722–2733. doi:10.1002/stem.226

Salathe M (2007) Regulation of mammalian ciliary beating. Annu Rev Physiol 69:401–422. doi:10.1146/annurev.physiol.69.040705.141253

Sanderson MJ, Sleigh MA (1981) Ciliary activity of cultured rabbit tracheal epithelium: beat pattern and metachrony. J Cell Sci 47:331–347

Sawamoto K, Wichterle H, Gonzalez-Perez O et al (2006) New neurons follow the flow of cerebrospinal fluid in the adult brain. Science 311:629–632

Schutzer SE, Liu T, Natelson BH et al (2010) Establishing the proteome of normal human cerebrospinal fluid. PLoS One 5:e10980. doi:10.1371/journal.pone.0010980.g003

Scott DE, Kozlowski GP, Sheridan MN (1974) Scanning electron microscopy in the ultrastructural analysis of the mammalian cerebral ventricular system. Int Rev Cytol 37:349–388

Segal MB (2000) The choroid plexuses and the barriers between the blood and the cerebrospinal fluid. Cell Mol Neurobiol 20:183–196. doi:10.1002/1097-0029(20010101)52:1<38::AID-JEMT6>3.0.CO;2-J

Segal MB (2001) Transport of nutrients across the choroid plexus. Microsc Res Tech 52(1):38–48

Shah AS, Ben-Shahar Y, Moninger TO et al (2009) Motile cilia of human airway epithelia are chemosensory. Science 325:1131–1134. doi:10.1126/science.1173869

Soria JM, Taglialatela P, Gil-Perotin S et al (2004) Defective postnatal neurogenesis and disorganization of the rostral migratory stream in absence of the Vax1 homeobox gene. J Neurosci 24:11171–11181. doi:10.1523/JNEUROSCI.3248-04.2004

Spassky N, Merkle FT, Flames N et al (2005) Adult ependymal cells are postmitotic and are derived from radial glial cells during embryogenesis. J Neurosci 25:10–18

Stubbs JL, Oishi I, Izpisua Belmonte JC, Kintner C (2008) The forkhead protein Foxj1 specifies node-like cilia in Xenopus and zebrafish embryos. Nat Genet 40:1454–1460. doi:10.1038/ng.267

Swoboda P, Adler HT, Thomas JH (2000) The RFX-type transcription factor DAF-19 regulates sensory neuron cilium formation in C. elegans. Mol Cell 5:411–421

Tissir F, Qu Y, Montcouquiol M et al (2010) Lack of cadherins Celsr2 and Celsr3 impairs ependymal ciliogenesis, leading to fatal hydrocephalus. Nat Neurosci 13:700–707. doi:10.1038/nn.2555

Town T, Breunig JJ, Sarkisian MR et al (2008) The stumpy gene is required for mammalian ciliogenesis. Proc Natl Acad Sci USA 105:2853–2858. doi:10.1073/pnas.0712385105

Tramontin AD, Garcia-Verdugo JM, Lim DA, Alvarez-Buylla A (2003) Postnatal development of radial glia and the ventricular zone (VZ): a continuum of the neural stem cell compartment. Cereb Cortex 13:580–587

Veening JG, Barendregt HP (2010) The regulation of brain states by neuroactive substances distributed via the cerebrospinal fluid; a review. Cerebrospinal Fluid Res 7:1. doi:10.1186/1743-8454-7-1

Wagshul ME, Chen JJ, Egnor MR et al (2006) Amplitude and phase of cerebrospinal fluid pulsations: experimental studies and review of the literature. J Neurosurg 104(5):810–819

Yu X, Ng CP, Habacher H, Roy S (2008) Foxj1 transcription factors are master regulators of the motile ciliogenic program. Nat Genet 40:1445–1453. doi:10.1038/ng.263

Zariwala MA, Knowles MR, Omran H (2007) Genetic defects in ciliary structure and function. Annu Rev Physiol 69:423–450. doi:10.1146/annurev.physiol.69.040705.141301

Zein El L, Ait-Lounis A, Morle L et al (2009) RFX3 governs growth and beating efficiency of motile cilia in mouse and controls the expression of genes involved in human ciliopathies. J Cell Sci 122:3180–3189. doi:10.1242/jcs.048348

Chapter 8
Primary Cilia and Brain Cancer

Shirui Hou and Young-Goo Han

Contents

Abstract Cell-extrinsic signals and intrinsic cell cycle regulators strictly control proliferation. Cancers develop from a cell that escapes these tight controls and proliferates unrestrictedly. The primary cilium critically controls proliferation by mediating cell-extrinsic signals and regulating cell cycle entry. Accordingly, recent studies showed that defective cilia can either promote or suppress cancers, depending on the cancer-initiating mutation, and that presence or absence of primary cilia is associated with specific cancer types. These novel findings suggest that primary cilia play central but distinct roles in different cancer types, opening up a completely new avenue of research to understand the biology and treatment of cancers.

Keywords Primary cilia • Cell cycle • Hedgehog • WNT • NOTCH • Receptor tyrosine kinase • Cancer

---

S. Hou • Y.-G. Han (✉)
Division of Brain Tumor Research, Department of Developmental Neurobiology, Neurobiology and Brain Tumor Program, St. Jude Children's Research Hospital, Memphis, TN, USA
e-mail: Young-Goo.Han@stjude.org

K.L. Tucker and T. Caspary (eds.), *Cilia and Nervous System Development and Function*, DOI 10.1007/978-94-007-5808-7_8, © Springer Science+Business Media Dordrecht 2013

Outline

In this chapter, we review evidence showing that primary cilia play important roles in cell cycle entry and multiple signaling pathways, and we discuss how these roles of primary cilia could contribute or inhibit brain cancer formation. First, we review the dual and opposing roles of primary cilia in medulloblastoma development, then discuss possible roles of primary cilia in other brain cancers focusing on diffuse intrinsic pontine glioma, glioblastoma multiforme, and cancer stem cell. Lastly, we discuss the potential of primary cilia in diagnosis and treatment for brain cancers.

8.1 Introduction

The primary cilium is at the crossroads of cell cycle progression and cellular signaling pathways. Cell cycle progression regulates assembly and disassembly of primary cilia, and primary cilia in turn regulate cell cycle entry (Rieder et al. 1979; Tucker et al. 1979; Kim et al. 2011b; Li et al. 2011). Primary cilia participate in multiple signaling pathways that control cell proliferation, differentiation, migration, polarity, and metabolism, deregulation of which are closely linked to oncogenesis (Huangfu et al. 2003; Schneider et al. 2005; Simons et al. 2005; Zhu et al. 2009; Boehlke et al. 2010; Berbari et al. 2011; Ezratty et al. 2011). Thus, it is reasonable to suspect that normal or abnormal functions of primary cilia may contribute to oncogenesis. Indeed, recent studies showed that primary cilia play fundamental roles in the development of basal cell carcinoma, the most common cancer in Caucasians, and medulloblastoma, the most common brain cancer in children (Han et al. 2009; Wong et al. 2009). Surprisingly, primary cilia are either required or suppressive for oncogenesis in mice, depending on the cancer-initiating mutation. Furthermore, the presence or absence of primary cilia is associated with specific cancer types in humans (Wheatley 1995; Han et al. 2009; Schraml et al. 2009; Seeley et al. 2009; Wong et al. 2009; Yuan et al. 2010; Kim et al. 2011a), suggesting that primary cilia can either promote or suppress human cancers as well. Therefore, understanding the mechanisms by which this intriguing organelle functions in normal and cancerous cells will reveal oncogenic mechanisms that were not apparent previously and will unmask new therapeutic approaches.

8.2 Primary Cilia and Cell Cycle Progression

The primary cilium contains a unique microtubule cytoskeleton called the axoneme, a ring of nine microtubule doublets (9+0) that runs longitudinally through the organelle. The axoneme grows from the basal body docked with the cell membrane. During cell division, the basal body detaches from membrane and transforms into the mother centriole to form the centrosome that organizes the mitotic spindle.

Thus, ciliogenesis is tightly regulated throughout the cell cycle; proliferating cells disassemble cilia before mitosis to release the basal body and reassemble them after mitosis (Rieder et al. 1979; Tucker et al. 1979). Some cells also resorb cilia upon entry into the S phase (Tucker et al. 1979). Consistent with this regulation, Aurora kinase A, a centrosomal kinase that regulates mitotic entry, becomes activated at the basal body as cells enter the S phase and mitosis and triggers ciliary disassembly (Pugacheva et al. 2007). As a cell exits mitosis, CP110, a distal centriolar protein that inhibits ciliogenesis, disappears from the mother centriole, allowing ciliogenesis to occur (Spektor et al. 2007). Cyclin-dependent kinase inhibitors also influence ciliogenesis in cultured cells; knockdown of $p16^{INK4a}$ positively influences ciliogenesis, and knockdown of $p15^{INK4b}$ negatively regulates it (Bishop et al. 2010; Kim et al. 2010a). However, mutant mice defective for $p16^{INK4a}$ and/or $p15^{INK4b}$do not display developmental defects associated with defective cilia. Remarkably, recent studies showed that this cell-cycle-dependent organelle is not passively linked with the cell cycle but actively regulates cell cycle progression (Bielas et al. 2009; Jacoby et al. 2009; Kim et al. 2011b; Li et al. 2011).

One of the first pieces of evidence supporting the role of primary cilia in cell cycle progression came from studies on the *inositol polyphosphate-5-phosphatase E* (*INPP5E*) gene mutated in two cilia-associated diseases, Joubert and MORM syndromes (Bielas et al. 2009; Jacoby et al. 2009). INPP5E, which hydrolyzes the 5-phosphate of phosphatidylinositol 3,4,5-triphosphate and phosphatidylinositol 4,5-bisphosphate, localized exclusively at primary cilia; thus, mutation in *INPP5E* caused the ciliary membrane to accumulate these lipids, which play important roles in signaling pathways and membrane trafficking. Notably, *INPP5E* mutation did not affect assembly of cilia but accelerated both ciliary disassembly and S-phase entry after serum stimulation. Previous work showed that overexpression of *INPP5E* caused cell cycle arrest (Kisseleva et al. 2002). These studies suggest that ciliary disassembly can affect S-phase entry. More direct evidence supporting this notion came from the following two studies.

In one study, Sung and colleagues (Li et al. 2011) found that Tctex-1, a light chain subunit of cytoplasmic dynein, plays critical roles in ciliary disassembly and cell cycle progression. Knockdown of Tctex-1 inhibited both ciliary disassembly and S-phase entry that occur after serum addition to serum-starved NIH3T3 and RPE cells. Importantly, Tctex-1 knockdown did not affect S-phase entry in cells lacking primary cilia, indicating that failure to disassemble cilia underlies the blockage of S-phase entry. This inhibition was independent of the function of Tctex-1 as a cytoplasmic dynein component. Serum stimulation induced phosphorylation of Tctex-1 at threonine 94 (T94), leading to its dissociation from the dynein complex and accumulation at the transition zone between the basal body and the ciliary axoneme. The knockdown effect was rescued by wild-type and phosphorylation-mimicking Tctex-1 (Tctex-1^{T94E}), but to a lesser extent by non-phosphorylatable Tctex-1^{T94A}, which binds dynein. Furthermore, Tctex-1^{T94E} accelerated ciliary disassembly and S-phase entry even in the absence of serum stimulation. *In vivo*, Tctex-1 is selectively enriched in proliferating neural progenitors. Remarkably, knockdown of Tctex-1 in radial glia, the neural progenitors in the embryonic brain, caused them to exit from the cell cycle and differentiate into neurons prematurely, whereas overexpression of Tctex-1^{T94E}

shortened the G1 phase, accelerated S-phase re-entry, and increased the proliferating progenitor cell population. These data indicate that the primary cilium blocks S-phase entry and cells use Tctex-1 to overcome this blockage.

In the other study, Tsiokas and colleagues (Kim et al. 2011b) found that Nde-1, a centrosomal protein, critically regulates cilia length and S-phase entry. Knockdown of Nde-1 in NIH3T3 and RPE cells lengthened cilia and delayed S-phase entry after serum stimulation. Importantly, similar to Tctex-1, the delay in S-phase entry upon Nde-1 knockdown was dependent on the presence of primary cilia; there was no delay in S-phase entry in the absence of primary cilia. Furthermore, expression of constitutively active form of Rab8a, which lengthens primary cilia independently of Nde-1, also inhibited S-phase entry, confirming that lengthened cilia delay S-phase entry. *In vivo*, knockdown of *Nde-1* in zebrafish caused lengthening of primary cilia and defective proliferation in Kuffer's vesicle. Interestingly, mutations in *Nde-1* cause microcephaly in both humans and mice (Feng and Walsh 2004; Alkuraya et al. 2011; Bakircioglu et al. 2011), which may be partly due to premature cell cycle exit and differentiation of neural progenitors that have abnormally long cilia. Taken together, these studies revealed that primary cilia regulate cell division as a barrier blocking S-phase entry raising the possibility that primary cilia negatively regulate oncogenesis.

8.3 Primary Cilia and Signaling Pathways

Studies over the past decade have established primary cilia as a signaling hub for multiple signaling pathways, including Hedgehog (Goetz and Anderson 2010), Wnt (Wallingford and Mitchell 2011), receptor tyrosine kinases (RTKs) (Christensen et al. 2012), and Notch signaling (Ezratty et al. 2011). These signaling pathways control a myriad of cellular processes, including proliferation, differentiation, migration, polarity, and metabolism, all of which play critical roles in development, homeostasis, and oncogenesis. Here we briefly discuss the role of primary cilia in these signaling pathways.

8.3.1 Hedgehog Signaling

Forward genetic studies in mice have shown that Hedgehog signaling requires the primary cilium for both activation and repression of the pathway (Huangfu et al. 2003; Haycraft et al. 2005; Huangfu and Anderson 2005; Liu et al. 2005; May et al. 2005) (detailed in Chap. 2). Ciliary mutant mice have phenotypes similar to those of mutant mice defective in Hedgehog signaling. Epistasis studies placed primary cilia downstream of Hedgehog receptor Patched1 (Ptch1) and Smoothened (Smo), but upstream of the GLI-Kruppel family transcription factors (Gli1-3) and their binding protein Suppressor of Fused (Sufu) (Huangfu et al. 2003; Haycraft et al.

2005; Huangfu and Anderson 2005; Liu et al. 2005; May et al. 2005; Han et al. 2008; Chen et al. 2009; Jia et al. 2009). In the absence of Hedgehog, Ptch1 localizes to primary cilia and inhibits Hedgehog signaling by preventing Smo from entering primary cilia (Rohatgi et al. 2007). Upon binding of Hedgehog, Ptch1 moves out of the cilia, leading to accumulation and activation of Smo in the primary cilium (Corbit et al. 2005; Rohatgi et al. 2007). Activated Smo induces Sufu-Gli complex accumulation in primary cilia and dissociation of the complex, leading to formation of Gli transcriptional activators (Humke et al. 2010; Tukachinsky et al. 2010; Zeng et al. 2010). Without activated Smo in primary cilia, Gli2 and Gli3 are truncated by the proteasome to become transcriptional repressors and repress Hedgehog target gene expression. In the absence of primary cilia, the processing of Gli3 (and probably Gli2 also) repressor is greatly reduced, resulting in derepression of Hedgehog target genes (Haycraft et al. 2005; Huangfu and Anderson 2005; Liu et al. 2005; May et al. 2005).

In contrast to complete loss of primary cilia, which results in complete irresponsiveness to Hedgehog and inefficient processing of Gli repressors, distinctive structural abnormalities in primary cilia cause a range of Hedgehog signaling defects. Mutations in components of intraflagellar transport (IFT) complex A, which functions in retrograde transport of ciliary components from cilia to the cell body, cause abnormally short and swollen cilia and constitutive activation of Hedgehog signaling (Tran et al. 2008; Cortellino et al. 2009). Loss of Arl13b, a ciliary small GTPase, causes opening of axonemal microtubules and constitutive activation of Gli activators at low levels without affecting Gli3 repressor activity (Caspary et al. 2007). Mutations in Broad-minded, a Rab-GAP–like protein, cause detachment of ciliary membrane from the axoneme and selective loss of responsiveness to high Hedgehog levels (Ko et al. 2010). Thus, primary cilia play complex and active roles in Hedgehog signaling rather than merely concentrating signaling molecules to facilitate their interactions.

8.3.2 Wnt Signaling

Secreted protein Wnt binds to Frizzled receptors to trigger the signaling activity of cytoplasmic protein Dishevelled (Dvl), where the signaling diverges into canonical and non-canonical Wnt signaling pathways (Logan and Nusse 2004). In the canonical pathway, activated Dvl leads to accumulation and nuclear localization of β-catenin and subsequent activation of Wnt target genes. The non-canonical pathway is independent of β-catenin and primarily controls cytoskeletons involved in planar cell polarity (PCP) and cell migration. One of the first pieces of evidence linking primary cilia to Wnt signaling came from a study on Inversin, a ciliary protein whose mutation causes cystic kidney diseases and *situs inversus* (Simons et al. 2005). The study showed that, in fish and frogs, Inversin functions as a switch from the canonical to the non-canonical Wnt signaling pathway by targeting Dvl for destruction. Moreover, mutant mice defective in genes mutated in Bardet-Biedl syndrome, a disease linked to ciliary dysfunction, have phenotypes associated with PCP mutants (Ross et al. 2005). Other studies from kidney, pancreas, cochlea, fish embryos, and cultured cells also showed that defective primary cilia increase canonical Wnt signaling activity and disrupt non-canonical Wnt signaling (Lin et al. 2003; Cano et al. 2004; Gerdes et al. 2007; Corbit et al. 2008; Jonassen et al. 2008; Jones et al. 2008). Notably, however, mutant mice defective for primary cilia do not have obvious developmental phenotypes associated with defective Wnt signaling. Furthermore, two recent studies showed that defective primary cilia do not affect Wnt signaling in fish, mice, and cultured cells and suggested that Wnt signaling requires the basal body rather than primary cilia *per se* (Huang and Schier 2009; Ocbina et al. 2009). Thus, the primary cilium and the basal body appear to have a subtle and cell type–specific roles in Wnt signaling (Wallingford and Mitchell 2011).

8.3.3 RTK Signaling

RTKs are activated by growth factors and initiate a series of signaling cascades, including mitogen-activated protein kinase pathways, phosphatidylinositol 3-kinase pathways, and phospholipase C pathways. The first connection between RTKs and primary cilia came from evidence that platelet-derived growth factor receptor α (PDGFRα) signaling requires primary cilia in NIH3T3 cells and mouse embryonic fibroblasts (MEFs) (Schneider et al. 2005). Serum starvation of confluent cells induced PDGFRα expression and its localization to primary cilia. Subsequent stimulation of cells with PDGF-AA ligand induced phosphorylation of PDGFRα and downstream dual specificity mitogen-activated protein kinase kinase 1/2 (MEK1/2) inside primary cilia, leading to phosphorylation of retinoblastoma-associated (RB) protein, which marks S-phase entry. The activation of PDGFRα in primary cilia was also required for directional migration mediated by Na+/H+exchanger NHE1 (Schneider et al. 2009). Importantly, PDGF-AA failed to activate PDGFRα to induce S-phase entry and directional migration in MEFs isolated from hypomorphic *IFT88* mutant mice (*IFT88*^{Tg737Rpw}) defective for ciliogenesis. These observations suggest that PDGF-AA and PDGFRα require primary cilia to transmit signals. Some of the observed defects, however, may be partly due to the low level of PDGFRα in *IFT88*^{Tg737Rpw} MEFs after serum starvation, which dramatically induced a higher PDGFRα level in wild-type but not in *IFT88*^{Tg737Rpw} MEFs (Schneider et al. 2005). Thus, it will be important to determine whether PDGF-AA can activate signaling in cells that lack primary cilia but have PDGFRα at a similar level as wild-type cells. It still remains to be determined whether primary cilia are required for PDGFRα signaling *in vivo*.

Another RTK, insulin-like growth factor 1 receptor (IGF1R) appears to function preferentially in primary cilia in MEF-adipose-like 3T3-L1 cells (Zhu et al. 2009), which can be differentiated into adipocyte after growth arrest at confluence. IGF1R activation by insulin is essential to induce differentiation. Interestingly, in 3T3-L1 cells, insulin activated IGF1R in primary cilia faster than IGF1R outside primary cilia and induced accumulation of activated downstream signaling molecules, phosphorylated insulin receptor substrate 1 (IRS1) and protein kinase B (Akt), at the basal body. Remarkably, knockdown of *IFT88* or *Kif3a* encoding a subunit of essential ciliogenic Kinesin-II motor disrupted IGF1R signaling and blocked differentiation of 3T3-L1, suggesting that IGF1R requires primary cilia to signal to induce differentiation. Like PDGFRα signaling, it is unknown whether primary cilia are required for IGF1R signaling *in vivo*.

In addition to PDGFRα and IGF1R, epidermal growth factor receptor (EGFR) and the angiopoietin receptors Tie-1 and Tie-2 localize to primary cilia (Ma et al. 2005; Teilmann and Christensen 2005; Danilov et al. 2009; Wu et al. 2009). Furthermore, recent studies showed that primary cilia negatively regulate the activity of mammalian target of rapamycin (mTOR) both *in vitro* and *in vivo* (DiBella et al. 2009; Boehlke et al. 2010; Berbari et al. 2011), whereas mTOR positively regulates the length of primary cilia (Yuan et al. 2012). mTOR is a key signaling molecule that integrates RTK signaling with cellular metabolism, a change in which is one of the hallmarks of cancers. Thus, primary cilia appear to participate in multiple RTK signaling pathways, providing them with a platform to crosstalk and integrate. Future work should investigate whether primary cilia participate in RTK signaling *in vivo* and in the crosstalk of multiple RTK signaling pathways.

8.3.4 Notch Signaling

A recent study showed that Notch signaling requires primary cilia during skin development (Ezratty et al. 2011). Notch is a transmembrane receptor protein that undergoes intramembrane proteolytic cleavage upon binding to its ligands, which are mainly transmembrane proteins as well (Kopan and Ilagan 2009). After cleavage, the Notch intracellular domain (NICD) enters the nucleus and activates transcription of target genes. Removing primary cilia in mouse embryonic skin cells by knockdown of *IFT74* or conditional ablation of either *IFT88* or *Kif3a* resulted in defective epidermal differentiation, a process dependent on Notch signaling. Consistently, expression of Notch responsive genes was disrupted in cells lacking primary cilia. Expression of NCID partially rescued expression of a Notch reporter gene and differentiation defects. The study also showed that Notch3 is selectively localized to primary cilia, and Presenilin-2, the catalytic subunit of γ-secretase that cleaves Notch receptor to generate NICD, is localized at the base of primary cilia in addition to intercellular membrane borders. These localizations were specific to suprabasal cells in the embryonic skin, where Notch signaling is active. Remarkably, nuclear NCID3, the processed Notch3, was observed only in ciliated suprabasal cells and not in *Kif3a* mutant cells. These findings raise interesting questions: Does Notch3 require primary cilia to signal in other tissues and animals? Does any other Notch require primary cilia to signal? Are membrane-bound ligands for Notch3 also exclusively localized to a specific domain of the signaling cell to be juxtaposed to the primary cilium of responding cells in developing skin? If so, what is the underlying mechanism?

Why do vertebrate cells require primary cilia for multiple signaling pathways? Simplistically, the concentration of signaling molecules in primary cilia whose large surface area relative to the small volume would provide high sensitivity for detection of low levels of extracellular signals. Subtle structural defects in primary cilia, however, cause unique Hh signaling defects, suggesting that the function of primary cilia are more than concentrating and sensitizing signaling molecules for extracellular signals. The primary cilium, although continuous from cytoplasm, is a distinct subcellular compartment, in which trafficking is restricted by IFT and the transition zone, a barrier at the base of cilia. This restriction may allow the coordinated spread or movement of second messengers or effecter molecules. In addition, the juxtaposition of the basal body and the Golgi complex at the base of cilia may facilitate rapid trafficking of molecules like Smo into the cilia upon receiving extracellular signals and could help coordinate the cell cycle.

8.4 Primary Cilia and Medulloblastoma

One of the first bits of direct evidence showing that primary cilia play salient roles in cancer came from a study on medulloblastoma (Han et al. 2009). Medulloblastoma is the most common malignant brain tumor in children, accounting for ~20% of childhood brain tumors. Medulloblastomas mostly arise in the cerebellum, but a recent study showed that a subgroup of medulloblastoma arises in the dorsal brain stem (Gibson et al. 2010). Several transcriptional profiling studies revealed that medulloblastoma comprises four principal subgroups, which have distinct demographic, clinical, transcriptional, and mutational characteristics (Thompson et al. 2006; Kool et al. 2008; Cho et al. 2011; Northcott et al. 2011; Taylor et al. 2012). These subgroups include Sonic Hedgehog (SHH, one of three mammalian Hedgehog proteins), WNT, subgroup 3, and subgroup 4. The SHH and WNT subgroups are named after the signaling pathways thought to drive tumorigenesis of that subgroup. Subgroup 3 often shows amplification of MYC. Molecular mechanisms that drive subgroup 4 have not been identified.

SHH subgroup medulloblastoma is characterized by aberrant activation of SHH signaling and constitutes about 25% of medulloblastoma cases. The SHH subgroup arises from granule neuron precursors (GNPs) in the cerebellum (Schuller et al. 2008; Yang et al. 2008). GNPs are produced from radial glia in the anterior roof of the fourth ventricle, known as the upper rhombic lip, and migrate rostrally to form the external granular layer (EGL) on the surface of the developing cerebellum (Altman and Bayer 1997). In the EGL, GNPs proliferate extensively to produce cerebellar granule neurons, the most abundant neurons that constitute more than half of the neurons in the central nervous system. Immature granule neurons produced from the EGL migrate inward, passing Purkinje neurons and forming the internal granule neuron layer. SHH secreted from Purkinje neurons is an essential mitogen for GNPs in the EGL (Dahmane and Ruiz i Altaba 1999; Wallace 1999; Wechsler-Reya and Scott 1999). Consistent with the critical role of primary cilia in

Hedgehog signaling, GNPs lacking primary cilia failed to proliferate, resulting in severe hypoplasia and underdevelopment of the cerebellum (Chizhikov et al. 2007; Spassky et al. 2008).

While SHH signaling is essential for the proliferation of GNPs, abnormal activation of SHH signaling leads to uncontrolled expansion of GNPs, resulting in medulloblastoma (Hatten and Roussel 2011). A recent study revealed surprising dual roles of primary cilia in medulloblastoma development driven by abnormal activation of SHH signaling (Han et al. 2009). Mice expressing a constitutively active form of Smo (SmoM2), which was identified in medulloblastoma and basal cell carcinoma (Lam et al. 1999), in GNPs develop medulloblastoma (Hallahan et al. 2004; Mao et al. 2006; Schuller et al. 2008; Han et al. 2009). In these mice, SmoM2 concentrated in the primary cilia of tumor cells and required this organelle to induce tumors; concomitant removal of primary cilia in SmoM2-expressing cells completely blocked medulloblastoma development (Han et al. 2009). Unlike SmoM2, GNPs expressing a constitutively active form of GLI2, a downstream transcription factor, that lacks an N-terminal repressor domain (GLI2ΔN) did not form medulloblastoma (Roessler et al. 2005; Pasca di Magliano et al. 2006; Han et al. 2009). Surprisingly, however, concomitant removal of primary cilia in GNPs expressing GLI2ΔN resulted in 100% medulloblastoma development, suggesting that primary cilia suppress medulloblastoma development when the oncogenic mutation is in the GLI2 transcription factor. Thus, the primary cilium plays opposing dual roles in medulloblastoma: it is required for SmoM2 but suppressive for GLI2ΔN to induce medulloblastoma (Fig. 8.1). The molecular mechanism by which primary cilia suppress GLI2ΔN-driven medulloblastoma development remains to be determined. In the presence of primary cilia, Gli3 repressors may counteract GLI2ΔN and inhibit medulloblastoma development, whereas in the absence of primary cilia, Gli3 repressors do not form (Haycraft et al. 2005; Huangfu and Anderson 2005; Liu et al. 2005; May et al. 2005), which may allow GLI2ΔN to induce medulloblastoma. Alternatively, primary cilia may be required for another signaling pathway to suppress tumorigenesis. Primary cilia may also function as a general barrier for cell cycle entry, as discussed above. Similar opposing dual functions of primary cilia were observed in basal cell carcinoma driven by SmoM2 and GLI2ΔN (Wong et al. 2009). Taken together, these suggest that, to induce cancer, some oncogenic mutations may require intact primary cilia but others may require losing them. Notably, in support of this hypothesis, the presence or absence of primary cilia is tightly associated with specific subgroups of medulloblastoma (Fig. 8.2). In humans, primary cilia are almost exclusively present in the SHH and WNT subgroups of medulloblastoma but absent in subgroups 3 and 4 (Han et al. 2009). Thus, primary cilia may be required for the SHH and WNT subgroups but suppressive for subgroups 3 and 4. The presence of primary cilia in the WNT subgroup is somewhat contradictory to several studies showing that primary cilia constrain canonical WNT signaling; however, as discussed above, the role of primary cilia in WNT signaling is still controversial and specific to cell type. Recently developed mouse models each representing WNT-subgroup medulloblastoma (Gibson et al. 2010) and subgroup 3 (Kawauchi et al. 2012; Pei et al. 2012) will provide an excellent opportunity

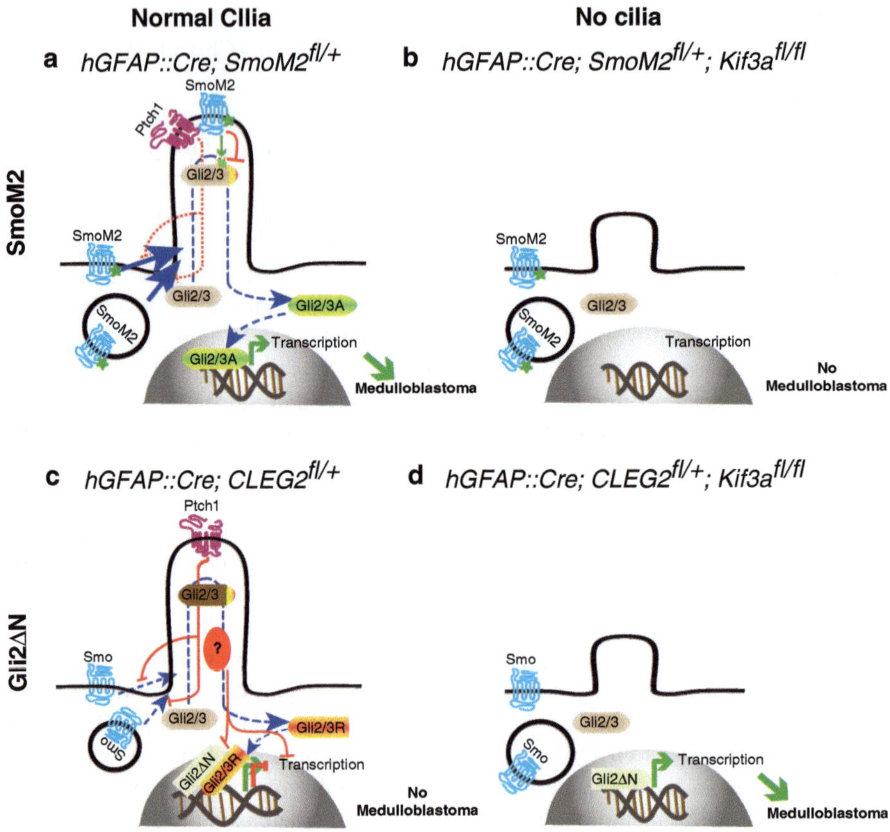

Fig. 8.1 Dual and opposing roles of primary cilia in medulloblastoma formation. (**a**) SmoM2 is insensitive to inhibition by Ptch1 and constitutively localizes to primary cilia, where it inhibits production of repressor forms of Gli2 and Gli3 and induces production of activator forms. Uncontrolled activation of the signaling leads to medulloblastoma formation in *hGFAP::Cre; SmoM2^fl/+^* mice. (**b**) Without primary cilia, SmoM2 cannot activate downstream signaling, thus loss of primary cilia in *hGFAP::Cre; SmoM2^fl/+^; Kif3a^fl/fl^* mice completely blocks medulloblastoma formation. (**c**, **d**) Constitutively active form of Gli2 (Gli2ΔN) is not sufficient to induce medulloblastoma in *hGFAP::Cre; CLEG2^fl/+^* mice. Loss of primary cilia in *hGFAP::Cre; CLEG2^fl/+^; Kif3a^fl/fl^* mice allows Gli2ΔN to induce medulloblastoma. Repressor forms of Gli2 and/or Gli3, whose formation requires primary cilia, may inhibits the tumorigenic activity of Gli2ΔN. Alternatively, unknown mechanism through primary cilia may inhibit the tumorigenic activity of Gli2ΔN

to investigate the role of primary cilia in these tumors. These investigations will also provide clues to whether the dual roles of primary cilia are generally applicable to cancers in addition to those driven by Hedgehog signaling.

Currently, SMO inhibitors are under clinical trials to treat SHH subgroup medulloblastoma and basal cell carcinoma. Although these clinical trials show promise for SMO inhibitors for treating medulloblastoma, resistance to a SMO inhibitor was observed in a patient who initially showed a dramatic response (Yauch

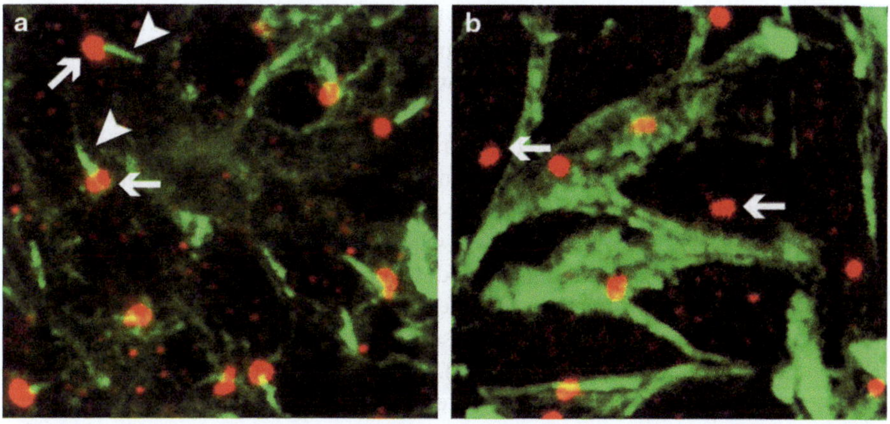

Fig. 8.2 Primary cilia are present in human medulloblastomas showing oncogenic activation of SHH or WNT signaling (**a**), but mostly absent in other molecular subgroups of medulloblastomas (**b**). *Arrows* indicate basal bodies stained with an antibody against pericentrin and *arrowheads* indicate primary cilia stained with an antibody against acetylated tubulin

et al. 2009). Subsequently, a point mutation in SMO conferring resistance to the SMO inhibitor was found in the medulloblastoma of this patient. Furthermore, *in vivo* and *in vitro* studies using the same SMO inhibitor revealed frequent appearance of resistance (Dijkgraaf et al. 2011). Currently, SMO is the only molecular target under clinical trials to treat medulloblastoma. Thus, it is necessary to develop strategies to overcome the resistance. Since Smo requires primary cilia to function and ciliogenesis requires a number of cellular processes involving a large number of proteins, primary cilia may provide multiple novel targets that can overcome resistance to SMO inhibitors.

8.5 Primary Cilia and Other Brain Cancers

Given that primary cilia play important roles in cell cycle progression and signaling pathways frequently involved in oncogenesis, they are also likely to play important roles in other brain cancers in addition to medulloblastoma. Here we will discuss two types of brain cancers, the deadliest brain cancers of children and adults, in which primary cilia may have important roles.

Diffuse intrinsic pontine gliomas (DIPGs) are diffusely infiltrative high-grade gliomas in the ventral pons. DIPGs affect mostly children, peaking at age 6–7 years, constituting 10–15% of pediatric brain cancer (Hawkins et al. 2011). DIPGs are extremely aggressive cancers that are almost universally fatal in less than a year. Yet, little is known about the biology of this tumor for which no effective therapy exists. The age- and region-specific natures of DIPGs suggest that these tumors arise from deregulation of a specific postnatal development process occurring in the ventral pons in children. Indeed, a recent study identified a putative neural precursor cell population positive for neural precursor markers, Nestin, Vimentin, and Olig2 in the ventral pons of both humans and mice, whose spatiotemporal distribution matches very closely to that of DIPGs (Monje et al. 2011). Interestingly, the Hedgehog signaling pathway was active in this precursor population in mice. Expression of SmoM2 increased proliferation of these cells in mice, leading to hypertrophy of the ventral pons. Furthermore, blocking Hedgehog signaling reduced self-renewal of neurospheres generated form human DIPGs, whereas addition of SHH increased self-renewal. Thus, aberrant activation of Hedgehog signaling may contribute to DIPG formation by driving proliferation of these precursor cells. Unlike medulloblastoma, however, expression of SmoM2 was not sufficient to induce DIPGs, suggesting that another hit is necessary for DIPGs to form. This is consistent with the fact that people having germline mutations of PTCH1 are predisposed to medulloblastoma but not DIPG (Johnson et al. 1996). Notably, recent genome-wide analyses identified frequent amplification or overexpression of *PDGFRα*, *PDGF-A*, and *IGF1R*, all of which encode proteins that may signal through primary cilia (Zarghooni et al. 2010; Paugh et al. 2011). Cancer cells that have amplification or overexpression of these oncogenes may rely on normal signaling mechanisms through primary cilia. Thus, primary cilia may play important roles in the development of this devastating disease. It would be interesting to find whether DIPGs are ciliated and co-activation of SHH and PDGFRα can cause DIPGs in mouse models, as well as to test the role of primary cilia in such mouse models. If primary cilia play a significant role, targeting ciliogenesis would be a valid treatment option.

Glioma is the most frequent brain tumor in adults, and malignant glioma (glioblastoma multiforme, GBM) comprises 80% of malignant tumors in the central nervous system (Chen et al. 2012). Currently, GBM patients' 5 year survival rate is less than 5%, and median survival is about 1 year. Two recent genome-wide studies including gene expression profiling, DNA copy number variation, protein-encoding gene sequencing, and DNA methylation status revealed three core pathways that commonly mutated in GBM: the p53 pathway, the RB pathway, and the RTK pathway (Parsons et al. 2008; TCGA 2008). The majority (74%) of GBMs had alterations in all three pathways, which enables cancer cells to proliferate unrestrictedly, escaping from cell-cycle checkpoints, senescence, and apoptosis. On the other hand, alterations affecting components in the same core pathway were mutually exclusive. Among RTKs, frequent aberrations were found in EGFR, ERBB2, PDGFRα, and MET. Similar to what was seen in medulloblastoma, several gene expression profiling studies identified distinct molecular subtypes in GBM (Vitucci et al. 2011). A recent study grouped GBM into four subtypes: proneural, neural, classical, and mesenchymal (Verhaak et al. 2010). By integrating gene expression profiles with the previous genome-wide analysis of GBM, this study showed that aberrant status of *EGFR*, *NF1*, and *PDGFRα/Isocitrate dehydrogenase 1* define the classic, mesenchymal, and proneural subtypes, respectively. Notably, the classic subtype expressed high levels of Hedgehog (SMO, GLI2, and GAS1) and NOTCH (NOTCH3, JAG1, and LFNG) signaling components. Given the important role of primary cilia in Hedgehog, Notch, and PDGFRα signaling pathways, the classic and proneural subtypes may require primary cilia for their growth. Thus, it would be interesting to determine whether the presence or absence of primary cilia is associated with specific GBM subtypes. Such associations may indicate distinct roles of primary cilia in different GBM subtypes. Primary cilia are absent in several GBM cell lines (Moser et al. 2009); however, the molecular subtypes to which the cell lines belong are unknown. If the presence or absence of primary cilia is associated with specific GBM subgroups and such an association is important for oncogenesis, primary cilia will be an important diagnostic tool and a treatment target for GBM.

Although the cancer stem cell (CSC) theory is controversial, it suggests that primary cilia may have important roles in GBM and possibly in other brain cancers. CSCs are a subpopulation of cells in a cancer that can self-renew and give rise to highly heterogeneous cancer cells that make up the bulk of cancer. GBM CSCs were one of the first CSCs isolated from solid cancers (Singh et al. 2004). Resistance to radiation and chemotherapies is thought to be partly due to CSCs, which have preferentially active DNA repair pathways (Bao et al. 2006) and high levels of ATP-binding cassette transporters to export chemotherapy agents (Bleau et al. 2009); thus, CSCs have important implications for cancer targeting strategy. CSCs are thought to have properties similar to those of somatic stem cells. A number of studies have shown that signaling pathways that critically regulate the behavior of normal somatic stem cells also regulate that of GBM CSCs (Clark et al. 2007; Takebe et al. 2011). These pathways include Hedgehog, Wnt, and Notch signaling pathways, for which primary cilia play important roles. Consistently, recent studies showed that expression of Gli1 and β-catenin are associated with recurrence after therapy and poor prognosis in GBM patients (Rossi et al. 2011; Kim et al. 2012). Therefore, targeting these pathways is a vital therapeutic approach to increase the efficacy of radiation and chemotherapies. Understanding the mechanism by which primary cilia function in these signaling pathways in normal stem cells and CSCs will be important for developing such a therapeutic intervention.

8.6 Conclusion

Brain cancer is a complex and heterogeneous disease. Its treatments, however, are largely similar, including surgical resection, radiation, and chemotherapy, thus resulting in individually different outcomes. Recent advances in genome-wide studies on large cohorts of brain cancer patients elucidated that cancers that otherwise appear identical are highly heterogeneous at the molecular level, with distinctive oncogenic mutations and gene expression profiles. These recent advances call for new treatment paradigms building on better understandings of the molecular and cellular processes involved in initiation and progression of particular brain cancer types. We envision that investigating the function of primary cilia together with oncogenic mutations specific to distinct cancer types will reveal oncogenic mechanisms that were not appreciated previously. We also envision that primary cilia hold a great therapeutic potential for treatment for brain cancer patients. Some cancers have primary cilia, but others do not (Wheatley 1995; Han et al. 2009; Schraml et al. 2009; Seeley et al. 2009; Wong et al. 2009; Yuan et al. 2010; Kim et al. 2011a). Most cells in our body have the primary cilium; thus, some cancers may have it as a default. Building and maintaining primary cilia requires complex processes involving a wide variety of proteins that function in cell cycle progression (Pugacheva et al. 2007; Spektor et al. 2007; Kim et al. 2011b; Li et al. 2011), cytoskeletal dynamics (Kim et al. 2010a), apicobasal polarity (Fan et al. 2004), planar cell polarity (Kim et al. 2010b; Wallingford 2010), intraflagellar transport (Rosenbaum and Witman 2002), vesicle trafficking (Nachury et al. 2007; Zuo et al. 2009; Knodler et al. 2010), and transcriptional regulation (Thomas et al. 2010). Thus, some cancers may have lost primary cilia secondarily as they progress and accumulate mutations. In cancers driven by Hedgehog signaling, however, the presence or absence of primary cilia directly controls oncogenesis (Han et al. 2009; Wong et al. 2009). Furthermore, the presence or absence of primary cilia is associated with specific cancer types. Thus, the status of primary cilia in a particular cancer may reflect the role of primary cilia in that cancer; some cancers may keep primary cilia and others may eliminate them for growth and progression. Primary cilia will be important targets for such cancers. Since many proteins are involved in ciliogenesis, inhibiting ciliogenesis or ciliary function may be a plausible strategy to treat cancers that require primary cilia for their growth. Indeed, a high-throughput screening for inhibitors of Hedgehog signaling discovered a small molecule that inhibits cytoplasmic dynein and ciliogenesis (Firestone et al. 2012). It will be challenging to restore primary cilia in cancers that have eliminated them for growth. Yet, recent studies showed that small molecules targeting signaling molecules or fatty acid synthesis can restore primary cilia even in cancer cells (Wang et al. 2009; Willemarck et al. 2010). Therefore, research in primary cilia will open up a completely new avenue of research to understand the biology and treatment of cancers.

Acknowledgments American Lebanese Syrian Associated Charities (Y.-G.H.) and the Cancer Center Support Grant CA021765 from the National Cancer Institute (Y.-G.H.) supported this work.

References

Alkuraya FS, Cai X, Emery C, Mochida GH, Al-Dosari MS, Felie JM, Hill RS, Barry BJ, Partlow JN, Gascon GG, Kentab A, Jan M, Shaheen R, Feng Y, Walsh CA (2011) Human mutations in NDE1 cause extreme microcephaly with lissencephaly [corrected]. Am J Hum Genet 88:536–547

Altman J, Bayer SA (1997) Development of the cerebellar system: in relation to its evolution, structure, and functions. CRC Press, Boca Raton

Bakircioglu M et al (2011) The essential role of centrosomal NDE1 in human cerebral cortex neurogenesis. Am J Hum Genet 88:523–535

Bao S, Wu Q, McLendon RE, Hao Y, Shi Q, Hjelmeland AB, Dewhirst MW, Bigner DD, Rich JN (2006) Glioma stem cells promote radioresistance by preferential activation of the DNA damage response. Nature 444:756–760

Berbari NF, Kin NW, Sharma N, Michaud EJ, Kesterson RA, Yoder BK (2011) Mutations in Traf3ip1 reveal defects in ciliogenesis, embryonic development, and altered cell size regulation. Dev Biol 360:66–76

Bielas SL et al (2009) Mutations in INPP5E, encoding inositol polyphosphate-5-phosphatase E, link phosphatidyl inositol signaling to the ciliopathies. Nat Genet 41:1032–1036

Bishop CL, Bergin AM, Fessart D, Borgdorff V, Hatzimasoura E, Garbe JC, Stampfer MR, Koh J, Beach DH (2010) Primary cilium-dependent and -independent Hedgehog signaling inhibits p16(INK4A). Mol Cell 40:533–547

Bleau AM, Hambardzumyan D, Ozawa T, Fomchenko EI, Huse JT, Brennan CW, Holland EC (2009) PTEN/PI3K/Akt pathway regulates the side population phenotype and ABCG2 activity in glioma tumor stem-like cells. Cell Stem Cell 4:226–235

Boehlke C, Kotsis F, Patel V, Braeg S, Voelker H, Bredt S, Beyer T, Janusch H, Hamann C, Godel M, Muller K, Herbst M, Hornung M, Doerken M, Kottgen M, Nitschke R, Igarashi P, Walz G, Kuehn EW (2010) Primary cilia regulate mTORC1 activity and cell size through Lkb1. Nat Cell Biol 12:1115–1122

Cano DA, Murcia NS, Pazour GJ, Hebrok M (2004) Orpk mouse model of polycystic kidney disease reveals essential role of primary cilia in pancreatic tissue organization. Development 131:3457–3467

Caspary T, Larkins CE, Anderson KV (2007) The graded response to Sonic Hedgehog depends on cilia architecture. Dev Cell 12:767–778

Chen MH, Wilson CW, Li YJ, Law KK, Lu CS, Gacayan R, Zhang X, Hui CC, Chuang PT (2009) Cilium-independent regulation of Gli protein function by Sufu in Hedgehog signaling is evolutionarily conserved. Genes Dev 23:1910–1928

Chen J, McKay RM, Parada LF (2012) Malignant glioma: lessons from genomics, mouse models, and stem cells. Cell 149:36–47

Chizhikov VV, Davenport J, Zhang Q, Shih EK, Cabello OA, Fuchs JL, Yoder BK, Millen KJ (2007) Cilia proteins control cerebellar morphogenesis by promoting expansion of the granule progenitor pool. J Neurosci 27:9780–9789

Cho YJ, Tsherniak A, Tamayo P, Santagata S, Ligon A, Greulich H, Berhoukim R, Amani V, Goumnerova L, Eberhart CG, Lau CC, Olson JM, Gilbertson RJ, Gajjar A, Delattre O, Kool M, Ligon K, Meyerson M, Mesirov JP, Pomeroy SL (2011) Integrative genomic analysis of medulloblastoma identifies a molecular subgroup that drives poor clinical outcome. J Clin Oncol 29:1424–1430

Christensen ST, Clement CA, Satir P, Pedersen LB (2012) Primary cilia and coordination of receptor tyrosine kinase (RTK) signalling. J Pathol 226:172–184

Clark PA, Treisman DM, Ebben J, Kuo JS (2007) Developmental signaling pathways in brain tumor-derived stem-like cells. Dev Dyn 236:3297–3308

Corbit KC, Aanstad P, Singla V, Norman AR, Stainier DY, Reiter JF (2005) Vertebrate Smoothened functions at the primary cilium. Nature 437:1018–1021

Corbit KC, Shyer AE, Dowdle WE, Gaulden J, Singla V, Chen MH, Chuang PT, Reiter JF (2008) Kif3a constrains beta-catenin-dependent Wnt signalling through dual ciliary and non-ciliary mechanisms. Nat Cell Biol 10:70–76

Cortellino S, Wang C, Wang B, Bassi MR, Caretti E, Champeval D, Calmont A, Jarnik M, Burch J, Zaret KS, Larue L, Bellacosa A (2009) Defective ciliogenesis, embryonic lethality and severe impairment of the Sonic Hedgehog pathway caused by inactivation of the mouse complex A intraflagellar transport gene Ift122/Wdr10, partially overlapping with the DNA repair gene Med1/Mbd4. Dev Biol 325:225–237

Dahmane N, Ruiz i Altaba A (1999) Sonic hedgehog regulates the growth and patterning of the cerebellum. Development 126:3089–3100

Danilov AI, Gomes-Leal W, Ahlenius H, Kokaia Z, Carlemalm E, Lindvall O (2009) Ultrastructural and antigenic properties of neural stem cells and their progeny in adult rat subventricular zone. Glia 57:136–152

DiBella LM, Park A, Sun Z (2009) Zebrafish Tsc1 reveals functional interactions between the cilium and the TOR pathway. Hum Mol Genet 18:595–606

Dijkgraaf GJ, Alicke B, Weinmann L, Januario T, West K, Modrusan Z, Burdick D, Goldsmith R, Robarge K, Sutherlin D, Scales SJ, Gould SE, Yauch RL, de Sauvage FJ (2011) Small molecule inhibition of GDC-0449 refractory smoothened mutants and downstream mechanisms of drug resistance. Cancer Res 71:435–444

Ezratty EJ, Stokes N, Chai S, Shah AS, Williams SE, Fuchs E (2011) A role for the primary cilium in Notch signaling and epidermal differentiation during skin development. Cell 145:1129–1141

Fan S, Hurd TW, Liu CJ, Straight SW, Weimbs T, Hurd EA, Domino SE, Margolis B (2004) Polarity proteins control ciliogenesis via kinesin motor interactions. Curr Biol 14:1451–1461

Feng Y, Walsh CA (2004) Mitotic spindle regulation by Nde1 controls cerebral cortical size. Neuron 44:279–293

Firestone AJ, Weinger JS, Maldonado M, Barlan K, Langston LD, O'Donnell M, Gelfand VI, Kapoor TM, Chen JK (2012) Small-molecule inhibitors of the AAA+ATPase motor cytoplasmic dynein. Nature 484:125–129

Gerdes JM, Liu Y, Zaghloul NA, Leitch CC, Lawson SS, Kato M, Beachy PA, Beales PL, Demartino GN, Fisher S, Badano JL, Katsanis N (2007) Disruption of the basal body compromises proteasomal function and perturbs intracellular Wnt response. Nat Genet 39(11):1350–1360

Gibson P et al (2010) Subtypes of medulloblastoma have distinct developmental origins. Nature 468:1095–1099

Goetz SC, Anderson KV (2010) The primary cilium: a signalling centre during vertebrate development. Nat Rev 11:331–344

Hallahan AR, Pritchard JI, Hansen S, Benson M, Stoeck J, Hatton BA, Russell TL, Ellenbogen RG, Bernstein ID, Beachy PA, Olson JM (2004) The SmoA1 mouse model reveals that notch signaling is critical for the growth and survival of sonic hedgehog-induced medulloblastomas. Cancer Res 64:7794–7800

Han YG, Spassky N, Romaguera-Ros M, Garcia-Verdugo JM, Aguilar A, Schneider-Maunoury S, Alvarez-Buylla A (2008) Hedgehog signaling and primary cilia are required for the formation of adult neural stem cells. Nat Neurosci 11:277–284

Han YG, Kim HJ, Dlugosz AA, Ellison DW, Gilbertson RJ, Alvarez-Buylla A (2009) Dual and opposing roles of primary cilia in medulloblastoma development. Nat Med 15:1062–1065

Hatten ME, Roussel MF (2011) Development and cancer of the cerebellum. Trends Neurosci 34:134–142

Hawkins CE, Bartels U, Bouffet E (2011) Molecular genetic approaches and potential new therapeutic strategies for pediatric diffuse intrinsic pontine glioma. J Clin Oncol 29:3956–3957

Haycraft CJ, Banizs B, Aydin-Son Y, Zhang Q, Michaud EJ, Yoder BK (2005) Gli2 and Gli3 localize to cilia and require the intraflagellar transport protein polaris for processing and function. PLoS Genet 1:e53

Huang P, Schier AF (2009) Dampened Hedgehog signaling but normal Wnt signaling in zebrafish without cilia. Development 136:3089–3098

Huangfu D, Anderson KV (2005) Cilia and Hedgehog responsiveness in the mouse. Proc Natl Acad Sci USA 102:11325–11330

Huangfu D, Liu A, Rakeman AS, Murcia NS, Niswander L, Anderson KV (2003) Hedgehog signalling in the mouse requires intraflagellar transport proteins. Nature 426:83–87

Humke EW, Dorn KV, Milenkovic L, Scott MP, Rohatgi R (2010) The output of Hedgehog signaling is controlled by the dynamic association between Suppressor of Fused and the Gli proteins. Genes Dev 24:670–682

Jacoby M, Cox JJ, Gayral S, Hampshire DJ, Ayub M, Blockmans M, Pernot E, Kisseleva MV, Compere P, Schiffmann SN, Gergely F, Riley JH, Perez-Morga D, Woods CG, Schurmans S (2009) INPP5E mutations cause primary cilium signaling defects, ciliary instability and ciliopathies in human and mouse. Nat Genet 41:1027–1031

Jia J, Kolterud A, Zeng H, Hoover A, Teglund S, Toftgard R, Liu A (2009) Suppressor of Fused inhibits mammalian Hedgehog signaling in the absence of cilia. Dev Biol 330:452–460

Johnson RL, Rothman AL, Xie J, Goodrich LV, Bare JW, Bonifas JM, Quinn AG, Myers RM, Cox DR, Epstein EH Jr, Scott MP (1996) Human homolog of patched, a candidate gene for the basal cell nevus syndrome. Science (NY) 272:1668–1671

Jonassen JA, San Agustin J, Follit JA, Pazour GJ (2008) Deletion of IFT20 in the mouse kidney causes misorientation of the mitotic spindle and cystic kidney disease. J Cell Biol 183:377–384

Jones C, Roper VC, Foucher I, Qian D, Banizs B, Petit C, Yoder BK, Chen P (2008) Ciliary proteins link basal body polarization to planar cell polarity regulation. Nat Genet 40:69–77

Kawauchi D, Robinson G, Uziel T, Gibson P, Rehg J, Gao C, Finkelstein D, Qu C, Pounds S, Ellison DW, Gilbertson RJ, Roussel MF (2012) A mouse model of the most aggressive subgroup of human medulloblastoma. Cancer Cell 21:168–180

Kim J, Lee JE, Heynen-Genel S, Suyama E, Ono K, Lee K, Ideker T, Aza-Blanc P, Gleeson JG (2010a) Functional genomic screen for modulators of ciliogenesis and cilium length. Nature 464:1048–1051

Kim SK, Shindo A, Park TJ, Oh EC, Ghosh S, Gray RS, Lewis RA, Johnson CA, Attie-Bittach T, Katsanis N, Wallingford JB (2010b) Planar cell polarity acts through septins to control collective cell movement and ciliogenesis. Science (NY) 329:1337–1340

Kim J, Dabiri S, Seeley ES (2011a) Primary cilium depletion typifies cutaneous melanoma in situ and malignant melanoma. PLoS One 6:e27410

Kim S, Zaghloul NA, Bubenshchikova E, Oh EC, Rankin S, Katsanis N, Obara T, Tsiokas L (2011b) Nde1-mediated inhibition of ciliogenesis affects cell cycle re-entry. Nat Cell Biol 13:351–360

Kim Y, Kim KH, Lee J, Lee YA, Kim M, Lee SJ, Park K, Yang H, Jin J, Joo KM, Nam DH (2012) Wnt activation is implicated in glioblastoma radioresistance. Lab Invest 92:466–473

Kisseleva MV, Cao L, Majerus PW (2002) Phosphoinositide-specific inositol polyphosphate 5-phosphatase IV inhibits Akt/protein kinase B phosphorylation and leads to apoptotic cell death. J Biol Chem 277:6266–6272

Knodler A, Feng S, Zhang J, Zhang X, Das A, Peranen J, Guo W (2010) Coordination of Rab8 and Rab11 in primary ciliogenesis. Proc Natl Acad Sci USA 107:6346–6351

Ko HW, Norman RX, Tran J, Fuller KP, Fukuda M, Eggenschwiler JT (2010) Broad-minded links cell cycle-related kinase to cilia assembly and hedgehog signal transduction. Dev Cell 18:237–247

Kool M, Koster J, Bunt J, Hasselt NE, Lakeman A, van Sluis P, Troost D, Meeteren NS, Caron HN, Cloos J, Mrsic A, Ylstra B, Grajkowska W, Hartmann W, Pietsch T, Ellison D, Clifford SC, Versteeg R (2008) Integrated genomics identifies five medulloblastoma subtypes with distinct genetic profiles, pathway signatures and clinicopathological features. PLoS One 3:e3088

Kopan R, Ilagan MX (2009) The canonical Notch signaling pathway: unfolding the activation mechanism. Cell 137:216–233

Lam CW, Xie J, To KF, Ng HK, Lee KC, Yuen NW, Lim PL, Chan LY, Tong SF, McCormick F (1999) A frequent activated smoothened mutation in sporadic basal cell carcinomas. Oncogene 18:833–836

Li A, Saito M, Chuang JZ, Tseng YY, Dedesma C, Tomizawa K, Kaitsuka T, Sung CH (2011) Ciliary transition zone activation of phosphorylated Tctex-1 controls ciliary resorption, S-phase entry and fate of neural progenitors. Nat Cell Biol 13:402–411

Lin F, Hiesberger T, Cordes K, Sinclair AM, Goldstein LS, Somlo S, Igarashi P (2003) Kidney-specific inactivation of the KIF3A subunit of kinesin-II inhibits renal ciliogenesis and produces polycystic kidney disease. Proc Natl Acad Sci USA 100:5286–5291

Liu A, Wang B, Niswander LA (2005) Mouse intraflagellar transport proteins regulate both the activator and repressor functions of Gli transcription factors. Development 132:3103–3111

Logan CY, Nusse R (2004) The Wnt signaling pathway in development and disease. Annu Rev Cell Dev Biol 20:781–810

Ma R, Li WP, Rundle D, Kong J, Akbarali HI, Tsiokas L (2005) PKD2 functions as an epidermal growth factor-activated plasma membrane channel. Mol Cell Biol 25:8285–8298

Mao J, Ligon KL, Rakhlin EY, Thayer SP, Bronson RT, Rowitch D, McMahon AP (2006) A novel somatic mouse model to survey tumorigenic potential applied to the Hedgehog pathway. Cancer Res 66:10171–10178

May SR, Ashique AM, Karlen M, Wang B, Shen Y, Zarbalis K, Reiter J, Ericson J, Peterson AS (2005) Loss of the retrograde motor for IFT disrupts localization of Smo to cilia and prevents the expression of both activator and repressor functions of Gli. Dev Biol 287:378–389

Monje M, Mitra SS, Freret ME, Raveh TB, Kim J, Masek M, Attema JL, Li G, Haddix T, Edwards MS, Fisher PG, Weissman IL, Rowitch DH, Vogel H, Wong AJ, Beachy PA (2011) Hedgehog-responsive candidate cell of origin for diffuse intrinsic pontine glioma. Proc Natl Acad Sci USA 108:4453–4458

Moser JJ, Fritzler MJ, Rattner JB (2009) Primary ciliogenesis defects are associated with human astrocytoma/glioblastoma cells. BMC Cancer 9:448

Nachury MV, Loktev AV, Zhang Q, Westlake CJ, Peranen J, Merdes A, Slusarski DC, Scheller RH, Bazan JF, Sheffield VC, Jackson PK (2007) A core complex of BBS proteins cooperates with the GTPase Rab8 to promote ciliary membrane biogenesis. Cell 129:1201–1213

Northcott PA, Korshunov A, Witt H, Hielscher T, Eberhart CG, Mack S, Bouffet E, Clifford SC, Hawkins CE, French P, Rutka JT, Pfister S, Taylor MD (2011) Medulloblastoma comprises four distinct molecular variants. J Clin Oncol 29:1408–1414

Ocbina PJ, Tuson M, Anderson KV (2009) Primary cilia are not required for normal canonical Wnt signaling in the mouse embryo. PLoS One 4:e6839

Parsons DW et al (2008) An integrated genomic analysis of human glioblastoma multiforme. Science (NY) 321:1807–1812

Pasca di Magliano M, Sekine S, Ermilov A, Ferris J, Dlugosz AA, Hebrok M (2006) Hedgehog/Ras interactions regulate early stages of pancreatic cancer. Genes Dev 20:3161–3173

Paugh BS, Broniscer A, Qu C, Miller CP, Zhang J, Tatevossian RG, Olson JM, Geyer JR, Chi SN, da Silva NS, Onar-Thomas A, Baker JN, Gajjar A, Ellison DW, Baker SJ (2011) Genome-wide analyses identify recurrent amplifications of receptor tyrosine kinases and cell-cycle regulatory genes in diffuse intrinsic pontine glioma. J Clin Oncol 29:3999–4006

Pei Y, Moore CE, Wang J, Tewari AK, Eroshkin A, Cho YJ, Witt H, Korshunov A, Read TA, Sun JL, Schmitt EM, Miller CR, Buckley AF, McLendon RE, Westbrook TF, Northcott PA, Taylor MD, Pfister SM, Febbo PG, Wechsler-Reya RJ (2012) An animal model of MYC-driven medulloblastoma. Cancer Cell 21:155–167

Pugacheva EN, Jablonski SA, Hartman TR, Henske EP, Golemis EA (2007) HEF1-dependent Aurora A activation induces disassembly of the primary cilium. Cell 129:1351–1363

Rieder CL, Jensen CG, Jensen LC (1979) The resorption of primary cilia during mitosis in a vertebrate (PtK1) cell line. J Ultrastruct Res 68:173–185

Roessler E, Ermilov AN, Grange DK, Wang A, Grachtchouk M, Dlugosz AA, Muenke M (2005) A previously unidentified amino-terminal domain regulates transcriptional activity of wild-type and disease-associated human GLI2. Hum Mol Genet 14:2181–2188

Rohatgi R, Milenkovic L, Scott MP (2007) Patched1 regulates hedgehog signaling at the primary cilium. Science (NY) 317:372–376

Rosenbaum JL, Witman GB (2002) Intraflagellar transport. Nat Rev Mol Cell Biol 3:813–825

Ross AJ et al (2005) Disruption of Bardet-Biedl syndrome ciliary proteins perturbs planar cell polarity in vertebrates. Nat Genet 37:1135–1140

Rossi M, Magnoni L, Miracco C, Mori E, Tosi P, Pirtoli L, Tini P, Oliveri G, Cosci E, Bakker A (2011) Beta-catenin and Gli1 are prognostic markers in glioblastoma. Cancer Biol Ther 11:753–761

Schneider L, Clement CA, Teilmann SC, Pazour GJ, Hoffmann EK, Satir P, Christensen ST (2005) PDGFRalphaalpha signaling is regulated through the primary cilium in fibroblasts. Curr Biol 15:1861–1866

Schneider L, Stock CM, Dieterich P, Jensen BH, Pedersen LB, Satir P, Schwab A, Christensen ST, Pedersen SF (2009) The Na+/H+ exchanger NHE1 is required for directional migration stimulated via PDGFR-alpha in the primary cilium. J Cell Biol 185:163–176

Schraml P, Frew IJ, Thoma CR, Boysen G, Struckmann K, Krek W, Moch H (2009) Sporadic clear cell renal cell carcinoma but not the papillary type is characterized by severely reduced frequency of primary cilia. Mod Pathol 22:31–36

Schuller U, Heine VM, Mao J, Kho AT, Dillon AK, Han YG, Huillard E, Sun T, Ligon AH, Qian Y, Ma Q, Alvarez-Buylla A, McMahon AP, Rowitch DH, Ligon KL (2008) Acquisition of granule neuron precursor identity is a critical determinant of progenitor cell competence to form Shh-induced medulloblastoma. Cancer Cell 14:123–134

Seeley ES, Carriere C, Goetze T, Longnecker DS, Korc M (2009) Pancreatic cancer and precursor pancreatic intraepithelial neoplasia lesions are devoid of primary cilia. Cancer Res 69:422–430

Simons M, Gloy J, Ganner A, Bullerkotte A, Bashkurov M, Kronig C, Schermer B, Benzing T, Cabello OA, Jenny A, Mlodzik M, Polok B, Driever W, Obara T, Walz G (2005) Inversin, the

gene product mutated in nephronophthisis type II, functions as a molecular switch between Wnt signaling pathways. Nat Genet 37:537–543
Singh SK, Hawkins C, Clarke ID, Squire JA, Bayani J, Hide T, Henkelman RM, Cusimano MD, Dirks PB (2004) Identification of human brain tumour initiating cells. Nature 432:396–401
Spassky N, Han YG, Aguilar A, Strehl L, Besse L, Laclef C, Ros MR, Garcia-Verdugo JM, Alvarez-Buylla A (2008) Primary cilia are required for cerebellar development and Shh-dependent expansion of progenitor pool. Dev Biol 317:246–259
Spektor A, Tsang WY, Khoo D, Dynlacht BD (2007) Cep97 and CP110 suppress a cilia assembly program. Cell 130:678–690
Takebe N, Harris PJ, Warren RQ, Ivy SP (2011) Targeting cancer stem cells by inhibiting Wnt, Notch, and Hedgehog pathways. Nat Rev Clin Oncol 8:97–106
Taylor MD, Northcott PA, Korshunov A, Remke M, Cho YJ, Clifford SC, Eberhart CG, Parsons DW, Rutkowski S, Gajjar A, Ellison DW, Lichter P, Gilbertson RJ, Pomeroy SL, Kool M, Pfister SM (2012) Molecular subgroups of medulloblastoma: the current consensus. Acta Neuropathol 123(4):465–472
TCGA (2008) Comprehensive genomic characterization defines human glioblastoma genes and core pathways. Nature 455:1061–1068
Teilmann SC, Christensen ST (2005) Localization of the angiopoietin receptors Tie-1 and Tie-2 on the primary cilia in the female reproductive organs. Cell Biol Int 29:340–346
Thomas J, Morle L, Soulavie F, Laurencon A, Sagnol S, Durand B (2010) Transcriptional control of genes involved in ciliogenesis: a first step in making cilia. Biol Cell 102:499–513
Thompson MC, Fuller C, Hogg TL, Dalton J, Finkelstein D, Lau CC, Chintagumpala M, Adesina A, Ashley DM, Kellie SJ, Taylor MD, Curran T, Gajjar A, Gilbertson RJ (2006) Genomics identifies medulloblastoma subgroups that are enriched for specific genetic alterations. J Clin Oncol 24:1924–1931
Tran PV, Haycraft CJ, Besschetnova TY, Turbe-Doan A, Stottmann RW, Herron BJ, Chesebro AL, Qiu H, Scherz PJ, Shah JV, Yoder BK, Beier DR (2008) THM1 negatively modulates mouse sonic hedgehog signal transduction and affects retrograde intraflagellar transport in cilia. Nat Genet 40:403–410
Tucker RW, Pardee AB, Fujiwara K (1979) Centriole ciliation is related to quiescence and DNA synthesis in 3 T3 cells. Cell 17:527–535
Tukachinsky H, Lopez LV, Salic A (2010) A mechanism for vertebrate Hedgehog signaling: recruitment to cilia and dissociation of SuFu-Gli protein complexes. J Cell Biol 191:415–428
Verhaak RG et al (2010) Integrated genomic analysis identifies clinically relevant subtypes of glioblastoma characterized by abnormalities in PDGFRA, IDH1, EGFR, and NF1. Cancer Cell 17:98–110
Vitucci M, Hayes DN, Miller CR (2011) Gene expression profiling of gliomas: merging genomic and histopathological classification for personalised therapy. Br J Cancer 104:545–553
Wallace VA (1999) Purkinje-cell-derived Sonic Hedgehog regulates granule neuron precursor cell proliferation in the developing mouse cerebellum. Curr Biol 9:445–448
Wallingford JB (2010) Planar cell polarity signaling, cilia and polarized ciliary beating. Curr Opin Cell Biol 22:597–604
Wallingford JB, Mitchell B (2011) Strange as it may seem: the many links between Wnt signaling, planar cell polarity, and cilia. Genes Dev 25:201–213
Wang G, Krishnamurthy K, Bieberich E (2009) Regulation of primary cilia formation by ceramide. J Lipid Res 50:2103–2110
Wechsler-Reya RJ, Scott MP (1999) Control of neuronal precursor proliferation in the cerebellum by Sonic Hedgehog. Neuron 22:103–114
Wheatley DN (1995) Primary cilia in normal and pathological tissues. Pathobiology: J Immunopathol Molecul Cell Biol 63:222–238
Willemarck N, Rysman E, Brusselmans K, Van Imschoot G, Vanderhoydonc F, Moerloose K, Lerut E, Verhoeven G, van Roy F, Vleminckx K, Swinnen JV (2010) Aberrant activation of fatty acid synthesis suppresses primary cilium formation and distorts tissue development. Cancer Res 70:9453–9462

Wong SY, Seol AD, So PL, Ermilov AN, Bichakjian CK, Epstein EH Jr, Dlugosz AA, Reiter JF (2009) Primary cilia can both mediate and suppress Hedgehog pathway-dependent tumorigenesis. Nat Med 15:1055–1061

Wu J, Du H, Wang X, Mei C, Sieck GC, Qian Q (2009) Characterization of primary cilia in human airway smooth muscle cells. Chest 136:561–570

Yang ZJ, Ellis T, Markant SL, Read TA, Kessler JD, Bourboulas M, Schuller U, Machold R, Fishell G, Rowitch DH, Wainwright BJ, Wechsler-Reya RJ (2008) Medulloblastoma can be initiated by deletion of Patched in lineage-restricted progenitors or stem cells. Cancer Cell 14:135–145

Yauch RL, Dijkgraaf GJ, Alicke B, Januario T, Ahn CP, Holcomb T, Pujara K, Stinson J, Callahan CA, Tang T, Bazan JF, Kan Z, Seshagiri S, Hann CL, Gould SE, Low JA, Rudin CM, de Sauvage FJ (2009) Smoothened mutation confers resistance to a Hedgehog pathway inhibitor in medulloblastoma. Science (NY 326:572–574

Yuan K, Frolova N, Xie Y, Wang D, Cook L, Kwon YJ, Steg AD, Serra R, Frost AR (2010) Primary cilia are decreased in breast cancer: analysis of a collection of human breast cancer cell lines and tissues. J Histochem Cytochem 58:857–870

Yuan S, Li J, Diener DR, Choma MA, Rosenbaum JL, Sun Z (2012) Target-of-rapamycin complex 1 (Torc1) signaling modulates cilia size and function through protein synthesis regulation. Proc Natl Acad Sci USA 109:2021–2026

Zarghooni M, Bartels U, Lee E, Buczkowicz P, Morrison A, Huang A, Bouffet E, Hawkins C (2010) Whole-genome profiling of pediatric diffuse intrinsic pontine gliomas highlights platelet-derived growth factor receptor alpha and poly (ADP-ribose) polymerase as potential therapeutic targets. J Clin Oncol 28:1337–1344

Zeng H, Jia J, Liu A (2010) Coordinated translocation of mammalian Gli proteins and suppressor of fused to the primary cilium. PLoS One 5:e15900

Zhu D, Shi S, Wang H, Liao K (2009) Growth arrest induces primary-cilium formation and sensitizes IGF-1-receptor signaling during differentiation induction of 3 T3-L1 preadipocytes. J Cell Sci 122:2760–2768

Zuo X, Guo W, Lipschutz JH (2009) The exocyst protein Sec10 is necessary for primary ciliogenesis and cystogenesis in vitro. Mol Biol Cell 20:2522–2529

Chapter 9
Abnormalities of the Central Nervous System Across the Ciliopathy Spectrum

Kate Baker and Philip L. Beales

Contents

K. Baker (✉)
Department of Medical Genetics, University of Cambridge, Cambridge, UK
e-mail: kb488@cam.ac.uk

P.L. Beales
Molecular Medicine Unit, UCL Institute of Child Health, London, UK

K.L. Tucker and T. Caspary (eds.), *Cilia and Nervous System Development and Function*,
DOI 10.1007/978-94-007-5808-7_9, © Springer Science+Business Media Dordrecht 2013

Abstract More than one hundred congenital anomaly syndromes have now been attributed to structural or functional disruption of the primary immotile cilium (ciliopathies). Whilst most are rare, collectively the ciliopathies may affect as many as 1 in 700 of the general population. The majority of ciliopathies are multi-system disorders. Neuroanatomical, neurological, cognitive, psychiatric and behavioural features (central nervous system phenotypes) are frequently but not universally observed. This chapter aims to systematically collate clinical information related to central nervous system (CNS) phenotypes across the ciliopathy spectrum. There are notable similarities in the types of structural brain abnormalities and functional deficits observed across the spectrum, albeit with variations in frequency of abnormalities between syndromes (in part reflected in diagnostic criteria). Considerable variation exists within syndromes in the extent and severity of CNS abnormalities. Underlying genetic and developmental mechanisms are emerging, and it is anticipated that integration of clinical observation with interdisciplinary basic science will reap benefits in the following years, both for understanding of these complex disorders and for improved patient outcomes.

Keywords Cilia • Ciliopathy • Brain • Neurology • Intellectual disability • Autism • Joubert • Bardet Biedl • MRI

Outline

This chapter will describe common and rare central nervous system (CNS) abnormalities observed in patients with ciliopathy syndromes. CNS abnormalities that cut across syndrome boundaries and that could reflect distinct ciliary functions during normal neurodevelopment will be discussed. We wish to emphasise that clinical observations can promote scientific progress and vice-versa, and may one day lead to patient benefit.

9.1 Introduction

Approximately 2–3% of singleton new-born babies have a major congenital anomaly, which can arise from local errors in morphogenesis, deformation, disruption, teratogen exposure or germline genetic alterations. When a particular set of developmental anomalies consistently occur together, the condition is referred to as a syndrome. Of the ~7,000 phenotypes described in OMIM (McKusick-Nathans Institute of Genetic Medicine 2011), approximately one third have a neurologic component, indicating the importance of dysmorphology and accurate diagnosis in the context of neurodevelopmental disorder. It is equally important to recognise that the combination of clinical signs that make up a syndrome can have several aetiologies. A classic example is non-syndromic retinitis pigmentosa with upward of 50 genes that when mutated give rise to often indistinguishable retinal phenotypes. Within this spectrum are represented every form of inheritance pattern possible, differing ages at onset and severity.

9.1.1 Ciliopathies

During the past decade we have witnessed the rapid emergence of knowledge surrounding the role of primary cilia in cellular sensing, signalling, human disease and development. This sessile appendage and its associated structures (basal body, centrosome) have been the focus of several recent protein inventories indicating that over 1,500 proteins are involved with both structure and function (Inglis 2006). This in turn has provided a rich resource for the identification of ciliopathy-related genes many of which are fundamentally important during embryogenesis. In spite of this plethora of information, we are only seeing the tip of the cilia iceberg as it emerges that many of the ciliome components assemble in to functional complexes and that the same proteins have possibly separate cytosolic functions including gate-keeping roles and vesicular transport (Sang et al. 2011).

Given the emerging level of complexity surrounding both ciliogenesis and the function of the primary cilium for cellular activities throughout the body, it is hardly surprising that dysfunction can have devastating effects on a large variety of tissues and organs. The pleiotropic presentation of the ciliopathies probably reflects the extent of underlying pathogenetic protein expression and/or the interactions with complex partners. The phenotypic manifestations range from isolated kidney disease or blindness to multi-organ involvement typified by the Bardet-Biedl syndrome (BBS) or Meckel syndrome (MKS). Amongst the hallmark features of ciliopathies are; retinitis pigmentosa (RP), cystic kidney disease, polydactyly, situs inversus, cognitive impairments, specific brain malformations, hepatic disease, deafness, skeletal anomalies and olfactory disturbances. In fact, these core presentations have been valuable for prediction of other ciliopathies, many of which have no prior known aetiology (Baker and Beales 2009). It has now been predicted that over 100 such conditions exist and whilst most are rare, collectively the ciliopathies may present in as many as 1 in 700 of the general population.

In addition to phenotypic variability, the ciliopathies also display extensive genetic heterogeneity. Mutations in a wide range of genes can give rise to similar phenotypes such as BBS, or conversely, mutations in a single gene can underlie several phenotypes with differing severity. For example, *CEP290* gene mutations are associated with organ-specific diseases such as RP or Leber's Congenital Amaurosis (LCA) as well as Senior-Löken syndrome (SLSN), Joubert syndrome, MKS and BBS. There is no pattern or clustering of the 112 mutations described so far in *CEP290* that might explain this phenomenon. Contrast this with the situation for *OFD1* where mutations residing before amino acid residue 631 that truncate the protein upstream of the fifth coiled-coil domain give rise to the male-lethal Oral

Facial Digital Type 1 syndrome, whilst mutations beyond residue 631 lead to the less severe Simpson-Golabi-Behmel type 2 syndrome or Joubert Syndrome.

9.1.2 CNS Abnormalities as Core Ciliopathic Features

Amongst the known and predicted ciliopathies, CNS system anomalies are present in over half of the syndromes described (Baker and Beales 2009). The core features that are associated with a ciliopathy include structural brain abnormalities such as hydrocephalus, hypoplasia or agenesis of the corpus callosum, cerebellar vermis hypoplasia, Dandy-Walker malformation and posterior encephalocoele formation (Table 9.1). However, many functional CNS abnormalities observed in ciliopathy patients do not have a clear structural correlate i.e. developmental delay and behavioural phenotypes. For example in BBS, some level of intellectual impairment is the norm, but even in the most severe cases, there are no consistent radiologically-apparent brain changes pathognomonic for the disorder. It is believed, but not proven, that ciliary dysfunction impacts neuronal function in the cerebral cortex. It has also been shown by us and others that BBS patients have abnormalities of the hippocampus, implicating defects in neurogenesis in intellectual deficits (see below). Structure-function relationships remain indistinct even for hard signs such as oculomotor apraxia and cerebellar malformations in Joubert Syndrome.

It remains to be discovered as to why neurological anomalies are associated with some ciliopathy syndromes but not others, and why frequency and severity of neurological features varies between individuals sharing the same ciliopathy diagnosis. Elucidation and integration of genetic and protein data into cilia systems networks will eventually reveal the source of phenotypic variability, including neurological involvement, so typical of ciliopathies.

9.2 Method of Review

The following sections describe CNS abnormalities reported for each of the currently-confirmed ciliopathies. There is an extensive and growing list of additional conditions which are associated with core ciliopathic phenotypes (including CNS phenotypes), but where evidence for definite ciliary pathology is lacking. In other conditions with ciliopathy-like phenotypes, disruption to signalling cascades downstream of the immotile primary cilium have been detected e.g. Gli3-associated disorders (Biesecker 2006) and Kif7-associated disorders (Putoux et al. 2011). These two latter groups of conditions have not been included in the current review.

The diagnostic features and genetic basis for each syndrome will be briefly presented, followed by (1) clinically-apparent neurological features such as sensory

Table 9.1 Structural brain abnormalities observed across the ciliopathy spectrum

Abnormality	Definition	Gestation[a]	Pathophysiology[b]	Functional consequence	Frequent or essential for diagnosis	Moderately frequent association	Occasionally reported but rare
Encephalocele	Protrusion of brain and meninges through an abnormal opening of skull	3–4 weeks	Non-separation of neural and surface ectoderm leading to defective bone formation	Often associated with abnormalities of posterior fossa structures and severe associated neurological disability	MKS	JS	
Holoprosencephaly	Impaired midline cleavage of the embryonic forebrain, varying from complete (alobar) to partial (semilobar) to minor (eg failure of thalamic cleavage)	4–6 weeks	Various mechanisms described in non-ciliopathic conditions	Very variable depending on extent		MKS	
Dandy Walker malformation	1. cystic dilatation of the 4th ventricle and enlarged posterior fossa with upward displacement of tentorium	4–6 weeks	Atresia of 4th ventricle outflow (aperture of Magendie) and persistence of rostral membrane a/w developmental arrest of hindbrain (sequence of events not known)	May require shunting to reduce pressure and secondary risk of seizures and developmental delay		JS, MKS	JATD, EVC, PKD
	2. cerebellar vermis hypoplasia 3. hydrocephalus						

(continued)

Table 9.1 (continued)

Abnormality	Definition	Gestation[a]	Pathophysiology[b]	Functional consequence	Frequent or essential for diagnosis	Moderately frequent association	Occasionally reported but rare
Agenesis of corpus callosum	Partial or complete malformation of interhemispheric commissures	11–13 weeks (formation of commissural plate) followed by extensive period of growth. Requires migration of a population of neurones to promote later axonal projections	Lack of development of callosal plate, or cortical agenesis (global or regional loss of layer three projection neurons), or failure of commissuration, or later white matter injury.	Spectrum from asymptomatic to severe mental retardation	OFD1	JS, MKS	CED
Cerebellar vermis hypoplasia	Complete or partial absence of the midline structures of the cerebellum, contributing to Molar Tooth Sign on MRI	Identifiable by 13 weeks, followed by inward migration and proliferation of granule cells (16–25 weeks)	Failure of growth and midline fusion of V-shaped tuberculum cerebelli, and/or failure of inward migration of granule cells from midbrain-hindbrain isthmus	Oculumotor and other apraxias. Volumetric reductions associated with non-syndromic autism and dyslexia in some studies	JS	OFD1, MKS	BBS, LCA

Abnormality	Definition	Gestation[a]	Pathophysiology[b]	Functional consequence	Frequent or essential for diagnosis	Moderately frequent association	Occasionally reported but rare
Heterotopia	Subependymal nodules of primitive neuronal tissue, often located in dorsolateral periventricular zones	12–24 weeks	Failure at onset of neuronal migration from ventricular zone	Often associated with seizure disorders and intellectual disability		JS, OFD1	
Polymicrogyria	Multiple thin small gyri with loss of layer V neurons	12–24 weeks	Abnormality of late phase of neuronal migration and start of cortical organisation	Often associated with seizure disorders and intellectual disability		JS, MKS	
Hippocampal dysgenesis	Abnormality of limbic lobe structure (precise cellular pathology not yet defined)	Extended process from late fetal structural organisation to postnatal proliferation and network organisation	Dysregulated neurogenesis and/or integration of neuronal progenitors into complex local and regional networks	May influence learning, memory and other higher cognitive functions (evidence in ciliopathies currently lacking)	BBS, JS		
Microcephaly	Reduction in cerebral volume, apparent either as reduced orbitofrontal circumference or brain tissue loss on imaging	Protracted	Likely diverse – abnormal prolifera-tion and organisation, or atrophy of brain tissue (grey and/or white matter)	Often associated with developmental delay/ID but unpredictable severity		OFD1, JS, BBS, CED	LCA

(continued)

Table 9.1 (continued)

Abnormality	Definition	Gestation[a]	Pathophysiology[b]	Functional consequence	Frequent or essential for diagnosis	Moderately frequent association	Occasionally reported but rare
Absent or reduced white matter decussations	Failure of formation of normal midline-crossing axonal structures (partial decussation is the normal situation)	Protracted	Failure of axonal growth cone pathfinding, or abnormalities of nuclei from which axons originate, or abnormality of chemotactic axon guidance cues	Where complete is presumed to underlie mirror movements	JS (superior cerebellar peduncles)	JS (corticospinal tracts)	
Hydrocephalus	Increased volume of CSF spaces varying from minor ventriculomegaly to extensive	Protracted	Heterogeneous	Dependent on whether normal or high pressure, and whether reflecting underlying cerebral dysgenesis		OFD1, BBS, JS	JATD
Intracerebral cysts including porencephaly or arachnoid cysts	Abnormal fluid-filled cavities in continuation with CSF circulation	Various	Can be developmental or destructive	Usually asymptomatic		OFD1, BBS, PKD	

[a]At which normal structure develops and/or abnormality occurs
[b]Unconfirmed for majority of features

abnormalities, seizures and motor abnormalities, (2) neurodevelopmental impairments such as developmental delay, learning disabilities, behavioural phenotypes and psychiatric illness, (3) neuroradiological features (qualitative and quantitative structural abnormalities) and (4) post-mortem neuropathology. To obtain this information on a semi-systematic basis, OMIM and GeneReviews entries for each syndrome were initially reviewed, followed by PubMed and bibliography literature searches.

There are several limitations to the existing published data which restrict comprehensive review. Clinical details for all diagnosed cases are not available, and published clinical descriptions are not always consistent in the use of terminology or in methods of investigation (notes review, questionnaire survey or direct examination). Publication biases exist, toward describing core diagnostic features for each syndrome and ignoring or not recognising non-core associations. There may be additional bias toward reporting exceptionally unusual clinical presentations.

Where syndromes are defined by, and cases ascertained for, non-CNS features (for example skeletal, retinal or renal abnormalities) neurodevelopmental assessment and related investigations such as neuroimaging may not have been carried out. For conditions that are lethal in early life, neurodevelopmental features may never become apparent. CNS assessment may not be a priority in the context of severe and life-limiting medical illness. For less severe conditions, it can be argued that neuroradiological investigations are not clinically indicated in the absence of focal neurological signs, epilepsy or regression. Indications for neuroimaging in the context of intellectual disability (ID) are controversial. Severe ID may preclude investigation without sedation or anaesthesia, and where a clear diagnosis has been achieved, imaging is unlikely to influence medical management or outcomes. Since there has been little systematic research of CNS abnormalities in these rare conditions, it is likely that the spectrum and frequency of CNS features has been incompletely described.

Interpretation of the existing clinical and research literature requires consideration of possible confounding factors, whereby non-CNS aspects of the ciliopathy syndrome may have an indirect impact on aspects of neurodevelopment. Importantly, visual impairment can confound assessment of developmental delay (motoric, cognitive and social), so requires specialist skills and methods. The impact of chronic medical illness, for example renal failure or cardiomyopathy, should also be taken into account. Several ciliopathy syndromes are associated with obesity and endocrine dysfunction, which may have indirect effects on brain development and function via metabolic and cardiovascular dysregulation. Interpretation of behavioural literature, for example on autistic spectrum disorders, attention problems and anxiety, should take into account the multiple physical, sensory and psychosocial challenges encountered by patients with ciliopathy syndromes and their families.

9.3 CNS-Predominant Ciliopathies

9.3.1 Joubert Syndrome and Joubert-Related Spectrum Disorders (JS/JRSD)

Joubert Syndrome (JS) is a ciliopathy defined according to specific neurological and neuroradiological features. No formal diagnostic criteria have been established, however it is generally agreed that diagnosis requires a radiological feature known as the Molar Tooth Sign (MTS, defined below), alongside clinical features of hypotonia, disordered breathing patterns during early life, abnormal eye moments and developmental delay. JS was initially considered an isolated CNS disorder, but recognition of associations with extra-CNS ciliopathy features (renal, ophthalmic, skeletal) has broadened the categorisation to include several Joubert-related conditions (JS-related spectrum disorders, JSRD). Features of JS in association with coloboma and hepatic fibrosis is referred to as COACH (Cerebellar vermis hypoplasia/aplasia, Oligophrenia, Ataxia, Coloboma, and Hepatic fibrosis), whilst association with ocular and renal signs is termed CORS (CerebelloOculoRenal Syndrome encompassing several previously defined conditions e.g. Arima syndrome). Phenotypic classifications into distinct JSRD sub-types have been proposed (Valente et al. 2008; Zaki et al. 2008), but it remains difficult to categorise many cases, therefore a spectrum rather than discrete conditions may be more realistic (Sattar and Gleeson 2011). CNS features characterising JS and JSRD will be considered collectively here.

Currently 14 genes have been associated with JS and JRSD, including genes previously associated with other CNS-phenotype ciliopathies (eg *MKS3, OFD1*) and genes previously associated with non-CNS isolated ciliopathies (eg *NPHP1, RPGRIP*) (reviewed by Sattar and Gleeson 2011). There is evidence for broad genotype-phenotype correlations, in that cerebello-renal subtypes and oculo-renal subtypes are more likely to be associated with genotypes that can also cause the relevant non-CNS isolated phenotypes (Parisi 2009). For example, *TMEM67* mutations are found at relatively high frequency in JSRD patients with hepatic fibrosis (Iannicelli et al. 2010). It has been argued that *AH1* is the JS genotype most likely to be associated with a CNS-only phenotype (Valente et al. 2006). Mutations in *CEP290*, on the other hand, seem capable of causing a very diverse range of ciliopathic disruptions including and in addition to JS, with no apparent mutation-to-phenotype predictability (Coppieters et al. 2010b). It appears likely that ascertainment of JS cases has been biased toward more severe cognitive and neurological abnormality, since siblings of severely affected cases have been observed to carry minor abnormalities such as isolated DWM without developmental delay which would not have otherwise come to diagnostic attention or been considered as part of the JS spectrum eg (Gunay-Aygun et al. 2009). Diverse presentations of ciliopathy within single families emphasises this complexity (Iannicelli et al. 2010; Zaki et al. 2011). A molecular-first approach to classification is therefore of limited clinical utility, because of variable penetrance for each CNS and non-CNS phenotype, and low predictive value of mutations either between or within families.

Structure-function relationships are explicable for some but not all of the key CNS features of JS. Moreover, the requirement for an imaging-based diagnostic criterion has facilitated collection of valuable imaging datasets for exploration of the wider structural and functional phenotype. The embryological and molecular origins of observed neuropathology and neuroradiological abnormalities are not fully understood and the primary developmental defects underlying regionally-specific CNS features remain unconfirmed.

9.3.1.1 Neurology

Neonatal hypotonia is a universal feature of JS, but is a very non-specific observation not strongly suggestive of a ciliopathic cause or JS. Apnoea and hyperpnoea in infancy are highly characteristic, although this feature usually improves with age therefore may only be contributory to diagnosis in retrospect depending on severity of other features. In an early series of 61cases, hyperpnoea was reported in 55% (with exacerbations during illness or excitement) and apnoea in 60% (Maria et al. 1999). Seizures were observed in 4% cases in the same study.

Eye movement abnormalities (oculomotor apraxia) in JS are diverse and include pendular or see-saw nystagmus, lack of visual orienting, inconsistent smooth pursuit, inability to initiate saccades or to relate saccades to a target. Perhaps surprisingly, only mild reduction in visual acuity is present in the majority of patients, where testing possible (Weiss et al. 2009). Related signs include compensatory head-thrusting, whilst the vestibulo-ocular reflex is usually intact (Khan et al. 2008). Additional ocular features are seen in some JS cases, most notably dystrophic retinal appearance but normal electroretinogram in around 20% of affected individuals (Maria et al. 1999; Khan et al. 2008). Ptosis and ophthalmic nerve palsies, and other cranial nerve abnormalities have also been reported (Burt et al. 2012). Interestingly, Khan et al. demonstrated asymmetric visual evoked potentials in all cases investigated. This asymmetry could be suggestive of abnormal decussation of the optic nerves, although this has not been directly visualised on neuroimaging to date. Abnormal chiasmatic decussation has been associated with see-saw nystagmus in other contexts.

Non-ocular motor control abnormalities are also characteristic of JS, although less systematically documented. Features include truncal ataxia, affecting acquisition of motor milestones and gait. Rhythmic, protruding tongue movements that may lead to tongue hypertrophy were reported in 45% by Maria et al. (1999). An interesting feature occurring at unknown frequency is the presence of mirror movements, where hands move synchronously instead of independently. This feature was particularly remarked-upon as being a feature of the first cases in whom *AHI1* mutations were identified (Ferland et al. 2004). This unusual motor abnormality is presumed to reflect lack of decussation of the pyramidal (corticospinal) tracts, although functional neuroimaging or electrophysiological evidence for this association has not been reported. More subtle neurobehavioural abnormalities reflecting abnormal interhemispheric connections, which could potentially include perceptual and cognitive disturbance, have not been investigated to date.

9.3.1.2 Neurodevelopment

Developmental delay and intellectual impairment are probably universal in JS, but the severity and extent of impairments is very variable. The nature, frequency and severity of impairments, and mechanisms underlying these aspects of the phenotype, have yet to be fully established.

Maria et al. (1999) reported that all patients had severe developmental delay. Seventy three percent could sit independently, and 50% walked independently (at average age 4 years). Only 60% acquired any verbal language. An independent follow-up study of 18 cases confirmed that motor development is affected in all cases (only 12 individuals were independently mobile at the time of follow-up, and age at which walking achieved was delayed in all ambulant cases, varying between 22 months and 10 years). However in contrast to the earlier study, one third of cases were attending mainstream education, with only mild or moderate impairments in communication and cognitive function (Hodgkins et al. 2004).

One potential explanation for the discrepancy between these studies in the severity of non-motor cognitive impairment could be that later-diagnosed cases are in general less severe than the first-reported JS cases. Another possibility however is that early estimates of cognitive function were hampered by visuomotor, visual and motor impairments. Braddock et al. (2006) objectively evaluated speech and language abilities and found some evidence for this methodological bias. Speech apraxia was a dominant feature, with severe difficulties in tongue movements and alternating lip/tongue sequences required for speech production. Expressive abilities were found to be much more severely impaired than receptive abilities which were only mildly impaired in most cases. Since receptive language abilities are closely correlated with general intelligence in the general population, historic assessments based primarily on observations and parental reporting may have led to underestimation of cognitive ability.

An association between JS and autism remains controversial. Given that visual fixation and control of pursuit are universally impaired in JS, it is not surprising that gaze-dependent social developmental interactions will be disrupted, with consequential effects for social and linguistic development. It remains unclear whether the resultant behavioural phenotype is truly autism, or a mimic thereof that has different origins and different consequences for social relationships and personality development. Autism was first reported in two JS cases by Holroyd et al. (1991). Ozonoff et al. (1999) investigated 11 JS cases at average age 7 years (seven cases being nonverbal at time of assessment), using standard diagnostic instruments (with unconfirmed validity for making diagnoses in children with both severe developmental delay and sensory impairments). In 4/11 cases an autism-spectrum disorder was diagnosed, but the two assessment tools did not yield convergent diagnoses in three of these four cases and, importantly, diagnostic assessments were not in agreement with parental reports of behavioural abilities and concerns.

Which aspects of the ASD triad (impaired language development, abnormalities of social interaction, and repetitive and stereotyped behaviours) may be most affected in JS? One of the two original cases reported by Holroyd was said to demonstrate repetitive and stereotyped behaviours but not deficits in reciprocal social interaction. Ozonoff et al. reported that stereotypies and restricted interests in addition to language delay were common amongst the total sample, whereas the children who were diagnosed with an ASD also demonstrated limited emotional responses and enjoyment of social interactions. Hence JS appears to be more strongly associated with repetitive and stereotyped behaviours than with social deficits. Braddock et al. (2006) emphasise that communication in JS is not typically autistic, in that children with JS use gesture to overcome limitations in verbal communication, especially when expressive ability is poor. Investigation of more subtle autistic phenomena, for example poor empathy and disturbance to theory of mind, has not yet been attempted.

Other behavioural difficulties reported in JS include temper tantrums, inattention and overactivity (Farmer et al. 2006). The extent to which these behaviours may be an understandable consequence of frustration due to motor control, visuomotor and communication impairments has not been explored. However, it is also possible that cognitive and behavioural features of impaired attention and inhibitory control could directly reflect focal pathology or disturbed connectivity.

In summary, the diversity and type of cognitive and behavioural abnormalities observed in JS is notable and surprising given the relatively isolated, subcortical neuroradiological and neuropathological features. Three non-exclusive explanations may be worth exploring. One possibility is that cerebellar vermis dysgenesis and resultant disruption to distributed cerebellar-cerebral circuitry has extensive developmental consequences for higher-cognitive functions, dispelling the previous view that the cerebellum is primarily a motor control centre. A second, linked possibility is that the cerebellum and associated network function are primarily involved in motor control, but that motor control is itself necessary for the emergence of higher order cognitive functions. Articulatory-sensory matching are recognised to be important for the development of language function, hence individuals with speech dyspraxia due to focal pathology can demonstrate associated linguistic impairments (Vargha-Khadem et al. 2005). It has been proposed that a more general deficit in matching observed and experienced behaviours (dependent on mirror neurons) could underlie abnormalities of social cognition, and the cerebellum could play a role within this system (Kana et al. 2011). A third possibility could be that cognitive and behavioural aspects of JS are independent of cerebellar vermis hypoplasia, and reflective of other structural abnormalities (hippocampal and temporal lobe disruptions, or disturbed white matter connectivity, or other abnormalities not yet described). Given interest in the role of the cerebellum (and vermis in particular) in autism-relevant cognitive functions in the general population, JS has been viewed as an important aetiologically-homogeneous population for investigation of developmental pathophysiology underlying an autistic phenotype. Association has been sought between *AHI1* variants and non-syndromic Autism Spectrum Disorders, with some evidence that common variation in this gene is associated with elevated risk of an ASD (Alvarez Retuerto et al. 2008).

9.3.1.3 Neuroradiology

The Molar Tooth Sign (MTS described by Quisling et al. 1999) is a neuroradiological observation required for diagnosis of JS. The MTS reflects the convergence of several co-localised anatomical abnormalities: a deepened interpeduncular fossa (reflecting abnormality at the pontomesencephalic junction), elongated, horizontally-orientated superior cerebellar peduncles, and cerebellar vermis hypoplasia or agenesis (Fig. 9.1, reproduced with permission of Mac Keith Press). Structures contributing to the MTS are derived from the primitive isthmus, but whether there is a primary problem with cerebellar patterning, or organisation of midbrain nuclei, or migration of neuronal precursors into both sets of structures, or a primary disturbance of brainstem-cerebellar connectivity is not clear. The severity of each of the component features of the MTS is variable between individuals with JS, leading investigators to speculate about corresponding variation in clinical features of JS. Gitten et al. (1998) found no correlation between a composite posterior fossa score and developmental delay, whereas subjective severity of pontine hypoplasia was predictive of severity of eye movement abnormalities (Weiss et al. 2009). No correlations have been detected between neuroradiological features and severity of respiratory dysfunction.

Whilst the MTS is a helpful objective diagnostic marker for JS, many additional structural abnormalities have been reported. Additional collections of cerebrospinal fluid in the posterior fossa are common, and the Dandy-Walker Malformation (DWM) is a relatively frequent observation (6 of 45 patients, Maria et al. 1999). It should be noted that DWM can sometimes obscure the MTS and that after decompression of posterior fossa, re-examination for MTS should be carried out (Sartori et al. 2010). Whilst extensive cerebral abnormalities are not common in JS, perisylvian polymicrogyria and heterotopias have been noted in at least six patients (Zaki et al. 2008; Bachmann-Gagescu et al. 2012). Ventriculomegaly has recently been reported to be characteristic of JS patients carrying *CC2D2A* mutations, accounting for around 8% JS cases overall (Bachmann-Gagescu et al. 2012). Whether this observation relates to obstructive hydrocephalus or compensatory ventricular enlargement occurring alongside cerebral dysgenesis or atrophy is not clear.

Two comprehensive reviews of neuroimaging abnormalities in addition to MTS have recently been published. Senocak et al. (2010) reviewed MRI of 20 JS patients (ages at scan 18 months to 17 years). Hippocampal malformations ("globular" in shape and medially displaced) were reported for 80% of patients, with temporal

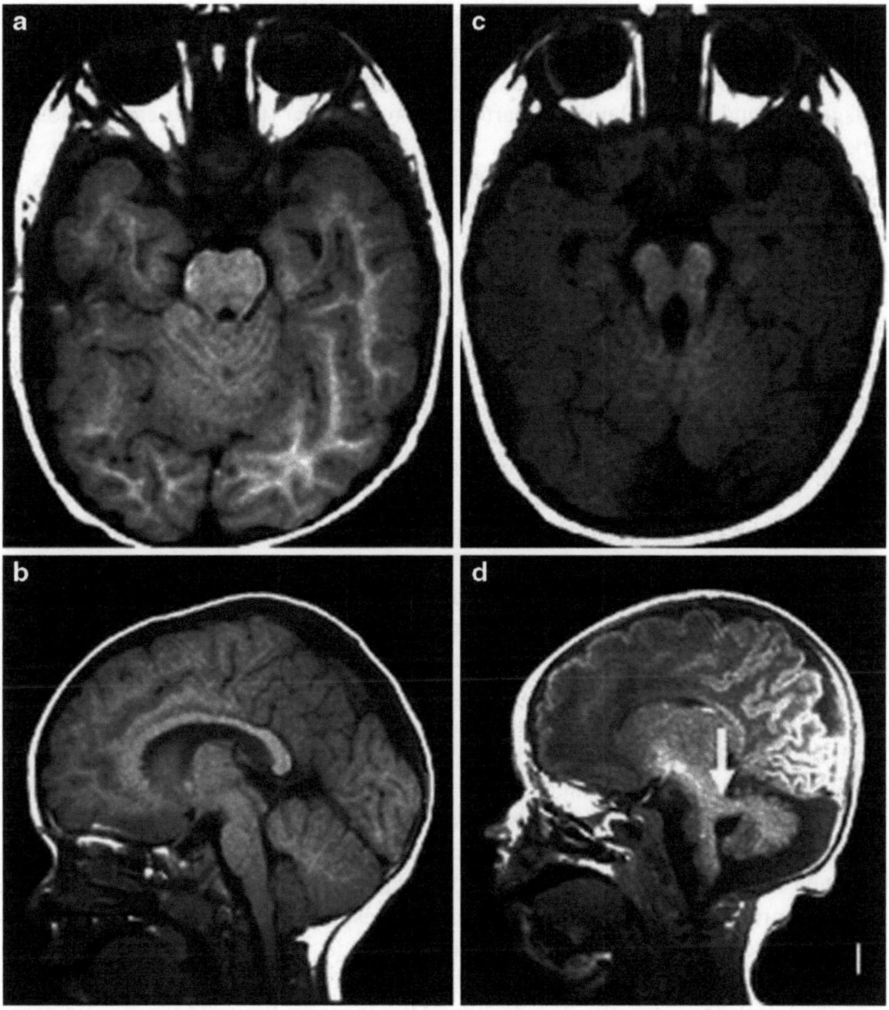

Fig. 9.1 Magnetic resonance imaging showing a normal cerebellar vermis and superior cerebellar peduncles on (**a**) axial and (**b**) sagittal images. (**c**) and (**d**) show cerebellar vermis hypoplasia and thickened superior cerebellar peduncles perpendicular to the brainstem (*arrow*), consistent with the molar tooth sign (*Figure 2* from Sattar and Gleeson (2011). Reproduced with permission of Mac Keith Press)

lobe hypoplasia in 25%. Callosal abnormalities were also very common, and co-existent with hippocampal abnormalities, suggesting that these reflect developmental disruption at a similar stage of neurodevelopment. Poretti et al. (2011) carried out an even larger survey of radiology in 75 JS patients (age at scan 2 days – 27 years). Occipital and atretic encephaloceles were detected in eight cases. The reported rate of hippocampal and corpus callosum abnormalities in this case series was much lower (11/71 and 6/71 cases respectively) than in the Senocak et al. review. An additional observation was abnormal cerebellar folia in a high proportion of cases.

Poretti et al. sought but found no evidence for genotype-neuroimaging correlations within their case series, except that amongst four patients with OFD-VI (presenting with severe developmental delay in addition to craniofacial features), hippocampal dysgenesis and hypothalamic hamartoma were frequent (two cases each), and cerebellar vermis dysplasia was severe.

An additional perplexing neuroradiological feature of JS (both in terms of developmental origin and functional correlates) is lack of decussation (either complete absence, or severe reduction) of both the Superior Cerebellar Peduncles (SCP) and Corticospinal Tracts (CST, pyramidal tracts). Absent SCP decussation was first recognised on postmortem examination of three early cases (discussed below). Lee et al. (2005) demonstrated that the same abnormality could be observed in vivo, applying diffusion-weighted imaging (DWI) to a single case. Poretti et al. (2007) confirmed the sensitivity of fibre orientation mapping of fractional anisotropy (FA), a measure of white matter organisation derived from DWI, for observing abnormal course and absent decussation of the SCP in six subjects (Fig. 9.2). Spampinato et al. (2008) demonstrated that the same abnormality can be observed on conventional imaging by seeking "an ill-defined oval area of lower T1 signal intensity in the region of the mesencephalic isthmus" (appearance of the normal decussation). Blind rating of this sign by two neuroradiologists indicated that the sign could not be identified in any JS cases or controls before the age of 30 months, but that after 30 months the sign was present in all 16 controls and absent in all six cases examined. Pyramidal tract decussation has not been systematically investigated, but it is interesting to note that similar abnormalities have been observed in individuals displaying mirror movements as an isolated abnormality (Nugent et al. 2012).

Parisi et al. (2004) explored the possible consequences of abnormal decussation for lateralised processing, using a finger-tapping task and functional MRI in a single adolescent JS subject. Atypical bilateral activations were observed in sensorimotor cortex and cerebellar hemispheres, an observation that has not (to our knowledge) been explored in further subjects to date. The one published visual-evoked potential study in JS (Khan et al. 2008) reported indirect evidence for abnormal optic chiasm decussation. This leads to the suggestion of a more generalised inter-hemispheric axonal routing problem in JS, affecting at least three spatially and functionally distinct pathways, with possible consequences for perception as well as motor control.

9.3.1.4 Neuropathology

Post-mortem evaluation is extremely valuable for delineating the type and extent of macroscopic and microscopic pathology in genetic disorders, but for many ciliopathies no such evaluations have been published. Whilst limitations such as cause and age of death, fixation methods and dissection artefacts should be recognised, integration of neuropathology with in vivo neuroimaging may give convergent perspectives on developmental anomalies.

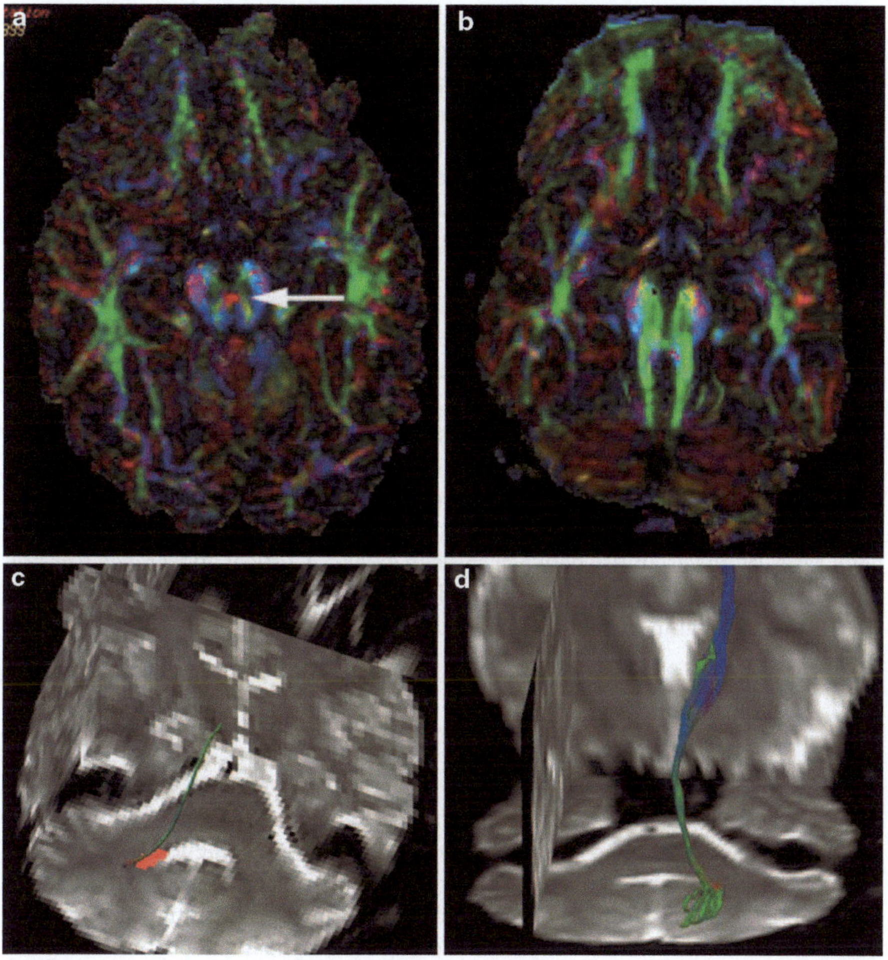

Fig. 9.2 Color-coded FA-maps at the level of the decussation of the superior cerebellar peduncles. (**a**) In a healthy subject, on the color-coded FA-maps the decussation of the superior cerebellar peduncles is identified as a "*red dot*" (*arrow*) at the level of the inferior colliculi of the midbrain. The decussating fibers have a transverse orientation and consequently show a "*red color* coding." (**b**) In JS, the absence of the "*red dot*" on color-coded FA-maps within the midbrain confirms the failure of the superior cerebellar peduncles to decussate. (**c**, **d**) Fiber tractography displays that, in JS, the fibers within the superior cerebellar peduncles that connect the dentate nucleus with the nucleus ruber do not cross and remain ipsilateral. Axial, coronal, and sagittal anatomic T2-weighted images are projected within the display for orientation purposes (*Figure 2* from Poretti et al. (2007))

Yachnis and Rorke (1999) reported the autopsy findings for a 30 year-old adult with JS. Vermal agenesis with enlargement and rostral extension of 4th ventricle were seen as expected, but additional features not visible on MRI include extensive abnormal development of medullary and cerebellar nuclei including hypoplasia of the inferior olives and islands of grey matter foci corresponding to dentate nuclei.

Disorganisation of the spinal cord posterior columns was also noted, and pigmented neurons corresponding to the locus coeruleus were dispersed from their typical location. No apparent decussation of pyramidal tracts could be traced. Within the cerebellum, reduction in Purkinje cell number was observed. Neuropathology for five individuals with diverse JS genotypes have recently been reported (Juric-Sekhar et al. 2012) (Figs. 9.3 and 9.4, reproduced with permission of Springer). Three cases were prenatal and carried out after diagnosis by fetal ultrasound and MRI and elective feticide (20–23 weeks gestation at autopsy). All three cases demonstrated cerebellar hypoplasia and enlargement of the 4th ventricle, and two cases had occipital encephalocoele (perhaps reflecting more severe presentations detectable at this gestation). Two cases showed dorsal enlargement of the medulla, but other macroscopic abnormalities of pons and midbrain were minimal. More extensive macroscopic abnormalities could be reported for the two postnatal cases (ages 13 months and 22 years), including fragmentation of the dentate gyri, abnormalities of the substantia nigra, and in one case hippocampal and parahippocampal hypoplasia. Histological examination confirmed extensive hypocellularity and fragmentation of cerebellar, midbrain and hindbrain nuclei in both fetal and postnatal cases (with variability between cases in extent). Cortical appearance was normal in all cases, but two cases showed hippocampal hypoplasia, and basal ganglia were also abnormal in two cases.

9.3.2 Meckel Syndrome (MKS)

MKS is a severe CNS-dominant ciliopathy that shares many features with both JS and with the multi-system ciliopathies described below. Diagnostic features of MKS are occipital encephalocele, polycystic kidney disease, polydactyly (more usually postaxial) and fibrosis of the liver with bile duct proliferation (Logan et al. 2011). Some authors have advocated widening the diagnostic net by including any major brain structural malformation in association with the extra-CNS features listed above (Baala et al. 2007b). Associated abnormalities include orofacial clefting, genitourinary anomalies, pulmonary hypoplasia, and microphthalmia/coloboma. MKS is genetically heterogeneous (7 loci to date: *MKS1, TMEM67, TMEM216, CEP290, CC2D2A, RPGRIP1L* and *B9D1*), and allelic to several others ciliopathies (JS, BBS, OFD1), with some evidence that more severe mutational mechanisms are more likely to lead to MKS (Tallila et al. 2009). For example, the mutational spectrum of *CCD2A* involves missense mutations in JS versus truncating mutations in MKS (Bachmann-Gagescu et al. 2012). Conversely, hypomorphic mutations at three loci associated with MKS can cause BBS but with some atypical BBS features notably seizures (Leitch et al. 2008). Hopp et al. (2011) deployed Next Generation Sequencing to investigate the complexity of MKS. In one family they identified a frameshift mutation in one allele and a deletion in the other allele of *B9D1*, and an additional likely-pathogenic difference in one copy of *CEP290*, suggesting oligogenic inheritance. It is possible that the

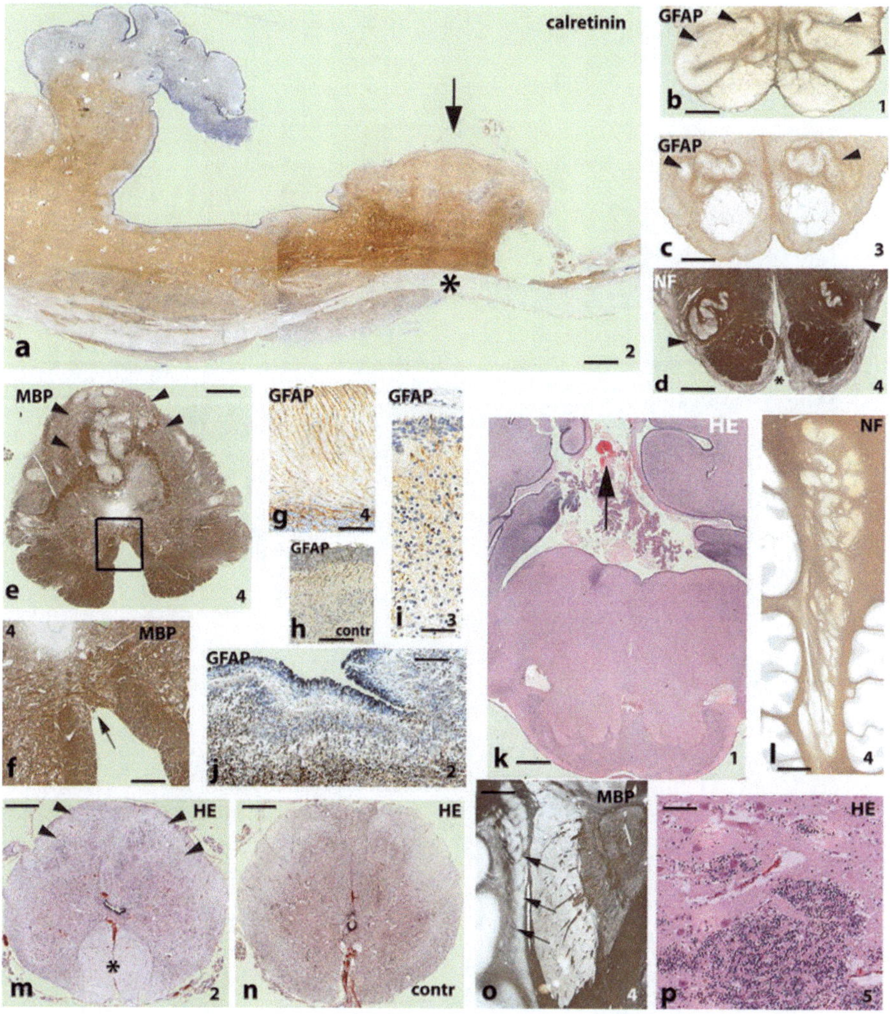

Fig. 9.3 Brain at postmortem examinations in subjects with Joubert syndrome. **a**, **b** Posterior views of the cerebellum and brainstem in Subjects 2 (**a**) and 3 (**b**) show enlarged wide-open 4th ventricle, hypoplastic vermis and markedly hypoplastic cerebellar hemispheres. Abnormal excess tissue is visible on the dorsal aspect of the cervicomedullary junction (*black arrows*). **c** Cerebellum and brainstem in an age-matched unaffected fetus. **d** The cerebellum and brainstem of Subject 2 (midline cut) demonstrate hypoplastic cerebellum (*black asterisk*), flat pons, and abnormal excess tissue on the dorsal aspect of the cervicomedullary junction (*black arrow*). **e** Subject 4 displays an elongated medulla lacking inferior olivary prominences (*black arrowhead*). No obvious pyramidal decussation is seen (*black arrow*). Cerebellar tissue wraps ventrally around the medulla to the pyramids (*black asterisk*). **f** Mid-pons and cerebellum (cut perpendicular to the long axis at the level of cranial nerve V exit) in Subject 4 show aplasia of the cerebellar vermis (*white arrow*) and enlarged 4th ventricle. Crossing fibers are seen in the hypoplastic basis pontis (*black arrowheads*), and fragmented dentate nuclei are seen bilaterally (*red arrowheads*). **g** The dentate nucleus of Subject 5 is also fragmented (*red arrowheads*). **h** The pons and cerebellum (midline cut, medial view) of Subject 5 demonstrate enlarged 4th ventricle, aplastic vermis (*black asterisk*), and hypoplastic cerebellum. *Scale bar* 1 cm (**a–h**) (*Figures 2 and 3* from Juric-Sekhar et al. (2012). Reproduced with permission of Springer)

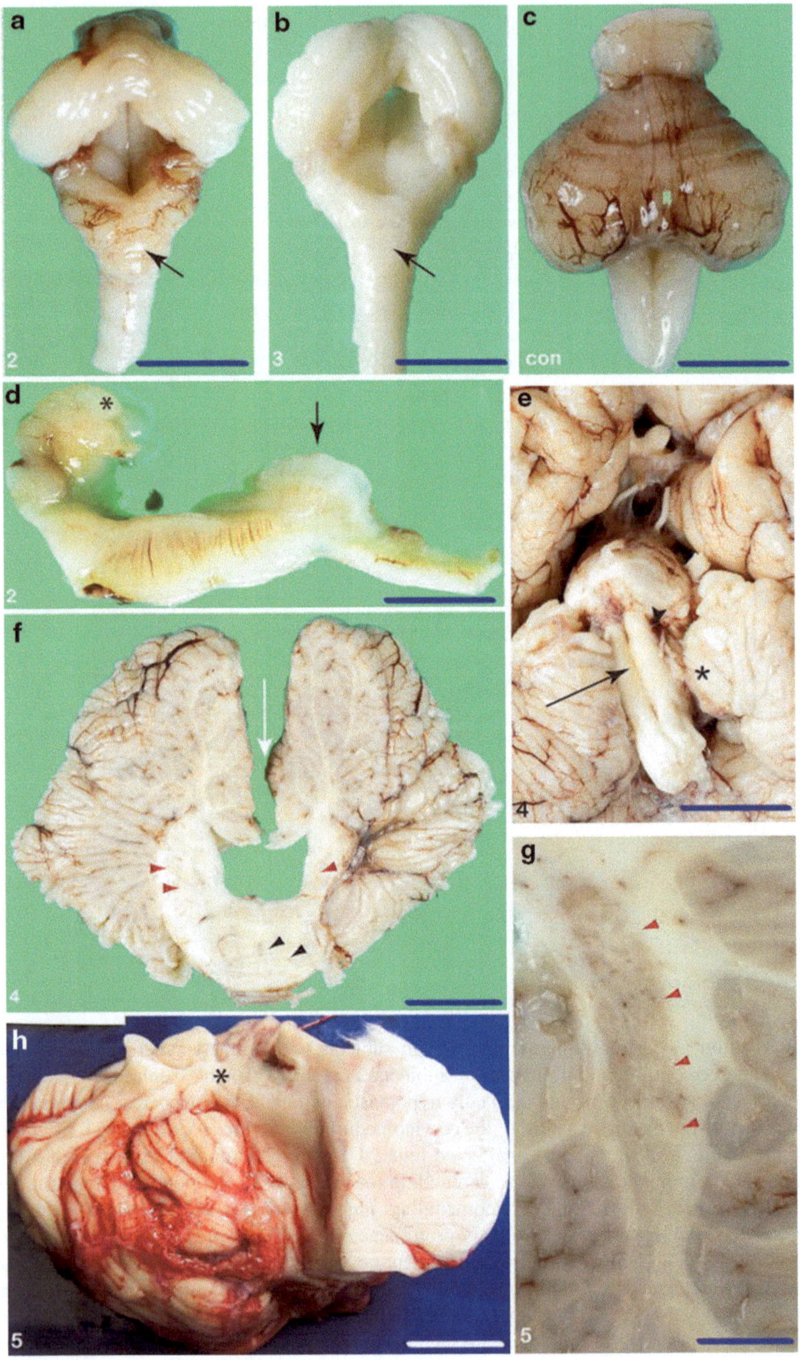
a
b
c
2
3
con
d
*
2
e
*
4
f
4
g
5
h
*
5

severity of MKS relates to mutational load, in addition to specific mutational mechanisms, converging on ciliogenesis and disrupting neural development at a relatively early stage.

9.3.2.1 Neurology and Neurodevelopment

To date, no case of classical MKS is known to have survived the perinatal period, with the majority of pregnancies resulting in intra-uterine demise, termination, stillbirth or premature delivery of a non-viable infant. Following the identification of MKS-associated genotypes, sequencing of the MKS genes has been carried out in patients with milder phenotypes overlapping with other ciliopathies in whom other attempts to identify causative mutations have been unsuccessful. For example, Gunay-Aygun et al. (2009) identified truncating (splice site) and missense MKS3 mutations in three individuals ascertained for cystic renal disease and minor CNS features of a JSRD. No case had an encephalocele, challenging the diagnostic specificity of *MKS3* mutations, and potentially broadening the MKS spectrum.

Fig. 9.4 Histopathology of the brain and spinal cord in Joubert Syndrome. **a** The cerebellum and brainstem of Subject 2 (sagittal section, rostral left) shows dorsal heterotopia with numerous calretinin-positive cells (*black arrow*). The pyramidal tract is appropriately positioned in the brainstem (*black asterisk*). **b, c** The medulla of Subjects 1 and 3, immunostained to detect GFAP, shows bilateral periolivary gliosis around the simplified inferior olivary nuclei (*black arrowheads*). **d** The rostral medulla of Subject 4, immunostained to detect *NF*, demonstrates hypoplastic inferior olivary nuclei (*black arrowheads*), and thickened arcuate nuclei (*black asterisk*). **e** Cervicomedullary junction of Subject 4 immunostained for *MBP* demonstrates absence of the posterior median sulcus, with disorganized dorsal column tracts and nuclei (*black arrowheads*). Due to nondecussation, the corticospinal tract is located mainly in anterior rather than lateral white matter. A few fascicles of *MBP*-positive corticospinal fibers cross the midline, as seen at higher magnification in **f** (corresponding to *boxed area* in **e**. **g** The cerebellar cortex in Subject 4 shows an increased GFAP-positive Bergmann glia. **h** GFAP-immunopositivity in the cerebellar cortex from an unaffected age-matched fetus. **i, j** The cerebellar cortex in Subjects 1 and 3 shows a paucity of GFAP-positive Bergmann glia, and decreased thickness of external and internal granular layers. **k** The pontomedullary junction in Subject 1 shows vermis aplasia (*HE*). **l** The cerebellar hemisphere of Subject 4 shows fragmented deep nuclei forming a chain of neuronal clusters in the white matter (*NF*). **m** The cervical spinal cord in Subject 2 shows disorganized, poorly defined posterior funiculi with nests of medium-sized neurons (*black arrowheads*), and a poorly formed posterior median sulcus. The anterior corticospinal tracts were greatly enlarged (*black asterisk*), consistent with nondecussation. **n** Axial section of cervical spinal cord from an unaffected age-matched fetus stained with HE. **o** The basal ganglia in Subject 4 demonstrate fragmentation of the claustrum (*black arrows*) with small neuronal heterotopia in the extreme capsule (*MBP*). **p** Cerebellar heterotopia in Subject 5 (stained with HE) consists of disorganized granule cells and Purkinje neurons. *Scale bars* 40 μm (**g–i**), 20 μm (**f, j, p**), 10 μm (**b–e, l**), 5 μm (**a, k, m–o**). Sections: sagittal (**a, l**), axial (**b–j, m–p**), coronal (**o**). *GFAP* glial fibrillary acidic protein, *HE* hematoxylin and eosin, *MBP* myelin basic protein, *NF* neurofilament protein (*Figures 2 and 3* from Juric-Sekhar et al. (2012). Reproduced with permission of Springer)

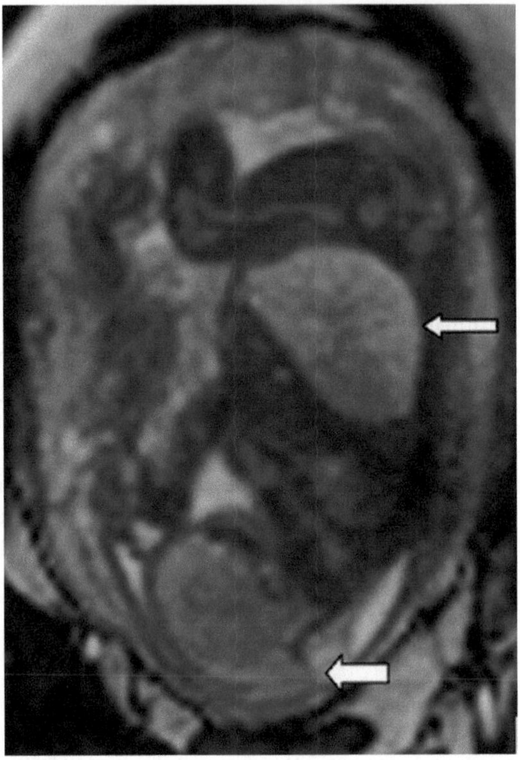

Fig. 9.5 A 24-week-fetus with Meckel Gruber syndrome. Fetal MRI in sagittal view showing marked oligohydramnios, bilateral enlarged dysplastic kidneys (*narrowed arrow*), small occipital cephalocele that was missed on US (*thick arrow*) (*Figure 5* from Behairy et al. (2010). Reproduced with permission of Elsevier Ireland Ltd.)

9.3.2.2 Neuroradiology

MKS can be detected by ultrasonography from as early as 11–14 weeks gestation, but is often not detected until anomaly scanning after 20 weeks (Chen 2007). Fetal MRI is now the modality of choice for delineating the extent of CNS and extra-CNS abnormalities to assist in definitive diagnosis and decision-making. Behairy et al. (2010) reported antenatal MRI findings in two MKS cases. Both cases had occipital encephalocoele (small in one case, not detected on ultrasonography), with vermian hypoplasia an additional finding in one case, and bilateral enlarged dysplastic kidneys in the other (Fig. 9.5, reproduced with permission of Elsevier Ireland Ltd.). MKS is sometimes described as being associated with neural tube defects. However, spinal dysraphisms in addition to occipital encephalocele (which may be mild, moderate or severe) have not been reported, and MKS mutations have not to date been identified in individuals with isolated neural tube defects.

9.3.2.3 Neuropathology

Unlike JS, the majority of information about the spectrum of neuro-embryological abnormalities in MTS comes from autopsy examination (Logan et al. 2011).

In addition to encephaloceles of variable location and extent, midline abnormalities predominate – holoprosencephaly, agenesis of the corpus callosum, fused thalami, small or absent pituitary gland. Somewhat in contrast to the dilated CSF systems seen in JS, hypoplasia of the third ventricle has been observed. Aqueductal stenosis leading to hydrocephalus has also been reported. Both common and rare features associated with JS are also seen in MKS (agenesis of the cerebellar vermis, elongated brain stem, Dandy–Walker malformation, microcephaly. polymicrogyria) leading to the suggestion that MKS represents the most severe end of the JSRD spectrum.

Recent studies exploring the frequency of different MKS genotypes and associated phenotypes has highlighted this overlap, and pointed toward genotype-phenotype correlations. Khaddour et al. (2007) reported the clinical characteristics of families mapped to the *MKS1* and *MKS3* loci (prior to gene identification). Baala et al. (2007a) provided detailed information about ten families (several with multiple affected siblings) in whom either homozygous or heterozygous *CEP290* mutations were identified. DWM was noted in nine cases, cerebellar vermis hypoplasia or agenesis in seven cases, and brain stem dysgenesis in 12. Histological observations included "chaotic organization of fibers and tracts" in the midbrain and involving the corticospinal tracts, enlarged and elongated superior cerebellar peduncles, as well as vermis hypoplasia.

9.3.3 Summary

Is there a unifying mechanism underlying the range of CNS features observed in patients with JS, JSRD and MKS? Barkovich et al. (2009), in their developmental classification of mid-brain hindbrain abnormalities, comment that vermian abnormalities in JS could reflect abnormal neuronal migration, whilst decussation abnormalities may be a sign of abnormal axonal growth. These discrete embryological anomalies could, however, be dependent on shared ciliary mechanisms: cilia functioning as chemotactic cue-guidance antennae, deployed by migrating neurons and by axonal growth cones, Abnormal migration would be consistent with the presence of polymicrogyria and heterotopias in some patients, but does not explain why the cerebellar-hindbrain pathology in JS is so region-specific. Subtle disturbances to corticogenesis and hippocampal organisation have also been observed and may relate to cognitive impairment. Whether one or multiple mechanisms can explain the neuropathological features of JS, and whether this mechanism is relevant to CNS features in other ciliopathies has yet to be determined.

Further insights into JS pathogenesis have been provided by recent investigation of an *Ahi1*-deficient mouse model (Lancaster et al. 2011) that recapitulates cerebellar vermis hypoplasia, but not fragmentation of nuclei or abnormal decussations. Examination of knock-down mouse embryos identified restricted proliferation of midline cerebellar granule neurons and a midline cerebellar fusion defect, similar to defects previously seen in Wnt-mutant mice. The investigators went on to demonstrate a specific decrease in Wnt activity in cerebellar cells surrounding the midline in mutant tissue, and showed that this phenotype could be partially rescued by lithium chloride injections. The authors propose that proliferation of specific cell populations, mediated via cilia-Wnt signalling, is responsible for vermian agenesis in JS. Embryological disruptions in MKS begin as early as neurulation, with disruption to neural tube formation and closure (Weatherbee et al. 2009). Whether these features reflect similar aspects of ciliary function and downstream signalling as JSRD has yet to be established.

9.4 Multi-System Ciliopathies

9.4.1 Alstrom Syndrome (AS)

AS is defined by cone-rod dystrophy before age 1, obesity and insulin resistance, dilated cardiomyopathy, recurrent pulmonary infections, progressive renal and hepatic disease, and absence of polydactyly (Marshall et al. 2011). *ALMS1* is the only gene in which mutations have been identified to date (Collin et al. 2002), with mutations detectable in 25–40% clinically diagnosed cases. Therefore AS appears to stand in contrast to CNS-dominant ciliopathies discussed above, with no genetic heterogeneity identified to date, and apparent absence of CNS features. However on closer inspection some similarities with the CNS-dominant ciliopathies may perhaps emerge, as well as neurological abnormalities that may be specific to this condition.

9.4.1.1 Neurology

Early visual impairment is a required criterion for diagnosis of AS. Features include pendular nystagmus, photophobia and reduced acuity, in addition to abnormal electroretinogram (ERG) characteristic of severe retinal dystrophy. Progressive bilateral sensorineural hearing impairment affects at least 70% in first decade of life (Van Den et al. 2001), perhaps due to requirement of ALMS1 for maintenance of planar cell polarity of cochlear hair cells (Jagger et al. 2011).

Marshall et al. (2005) surveyed clinical features in 150 patients with AS, and reported a number of unexplained neurological or neuromuscular complaints – partial unilateral paralysis, unexplained joint or muscle pain and muscle dystonia. One possibility is that these features could reflect peripheral nervous system abnormalities, secondary to metabolic syndrome and disturbed glycaemic control. One interesting observation, however, is the relative lack of neuropathic signs and apparent protection from neuropathic ulcers in AS patients with diabetes compared to other young type 2 DM patients (Paisey et al. 2009). Marshall et al. reported that hyporeflexia was present in 33 patients (20%; age range, 9 months to 41 years). Twenty of these same patients reported absence seizures, although supporting EEG evidence is lacking.

9.4.1.2 Neurodevelopment

Although developmental delay is not a diagnostic criterion for AS, Marshall et al. (2005) reported that developmental milestones were delayed in around half of their cohort of patients. In around half of patients who demonstrated delay, all domains of development were affected, whereas in the remainder either fine motor function or language development were predominantly affected. Later cognitive functions were in the learning disability range (IQ<70) in 16% of cases, although it should be noted that the availability of assessments was variable, and necessarily limited by visual impairment, which is severe in all cases. Severe medical problems such as cardiomyopathy also affect a significant proportion of AS children from an early age, with likely impact on global abilities. Mild autistic spectrum behaviours were reported to affect a minority of children but have not been systematically evaluated or compared to other children with visual impairment, including JS.

9.4.1.3 Neuroradiology

No publications found

9.4.2 Oro-Facial-Digital Syndrome (OFD)

OFD is a large group of conditions. Shared diagnostic criteria for OFD are the presence of a lobed tongue with hamartomatous or lipomatous oral lesions, cleft palate (50%) and variable facial and digital abnormalities. OFD-1 is an X-linked dominant condition that demonstrates male lethality, caused by mutations in *OFD1*. OFD-6 is generally considered to be part of the JSRD spectrum, indicating that multi-system and CNS-dominant ciliopathies may overlap to some degree.

9.4.2.1 Neurology

Epilepsy is a recognised feature of OFD-1 but the frequency and types of seizure disorders have not been systematically documented. Sensory abnormalities appear not to be frequent in OFD.

9.4.2.2 Neurodevelopment

At least 50% of OFD-1 individuals demonstrate developmental delay and learning disabilities. Mild to moderate intellectual disability was reported in 11 of 25 cases in the largest study to date (Thauvin-Robinet et al. 2006). No domain specificity of deficits has been described.

9.4.2.3 Neuroradiology

Structural brain abnormalities are common and diverse in OFD-1, and show many similarities with JS/JSRD. Odent et al. (1998) reported imaging in two families and three sporadic cases. Abnormalities included porencephaly or arachnoid cysts, periventricular heterotopia and cerebellar vermis hypoplasia. In the larger case series of Thauvin-Robinet et al. (2006), imaging was mainly available for patients with significant developmental delay. The most common abnormality was agenesis of corpus callosum common (11/14), followed by cerebellar hypoplasia, arachnoid cysts, hydrocephalus and porencephaly. OFD-1 mutations can also be associated with severe cerebral dysgenesis (affecting all midline structures), and may underlie the observed male lethality. Recent molecular testing of post-mortem material (from a severely affected female fetus) identified a de novo OFD1 mutation. This same mutation had previously been found in two sporadic cases, one with normal intelligence and one with learning disability, highlighting possible genetic complexity in determining phenotype (Thauvin-Robinet et al. 2011).

9.4.3 Bardet-Biedl Syndrome (BBS)

BBS is a multi-system ciliopathy in which CNS features are core but not absolutely required for diagnosis (Beales et al. 1999). Primary diagnostic features are rod-cone dystrophy, postaxial polydactyly, truncal obesity, hypogonadism, developmental delay, and cystic renal disease. Secondary features that add support to the diagnosis in the absence of sufficient primary features include neurological abnormalities and learning disability. Historically, two syndromes were described (BBS and Laurence Moon Syndrome, LMS) which could not be consistently discriminated from each other, hence were subsumed into (LM)BBS (Moore et al. 2005). CNS features were recognised in both BBS and LMS, but it is notable that the original description of LMS involved more extensive and frequent expression of neurological abnormalities than BBS.

BBS demonstrates considerable genetic heterogeneity, with 16 loci identified to date, including genes implicated in other ciliopathy syndromes (Forsythe 2012). Novel diagnostic strategies have recently been developed to improve diagnostic efficiency (Janssen et al. 2011). The number of patients diagnosed with BBS is likely to increase over time, leading to renewed focus on understanding each component of the phenotype, in the hope that interventions can be directed at each aspect of morbidity, and long-term health outcomes improved.

9.4.3.1 Neurology

Information about the range of clinical difficulties experienced across the lifespan in BBS was collated in a questionnaire-based study by Beales et al. (1999). Sensory impairments not directly related to cone-rod dysytrophy were frequent. Seven percent reported ocular abnormalities (strabisumus, cataracts, astigmatism), whilst 21% (23/109) had a history of conductive hearing loss. Mixed sensorineural and conductive hearing loss was reported rarely (3%). An additional sensory deficit of note in BBS is anosmia – objective evidence for smell identification impairment was reported for two members of an extended family carrying homozygous *BBS4* deletion (Iannaccone et al. 2005), and replicated in humans and in a *Bbs4* mouse knockout model (Kulaga et al. 2004).

The frequency of seizures reported in BBS cohorts has varied from 11% (Moore et al. 2005) to 4% (Beales et al. 1999). This may reflect population differences in BBS genotype frequencies, since the Moore et al. study was carried out in a genetically-isolated population (Newfoundland) with an increased BBS prevalence. As discussed above, BBS caused by mutations in the MKS genes appears to demonstrate a specific association with seizures (Leitch et al. 2008). The neurobiological basis for seizures in BBS, and in association with MKS genotypes, has not been elucidated.

Motor co-ordination impairment appears to be relatively frequent problem in BBS. Abnormal gait was a feature of the first sib pair reported by Laurence and Moon. Beales et al. (1999) reported the presence of ataxia with poor coordination in 40% (43/109). Moore et al. (2005) reported that ataxia and impaired co-ordination were present in 86% (18/21) of patients directly examined. The temporal relationship between gait co-ordination difficulties and slowly-progressive retinal dystrophy from late childhood has not been established, but it appears that ataxia may be independent of visual impairment in BBS. Examination of cerebellar function after the onset of visual impairment is challenging, but upper limb dysmetria is not usually observed.

Hypertonia was historically considered to be a core feature in LMS. Moore et al. (2005) reported that 21% (5/24) of BBS patients had spasticity, involving all limbs in 4/5 patients and only lower limbs in one. However, hypotonia was also reported in two families. Progressive spastic paraplegia was reported in two cases by Beales et al. (1999), and quadraparesis had been observed in LMS previously. Genotypes and mechanisms underlying this rare severe neurological presentation of BBS is not known. One possible link between progressive hypertonia and BBS could be that *TRIM32* (an E3 ligase which specifically binds to and ubiquitinates dysbindin, and is a cause of limb-girdle muscular dystrophy) is synonymous with *BBS11* (Chiang et al. 2006).

Hirschprungs disease (colonic aganglionosis) is a rare feature of BBS, present in 2% of the Beales et al. cohort. Moore et al. (2005) reported colonic dysmotility in a much higher proportion (15%) including one child with confirmed Hirschprungs. Tobin et al. (2008) showed that inhibition of neural crest cell migration underlies this rare manifestation of BBS, and it is not clear whether a minor degree of dysmotility, not of sufficient severity to necessitate rectal biopsy or surgical management of bowel obstruction, may be a more frequent peripheral neurological manifestation of BBS. Hirschprungs has also, rarely, been reported in Joubert Syndrome (Maria et al. 1999).

9.4.3.2 Neurodevelopment

Parental survey of 109 BBS patients (Beales et al. 1999) confirmed previous clinical observation that developmental delay evolving into learning disability is a frequent but variable component of the BBS phenotype. Some degree of delay had been recognised in 50% of cases. Sixty two percent were reported by parents to have learning disabilities, although lack of formal evaluation is likely to underestimate the true prevalence of mild or domain-specific deficits. Neuropsychological evaluation of 21 children with BBS reported that IQ ranged from 42 to 108 (Barnett et al. 2002). A recent investigation of adults with BBS (Bennouna-Greene et al. 2011) classified 26% of individuals as having moderate intellectual disability, 35% mild or borderline impairments and the remainder performing within the average range of abilities (one above average). Moore et al. (2005) reported that mean verbal IQ (VIQ) in 24 children and adults with BBS was 75 (range 53–102, n = 24). The Haptic Intelligence Scale, designed to estimate cognitive abilities in individuals with severe visual impairment, was used to obtain an estimate of performance IQ (PIQ). Mean PIQ was 83 (range 44–105, n = 14), indicating no apparent difference between verbal and performance abilities. Population-normed scores of less than 70 (cut-off generally used to define significant learning disability) were recorded in 33% (8/24) of cases for VIQ and 21% (3/14) for PIQ, again indicating the lack of apparent domain-specificity of learning disability in BBS.

Delay and disorder to language development has been noted as a predominant concern, with 54% of the Beales et al. (1999) cohort manifesting speech and language impairments requiring therapeutic intervention. Specific abnormalities of speech and language reported (based on parental report and SALT examination in four cases) included high pitched voice, hypernasal speech, poor articulation, palatal incoordination, consonant omissions/substitutions, and poor comprehension. Moore et al. (2005) reported speech assessments for 19 patients (18 adults, one child aged 13 years), including a standardised consonant and syllable articulation test. Syllable repetition times for patients were markedly prolonged, suggesting a subtle impairment in oromotor co-ordination.

In addition to concerns about developmental and educational progress, families frequently report that behavioural, emotional and social difficulties can be equally problematic for children with BBS. Parental survey highlighted frequent concerns about social development, anxieties and obsessions in children with BBS (Beales et al. 1999). Barnett et al. (2002) deployed standardised questionnaires to explore behavioural features further. One quarter of children were assessed to have clinically significant internalising behaviours (for example anxiety and somatic complaints). Social problems and attention problems were also frequently reported and to be of a severity suggestive of clinically-significant difficulties. Rating scales for autistic spectrum disorders (ASD) were completed for around half of cases. Two children demonstrated mild to moderate features suggestive of an ASD, and a further two children demonstrated severe features that may correspond to a diagnosis of autism on more thorough evaluation. Moore et al. (2005) reported that one child with mutations in the *MKKS/BBS6* gene had an ASD, although systematic evaluations were unavailable for the majority of cases.

During adulthood, mental health difficulties are a relatively frequent and disabling feature of BBS. Anxiety and depression are common, and relatively severe. Psychiatric chart reviews for 14 adults reported by Moore et al. (2005) indicated that 30% of individuals met criteria for a lifetime diagnosis of a major psychiatric disorder. Anxiety disorders occurred in 20%, a common presentation being recurrent psychosomatic symptoms at ages 20–40 years. Mood disorders occurred in 9%, including three with major depression and one with bipolar disorder. All except one of these cases had required at least one inpatient admission to a psychiatric facility i.e. symptoms were severe and disruptive. Neuropsychiatric evaluation of 34 adults with BBS (mean age = 25 years old, ranging from 17 to 53 years old) confirmed that significant psychopathology was present in 35%, with diverse difficulties including depression, anxiety disorders, obsessive compulsive disorder, psychosis and eating disorders (Bennouna-Greene et al. 2011). Whether mental health difficulties reflects the understandable challenge of adjustment to progressive visual loss during adolescence and other medical and social difficulties, or a more direct neurobiological susceptibility to mood disorder has yet to be determined. Neuroanatomical abnormalities discussed below, together with genetic and cellular evidence for convergence between disruptions in BBS and in idiopathic mental illnesses (Kamiya et al. 2008; Han et al. 2008), led to the speculation that mood disorder may be a direct reflection of neuronal ciliopathy in BBS.

9.4.3.3 Neuroradiology

Structural brain abnormalities apparent on qualitative radiological assessment are not frequent in BBS. Cerebellar abnormalities, including cerebellar vermis hypoplasia, have been reported in single case studies (Baskin et al. 2002; Rooryck et al. 2007). Recently, three independent neuroimaging studies of BBS have been carried out, aiming to describe qualitative and quantitative neuroanatomical abnormalities, that may correlate with neurodevelopmental aspects of the BBS phenotype. These three studies differ somewhat in methodology, and not all findings are consistent.

Global grey matter (GM) volume reduction was reported by (Baker et al. 2011), but not replicated by Keppler-Noreuil et al. (2011b). Global white matter (WM) volume reduction was reported by Keppler-Noreuil et al. (2011b), with a trend toward similar reduction in the Baker et al. (2011) study. Bennouna-Greene et al. 2011 reported a radiological appearance of hippocampal dysgenesis in 11 out of 34 cases (Fig. 9.6, reproduced with permission of Blackwell Munksgaard). Significant reduction in hippocampal volumes has been quantified by region-of-interest morphometry and temporal lobe grey matter volume differences quantified by voxel-based morphometry (Baker et al. 2011) (Fig. 9.7, reproduced with permission of Wiley). Caudate volumes were found to be reduced in both quantitative studies

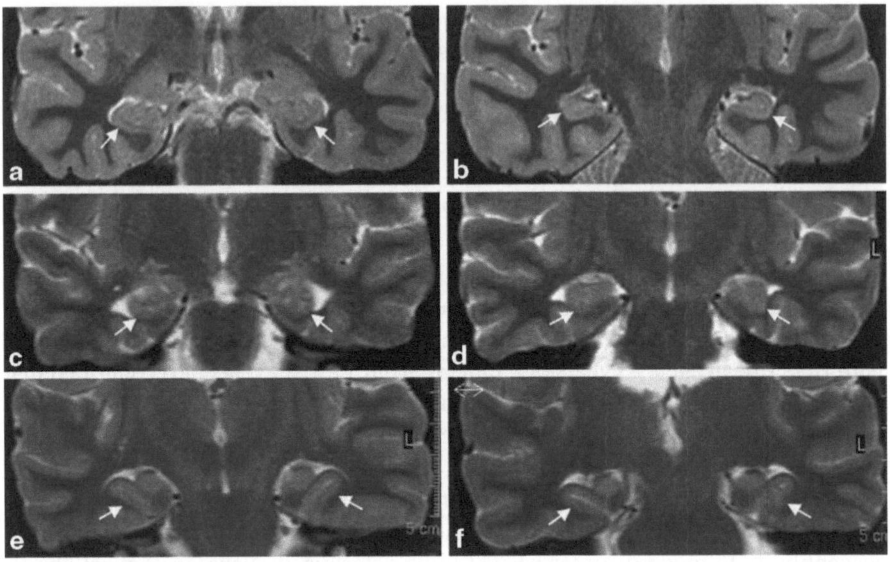

Fig. 9.6 Magnetic resonance imaging of normal hippocampus and hippocampal dysgenesis in Bardet Biedl Syndrome. (**a** and **b**) Coronal T2 weighted image centered on the hippocampus showing normal anatomy. The hippocampus presents a lateral position, an ovoid form, a transversal orientation, its internal structures are distinguishable and the collateral sulcus is short and transversally orientated. (**c** and **d**) Coronal T2 weighted images of patient 4.02.02. Bilateral hippocampal dysgenesis extending from the body to the head, partial on the right side (medial position of the hippocampus, pyramidal form, blurring of hippocampal internal structures, deep and vertically oriented collateral sulcus, but the hippocampus is not vertically oriented) and complete on the left side (medial position of the hippocampus, a round vertically oriented hippocampus, blurring of hippocampal internal structures, deep and vertically oriented collateral sulcus). (**e** and **f**) Coronal T2 weighted images of the hippocampus of patient 1.08.12. Bilateral complete hippocampal dysgenesis extending from the tail to the body and the head (medial position of the hippocampi, round vertically oriented, blurring of hippocampal internal structures, deep and vertically oriented collateral sulci) (*Figure 1* from Bennouna-Greene et al. (2011). Reproduced with permission of Blackwell Munksgaard)

(Baker et al. 2011; Keppler-Noreuil et al. 2011a, b). Studies are in agreement that cerebellar volumetric abnormalities are not a consistent feature of BBS.

Keppler-Noreuil et al. (2011b) reported that occipital lobe WM volumes were disproportionately reduced in comparison to volumes in the temporal, parietal and frontal lobes. Baker et al. (2011) applied voxel-based morphometry to identify

Fig. 9.7 (**a**) Neuroradiological abnormalities—T1-weighted scans illustrating qualitative abnormalities detected in *BBS* patients (presented in comparison to age-matched and gender-matched comparison subjects). (**b**) Voxel-based morphometry (*gray matter*). Background: averaged normalized GM segment for all patients and controls; *Red areas*: relative gray matter deficit in *BBS* patients (P<0.001 uncorrected); presented in neurological orientation. (**c**) *Total gray matter and hippocampal volume* measurements (*Figures 1, 2 and 4* from Baker et al. (2011). Reproduced with permission of Wiley)

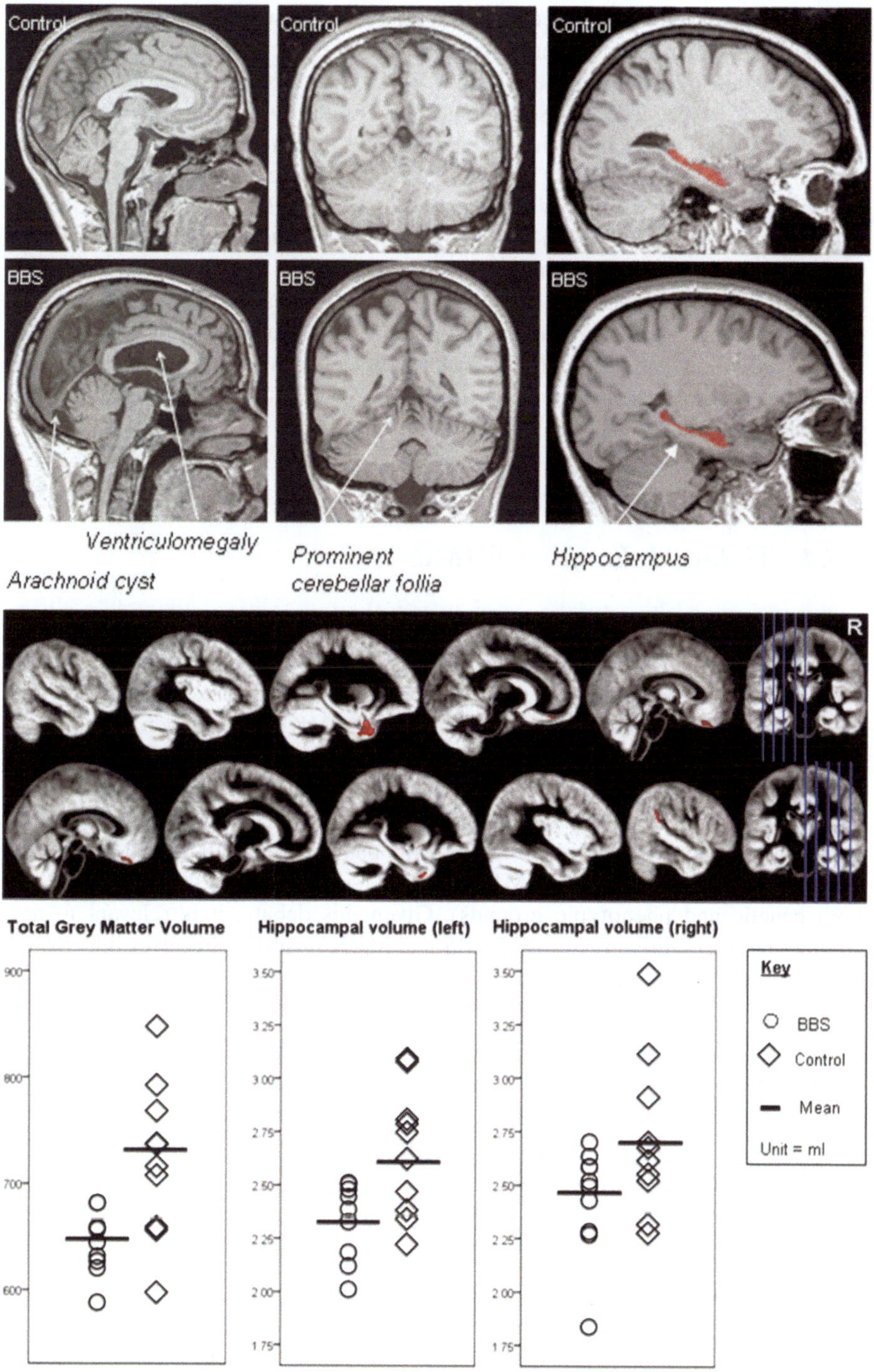
Control
Control
Control
BBS
BBS
BBS
Ventriculomegaly
Arachnoid cyst
Prominent cerebellar follia
Hippocampus
R
Total Grey Matter Volume
Hippocampal volume (left)
Hippocampal volume (right)
900
800
700
600
3.50
3.25
3.00
2.75
2.50
2.25
2.00
1.75
Key
BBS
Control
Mean
Unit = ml

regions of WM volume difference. Reductions in WM volume within the inferior longitudinal fasciculus and inferior fronto-occipital fasciculus were detected, and at lower statistical thresholds more extensive WM reductions within the occipital lobes were also observed. These observations show similarity with previous reports of WM tract abnormalities involving visual pathways in individuals with congenital or early onset visual impairment (Bridge et al. 2009; Ptito et al. 2008), leading to the possibility that these specific WM tract abnormalities in BBS may be consequential to visual impairment. Diffusion-weighted imaging and tractography is supportive of this proposal (Baker et al. unpublished data).

Hence it appears that there are several discrete neuroanatomical characteristics of BBS, including hippocampal and temporal lobe GM abnormalities, global WM volume reduction, and regional WM abnormalities involving the occipital lobe and associated pathways. Developmental mechanisms and functional associations are yet to be resolved.

9.4.4 McKusick-Kaufman (MKKS)

MKKS is a rare ciliopathy defined by postaxial polydactyly, congenital heart disease, and abnormalities of the internal genitalia (hydrometrocolpos in females), associated with poor prognosis. Absence of additional features consistent with a diagnosis of BBS is a requirement for diagnosis of MKKS, and given the considerable overlap in physical features, individuals with features suggestive of MKKS require observation until at least age 5 to determine which syndromic phenotype will evolve. MKKS is likely to be genetically heterogeneous, but it is notable that *BBS6* mutations have been identified in some patients with MKKS, adding further fuel to the debate about whether MKKS should be considered a subtype of BBS (on both genetic and phenotypic grounds). Given this debate, it is relevant to note whether neurocognitive features of BBS and MKKS overlap, however information is lacking that would enable this issue to be addressed.

9.4.4.1 Neurology

By definition, sensory impairments including retinal dystrophy would lead to a diagnosis of BBS rather than MKKS.

9.4.4.2 Neurodevelopment

Developmental delay was reported for 3 of 37 patients (14%) in review of reported cases (Slavotinek and Biesecker 2000). Where significant learning disability is present, MKKS is likely to be reclassified as BBS.

9.4.4.3 Neuroradiology

No studies identified.

9.4.5 Summary

Multi-system ciliopathies are associated with neurological signs, developmental delay, learning disability, childhood behavioural features and adult psychiatric illness in a significant proportion of cases. By definition, the posterior fossa abnormalities and associated oculomotor apraxia characteristic of JSRD are absent. The observation that BBS is associated with hippocampal and temporal lobe structural abnormalities, which have also been seen in JS, suggests that there could be common mechanisms underlying these neurodevelopmental features across the ciliopathy spectrum. However, there is a high degree of intra-syndrome and inter-syndrome variability in CNS features, and mechanisms of pathology and modifying factors, including the impact of sensory impairment and physical illness remains to be investigated. Addressing these complex aspects of the ciliopathy phenotype is a high priority for improving quality-of-life for families and patients.

9.5 Skeletal-Predominant Ciliopathies

9.5.1 Jeune Asphyxiating Thoracic Dystrophy (JATD)

JATD is a severe abnormality of the development of the thoracic skeleton, associated with polydactyly. Diagnosis depends on presence of characteristic chest wall deformity, specific signs of osteochondrogenesis on skeletal survey, and supporting evidence of relevant pathology in other organ systems. It is usually a lethal disorder with prenatal or early postnatal demise. Survivors into infancy develop severe kidney and liver disease, and some have manifested retinal dystrophy.

9.5.1.1 Neurology and Neurodevelopment

CNS features have not been extensively documented in view of the severity of the somatic phenotype in the majority of diagnosed cases. However, in a recent review (Keppler-Noreuil et al. 2011a) 140 previously published JATD cases were discussed alongside description of eight new cases, six of whom survived beyond infancy, indicating that the spectrum of milder clinical presentations may not previously have been appreciated. Neurological features were not prominent, however one case was reported to have generalised hypotonia and oculomotor dyspraxia (with no evidence of MTS on MRI). Developmental delay of moderate severity was reported in one out of eight new cases. In one additional case mild motor delay was apparent, but was attributed to severity of respiratory compromise and skeletal dysplasia. Interestingly, Hirschprungs disease was reported in two cases from the reviewed literature – this phenotype appears to be present, consistently in a very small number of cases across the ciliopathy spectrum.

9.5.1.2 Neuroradiology

Evidence for structural brain abnormalities in JATD is very limited – two cases in the literature were reported to have CNS defects, specifically Dandy–Walker malformation and hydrocephalus. Cranial MRI was abnormal in one case in the Keppler-Noreuil 2011 series, features being mild ventriculomegaly including 4th and lateral ventricles, hypoplastic corpus callosum, and diffuse spinal stenosis (development not documented, alive at 9 years of age).

9.5.2 Ellis-van Creveld/Short-Rib Polydactyly/ ChondroEctodermal Dysplasia

This group of conditions is characterised by short limbs, short ribs, postaxial polydactyly, and dysplastic nails and teeth. Congenital heart defects are present in around 60%, typically atrio-ventricular canal defects including common atrium, frequently requiring surgery in the early months of life with variable prognosis (Hills et al. 2011). Mutations in two genes (*EVC* and *EVC2* that encode centrosomal proteins) are detectable in two thirds of cases (Tompson et al. 2007). OMIM lists intellectual disability and Dandy-Walker malformation as part of the phenotypic spectrum for this syndrome, however the frequency, nature and severity of these deficits have not been surveyed. Stevens and Lachman (2010) reported a severely affected infant with a skeletal dysplasia, Dandy-Walker malformation and congenital heart defect, reminiscent of EVC but without confirmed genetic diagnosis, suggesting that CNS involvement in EVC may be a genetically distinct condition.

9.5.3 Cranioectodermal Dysplasia/Sensenbrenner Syndrome

This rare condition is unusual in that sagittal craniosynostosis (not reported in other ciliopathies) is a core diagnostic criterion, in addition to facial, ectodermal and skeletal abnormalities. The first gene mutated in cases of CED was *IFT122*, the encoded protein of which is a component of the intra-flagellar transport complex B important for ciliogenesis and hedgehog signal transduction (Walczak-Sztulpa et al. 2010). Exome sequencing in CED patients recently identified mutations in

WDR35 (Gilissen et al. 2010) and in *IFT43* (Arts et al. 2011) and *WDR19* (Bredrup et al. 2011), genes that encode presumed ciliary transport proteins.

Amar et al. (1997) presented a case report and review of 12 previously published cases. The detailed case report involved a female child with delayed psychomotor development and hypoplasia of corpus callosum, in addition to skeletal findings and abnormal calcium homeostasis. Three of twelve previously reported cases had CNS abnormalities (unspecified), and developmental delay was not surveyed. In both cases with *WDR35* mutations intelligence was normal with no behavioural concerns and therefore no clinical indications for neuroimaging (Gilissen et al. 2010). Therefore the spectrum of CNS abnormality, and possibly genotype-phenotype associations remains to be confirmed.

9.6 Retinal-Predominant Ciliopathies

9.6.1 Leber Congenital Amaurosis (LCA)

LCA is a severe retinal dystrophy, presenting with early-onset blindness, variable retinal appearances and severely abnormal ERG. LCA is associated with roving-eye movements and nystagmus, plus a characteristic behavioural sign: eye-poking, rubbing and pressing (Franceschetti's oculo-digital sign) in order to stimulate scotopic sensation, resulting in enophthalmos. LCA is genetically heterogeneous – chip-based testing for 14 genes identified mutations in 70% of 90 probands, with *CEP290* mutations contributing in 30% (Coppieters et al. 2010a). LCA gene functions are diverse – in addition to ciliary proteins, three LCA genes are involved in Vitamin A metabolism. It has been proposed that oligogenic inheritance and mutational load may be a contributory factor in some cases (Wiszniewski et al. 2011). Whether LCA is associated with extraocular features, or whether in the presence of extraocular features an alternative diagnosis is appropriate (for CNS feature, reclassification as a JSRD is likely to be suggested), remains a matter of debate.

9.6.1.1 Neurology

Fazzi et al. (2005) carried out neurological examinations in 40 children with LCA (ages 8–50 months), with abnormal findings in 31 cases. Hypotonia was observed in 27 cases (67.5%), and cerebellar ataxia in 4 (10%). In two of these cases, ataxia was associated with other cerebellar signs (dysmetria and intentional tremor). Yzer et al. (2012) reported epilepsy in one patient, and abnormal EEG in an additional patient. Abnormal proprioception was also a feature in two patients. Anosmia may be a feature in a proportion of cases – objective smell testing revealed impairment in LCA patients with *CEP290* mutations only (McEwen et al. 2007).

9.6.1.2 Neurodevelopment

A relatively early study of LCA reported the presence of intellectual disability in 20% of over 200 cases (Schuil et al. 1998). Limited neuroimaging and genetic testing was available, and methods of assessing visual impairment are unclear. Fazzi et al. (2005) carried out neuropsychological assessments using the Reynell–Zinkin Scales for visually impaired children (assessing social adaptation; sensorimotor understanding; exploration of the environment; response to sound and verbal comprehension; expressive language): performance was normal in 24 cases (60%), borderline in 5 (12.5%), while in 11 cases (27.5%) definite impairments of variable severity were detected. Perhaps unsurprisingly, predominant difficulties related to sensorimotor understanding and exploration of the environment. Perrault et al. (2007) investigated genotype-phenotype correlations and reported that *CEP290*-associated LCA was not linked to intellectual disability, unless neuroimaging findings suggestive of JSRD were present.

9.6.1.3 Neuroradiology

Fazzi et al. (2005) reported MRI to be normal in 22 (55%) of subjects and pathological in 13 (32.5%). In two cases (5%) a slight reduction in the dimensions of the chiasma and/or optic nerve were observed, whilst seven cases (17.5%) demonstrated mild and nonspecific alterations on MRI—mild atrophy in four subjects (10%) and white matter alterations in three (7.5%). Specific WM imaging (DWI and tractography) would assist in determining whether these abnormalities are most likely to reflect the impact of early severe visual loss on maturation of visual and non-visual pathways. The remaining four (10%) subjects (two of whom were siblings) had clearcut evidence of MTS on neuroimaing. All four cases were ataxic with developmental delay, but a diagnosis of JS was not thought to be appropriate because the ERG was too severely abnormal, and other supporting features such as disordered breathing were absent. Yang et al. (2010) reported that 6 of 31 patients (19%) with LCA in whom MRI was clinically indicated had radiologically documented brain abnormalities. Two patients had cerebellar vermis hypoplasia, 1 patient showed an absence of septum pellucidum, two subjects showed mild external hydrocephalus, and one patient was found to have a small cerebellum. Therefore in patients with LCA in whom neurological or neurodevelopmental concerns are highlighted, neuroradiological features similar to JS are likely.

9.7 Renal/Hepatic Predominant Ciliopathies

9.7.1 Nephronophthisis (NPHP)

NPHP pathology can be isolated or feature as part of multisystem ciliopathies. Key aspects of renal pathology include tubulointerstitial fibrosis, tubular dilatation and cyst formation and tubular atrophy, in most cases leading to end-stage renal failure by the teenage years. Genotypes associated with NPHP have been shown to be associated with JS as well as Senior-Loken syndrome, challenging previous classification systems. Moreover, it is unclear whether patients ascertained for NPHP have always been fully evaluated for neurodevelopmental features. Salomon et al. (2009) reported that 10–20% of individuals with a primary diagnosis of NPHP demonstrate extra-renal features including cerebellar signs and learning difficulties (in some cases suggestive of the milder end of the JS spectrum). An example of the genetic and phenotypic complexity inherent to the ciliopathy spectrum was provided by Tory et al. (2007), who identified a family expressing NPHP and JS-related neurological symptoms carrying homozygous *NPHP1* deletions in combination with a heterozygous truncation mutation in *CEP290* and a heterozygous missense mutation in *AHI1*.

9.7.2 *Polycystic Kidney Disease (PKD)*

PKD (autosomal dominant or recessive forms) is defined by the presence of cystic kidney disease, sometimes in association with hepatic cysts, and absence of features suggestive of a multi-system ciliopathy. Rarely, PKD has been reported to be associated with DWM (Goldston syndrome). The presence of more severe CNS structural abnormalities with functional consequences is likely to prompt reclassification as a cerebro-renal syndrome (for example a Meckel syndrome subtype). One exception to this rule is that a small proportion of ADPKD cases (8% of almost 250 cases), have been found to have arachnoid cysts. Arachnoid cysts are usually asymptomatic but sometimes associated with chronic subdural haematoma (Leung and Fan 2005), (Schievink et al. 1995).

9.8 Discussion

We have described the broad panoply of CNS abnormalities accompanying many ciliopathies. The degree of variability that exists between and within diseases is striking, despite common genetic aetiologies. Moreover, many CNS features cut across syndromal boundaries, and are seen at variable frequency across the entire ciliopathy spectrum. In spite of the recent rapid growth in our understanding of the causation of ciliopathies, we are only just scratching the surface of the knowledge required to understand how the cilium functions in typical brain development and dysfunctions in neurological disease, including potential involvement of the primary cilium in the pathogenesis of acquired or congenital isolated CNS disease.

A greater understanding of the genetic basis of ciliopathies will provide insight into phenotypic determinants. With the advent of new and rapid sequencing technologies, diagnosis of most rare, monogenic diseases will be achieved within the next 5 years. For us to make any correlation between genotype and phenotype amongst the ciliopathies will require more extensive investigation of the genetic mechanisms underlying disease in general. One explanation may be afforded by evidence of the influence of the total mutational load on final phenotypic outcomes. Here, both oligogenic inheritance patterns and modifier mutations have an important role to play in generating the clinical variability observed even within families with ciliopathies. Such examples of published modifier genes already include *TTC21B* (Davis et al. 2011), *AHI1* (Louie et al. 2010) and *RPGRIP1L* (Khanna et al. 2009) each of which codes for a protein within the cilia proteome. Oligogenic inheritance has been described for BBS and nephronophthisis whereby more than two mutations in two different genes are necessary and required for manifestation of the phenotype (Katsanis et al. 2001).

The cilium should be viewed as a complex organelle, rather than an isolated cellular appendage, that forms an integral part of the cell. Despite periods of absence at cell division, the primary cilium is clearly important for many and diverse cellular activities. Research in the coming years will unravel the extent of networks of protein complexes that underlie the role of the primary cilium and this is likely to differ from cell to cell and tissue to tissue. Loss of function of a given cilia protein in the epidermis, for example, may be fully tolerated whereas loss within the developing brain of the same protein may have a profound effect on neuronal migration or synapse formation. Recently, a network of crucial protein interactions was discovered linking nephronophthisis, Joubert, and Meckel-Gruber syndromes, ciliopathies presenting with cystic kidneys, retinal degeneration, and cerebellar/neural tube malformation (Sang et al. 2011). Using proteomics, 850 interactors copurifying with nine NPHP/JBTS/MKS proteins were discovered. These revealed three connected modules: one of which functions predominantly at the apical surface, another at centrosomes, and a third linked to Hedgehog signalling. This study showed that perturbation of the first, apical module leads to predominantly renal cystic disease whereas the other two modules are linked to retinal or neural deficits. This illustrates the power of combining proteomic networks with human genetics to uncover critical disease pathways.

Through the study of human diseases and animal models we are beginning to piece together the mechanisms by which cilia impact development of the cerebellum, hippocampus and the retina. The cerebellum comprises numerous granule cells, which make up more than 50% of all the neurons in the brain. It also contains Purkinje cells – large, complex neurons that control motor coordination. Normally the cerebellum undergoes massive proliferation of granule cell progenitors during development but this is interrupted in the absence of cilia (Chizhikov et al. 2007). Other studies have shown that the loss of cilia genes prevented granule cell progenitors from responding to sonic hedgehog signals important for guiding axons and controlling division of stem cells (Breunig 2008; Spassky 2008). More recently Han et al. (2008) identified a requirement for cilia in the dentate gyrus progenitor cells of the hippocampus, and linked this to volume loss and memory deficits in rodents.

Importantly, ciliary function could be partially restored resulting in improvements in recall and maze navigation (Amador-Arjona et al. 2011).

Already, the combined of use of the cilia proteome (Gherman et al. 2006) with syndrome categorisation (Baker and Beales 2009) has proved invaluable in predicting and defining many ciliopathies. It remains a challenge to extend their benefits to clinical diagnosis (targeted capture/whole exome analysis), prognosis (genotype-phenotype correlates), recurrence risks (confounded by intrafamilial variability of phenotypes), and ultimately clinical management of disease, including amelioration of CNS disruptions. The latter is now being actively explored by several labs whereby the approaches fall into three main categories (1) gene therapy (retinal disease) (2) cell therapies (retinal disease) and (3) drug therapy. Drug therapies hold great promise as it may be possible to administer compounds systemically to either limit damage to cells resulting from protein misfolding events, or to genetically modify the translation of damaged proteins in a subset of patients. Other considerations for drug therapy especially for neuronal disease will be the ability of test compounds to cross the blood–brain barrier. Biomarkers to monitor the impact of any interventions targeted at CNS disease are lacking, and must reflect clinically-meaningful outcomes. What is certain is that it will take the combined efforts of several disciplines to come up with effective therapeutic interventions for ciliopathies, a promise that may be realised sooner than expected.

References

Alvarez Retuerto AI, Cantor RM, Gleeson JG, Ustaszewska A, Schackwitz WS, Pennacchio LA, Geschwind DH (2008) Association of common variants in the Joubert syndrome gene (AHI1) with autism. Hum Mol Genet 17:3887–3896

Amador-Arjona A, Elliott J, Miller A, Ginbey A, Pazour GJ, Enikolopov G, Roberts AJ, Terskikh AV (2011) Primary cilia regulate proliferation of amplifying progenitors in adult hippocampus: implications for learning and memory. J Neurosci 31:9933–9944

Amar MJ, Sutphen R, Kousseff BG (1997) Expanded phenotype of cranioectodermal dysplasia (Sensenbrenner syndrome). Am J Med Genet 70:349–352

Arts HH, Bongers EM, Mans DA, van Beersum SE, Oud MM, Bolat E, Spruijt L, Cornelissen EA, Schuurs-Hoeijmakers JH, de Leeuw N, Cormier-Daire V, Brunner HG, Knoers NV, Roepman R (2011) C14ORF179 encoding IFT43 is mutated in Sensenbrenner syndrome. J Med Genet 48:390–395

Baala L, Audollent S, Martinovic J, Ozilou C, Babron MC, Sivanandamoorthy S, Saunier S, Salomon R, Gonzales M, Rattenberry E, Esculpavit C, Toutain A, Moraine C, Parent P, Marcorelles P, Dauge MC, Roume J, Le MM, Meiner V, Meir K, Menez F, Beaufrere AM, Francannet C, Tantau J, Sinico M, Dumez Y, MacDonald F, Munnich A, Lyonnet S, Gubler MC, Genin E, Johnson CA, Vekemans M, Encha-Razavi F, Attie-Bitach T (2007a) Pleiotropic effects of CEP290 (NPHP6) mutations extend to Meckel syndrome. Am J Hum Genet 81:170–179

Baala L, Romano S, Khaddour R, Saunier S, Smith UM, Audollent S, Ozilou C, Faivre L, Laurent N, Foliguet B, Munnich A, Lyonnet S, Salomon R, Encha-Razavi F, Gubler MC, Boddaert N, de Lonlay P, Johnson CA, Vekemans M, Antignac C, Attie-Bitach T (2007b) The Meckel-Gruber syndrome gene, MKS3, is mutated in Joubert syndrome. Am J Hum Genet 80:186–194

Bachmann-Gagescu R, Ishak GE, Dempsey JC, Adkins J, O'Day D, Phelps IG, Gunay-Aygun M, Kline AD, Szczaluba K, Martorell L, Alswaid A, Alrasheed S, Pai S, Izatt L, Ronan A, Parisi MA, Mefford H, Glass I, Doherty D (2012) Genotype-phenotype correlation in CC2D2A-related Joubert syndrome reveals an association with ventriculomegaly and seizures. J Med Genet 49:126–137

Baker K, Beales PL (2009) Making sense of cilia in disease: the human ciliopathies. Am J Med Genet C Semin Med Genet 151C:281–295

Baker K, Northam GB, Chong WK, Banks T, Beales P, Baldeweg T (2011) Neocortical and hippocampal volume loss in a human ciliopathy: a quantitative MRI study in Bardet-Biedl syndrome. Am J Med Genet A 155:1–8

Barkovich AJ, Millen KJ, Dobyns WB (2009) A developmental and genetic classification for midbrain-hindbrain malformations. Brain 132:3199–3230

Barnett S, Reilly S, Carr L, Ojo I, Beales PL, Charman T (2002) Behavioural phenotype of Bardet-Biedl syndrome. J Med Genet 39:e76

Baskin E, Kayiran SM, Oto S, Alehan F, Agildere AM, Saatci U (2002) Cerebellar vermis hypoplasia in a patient with Bardet-Biedl syndrome. J Child Neurol 17:385–387

Beales PL, Elcioglu N, Woolf AS, Parker D, Flinter FA (1999) New criteria for improved diagnosis of Bardet-Biedl syndrome: results of a population survey. J Med Genet 36:437–446

Behairy NH, Talaat S, Saleem SN, El-Raouf MA (2010) Magnetic resonance imaging in fetal anomalies: what does it add to 3D and 4D US? Eur J Radiol 74:250–255

Bennouna-Greene V, Kremer S, Stoetzel C, Christmann D, Schuster C, Durand M, Verloes A, Sigaudy S, Holder-Espinasse M, Godet J, Brandt C, Marion V, Danion A, Dietemann JL, Dollfus H (2011) Hippocampal dysgenesis and variable neuropsychiatric phenotypes in patients with Bardet-Biedl syndrome underline complex CNS impact of primary cilia. Clin Genet 80:523–531

Biesecker LG (2006) What you can learn from one gene: GLI3. J Med Genet 43:465–469

Braddock BA, Farmer JE, Deidrick KM, Iverson JM, Maria BL (2006) Oromotor and communication findings in joubert syndrome: further evidence of multisystem apraxia. J Child Neurol 21:160–163

Bredrup C, Saunier S, Oud MM, Fiskerstrand T, Hoischen A, Brackman D, Leh SM, Midtbo M, Filhol E, Bole-Feysot C, Nitschke P, Gilissen C, Haugen OH, Sanders JS, Stolte-Dijkstra I, Mans DA, Steenbergen EJ, Hamel BC, Matignon M, Pfundt R, Jeanpierre C, Boman H, Rodahl E, Veltman JA, Knappskog PM, Knoers NV, Roepman R, Arts HH (2011) Ciliopathies with skeletal anomalies and renal insufficiency due to mutations in the IFT-A gene WDR19. Am J Hum Genet 89:634–643

Breunig JJ, Sarkisian MR, Arellano JI, Morozov YM, Ayoub AE, Sojitra S, Wang B, Flavell RA, Rakic P, Town T (2008) Primary cilia regulate hippocampal neurogenesis by mediating sonic hedgehog signaling. Proc Natl Acad Sci USA 105:13127–13132

Bridge H, Cowey A, Ragge N, Watkins K (2009) Imaging studies in congenital anophthalmia reveal preservation of brain architecture in 'visual' cortex. Brain 132:3467–3480

Burt B, Levine J, Le K (2012) Adult-onset bulbar ptosis in Joubert syndrome. Clin Ophthal 6:151–153

Chen CP (2007) Meckel syndrome: genetics, perinatal findings, and differential diagnosis. Taiwan J Obstet Gynecol 46:9–14

Chiang AP, Beck JS, Yen HJ, Tayeh MK, Scheetz TE, Swiderski RE, Nishimura DY, Braun TA, Kim KY, Huang J, Elbedour K, Carmi R, Slusarski DC, Casavant TL, Stone EM, Sheffield VC (2006) Homozygosity mapping with SNP arrays identifies TRIM32, an E3 ubiquitin ligase, as a Bardet-Biedl syndrome gene (BBS11). Proc Natl Acad Sci U S A 103:6287–6292

Chizhikov VV, Davenport J, Zhang Q, Shih EK, Cabello OA, Fuchs JL, Yoder BK, Millen KJ (2007) Cilia proteins control cerebellar morphogenesis by promoting expansion of the granule progenitor pool. J Neurosci 27:9780–9789

Collin GB, Marshall JD, Ikeda A, So WV, Russell-Eggitt I, Maffei P, Beck S, Boerkoel CF, Sicolo N, Martin M, Nishina PM, Naggert JK (2002) Mutations in ALMS1 cause obesity, type 2 diabetes and neurosensory degeneration in Alstrom syndrome. Nat Genet 31:74–78

Coppieters F, Casteels I, Meire F, De Jaegere S, Hooghe S, van Regemorter N, Van Esch H, Matuleviciene A, Nunes L, Meersschaut V, Walraedt S, Standaert L, Coucke P, Hoeben H, Kroes HY, Vande WJ, de Ravel T, Leroy BP, De Baere E (2010a) Genetic screening of LCA in Belgium: predominance of CEP290 and identification of potential modifier alleles in AHI1 of CEP290-related phenotypes. Hum Mutat 31:E1709–E1766

Coppieters F, Lefever S, Leroy BP, De BE (2010b) CEP290, a gene with many faces: mutation overview and presentation of CEP290base. Hum Mutat 31:1097–1108

Davis EE, Zhang Q, Liu Q, Diplas BH, Davey LM, Hartley J, Stoetzel C, Szymanska K, Ramaswami G, Logan CV, Muzny DM, Young AC, Wheeler DA, Cruz P, Morgan M, Lewis LR, Cherukuri P, Maskeri B, Hansen NF, Mullikin JC, Blakesley RW, Bouffard GG, Gyapay G, Rieger S, Tonshoff B, Kern I, Soliman NA, Neuhaus TJ, Swoboda KJ, Kayserili H, Gallagher TE, Lewis RA, Bergmann C, Otto EA, Saunier S, Scambler PJ, Beales PL, Gleeson JG, Maher ER, Attie-Bitach T, Dollfus H, Johnson CA, Green ED, Gibbs RA, Hildebrandt F, Pierce EA, Katsanis N (2011) TTC21B contributes both causal and modifying alleles across the ciliopathy spectrum. Nat Genet 43:189–196

Farmer JE, Deidrick KM, Gitten JC, Fennell EB, Maria BL (2006) Parenting stress and its relationship to the behavior of children with Joubert syndrome. J Child Neurol 21:163–167

Fazzi E, Signorini SG, Uggetti C, Bianchi PE, Lanners J, Lanzi G (2005) Towards improved clinical characterization of Leber congenital amaurosis: neurological and systemic findings. Am J Med Genet A 132A:13–19

Ferland RJ, Eyaid W, Collura RV, Tully LD, Hill RS, Al-Nouri D, Al-Rumayyan A, Topcu M, Gascon G, Bodell A, Shugart YY, Ruvolo M, Walsh CA (2004) Abnormal cerebellar development and axonal decussation due to mutations in AHI1 in Joubert syndrome. Nat Genet 36:1008–1013

Forsythe E, Beales PL (2012) Bardet-Biedl syndrome

Gherman A, Davis EE, Katsanis N (2006) The ciliary proteome database: an integrated community resource for the genetic and functional dissection of cilia. Nat Genet 38:961–962

Gilissen C, Arts HH, Hoischen A, Spruijt L, Mans DA, Arts P, van Lier B, Steehouwer M, van Reeuwijk J, Kant SG, Roepman R, Knoers NV, Veltman JA, Brunner HG (2010) Exome sequencing identifies WDR35 variants involved in Sensenbrenner syndrome. Am J Hum Genet 87:418–423

Gitten J, Dede D, Fennell E, Quisling R, Maria BL (1998) Neurobehavioral development in Joubert syndrome. J Child Neurol 13:391–397

Gunay-Aygun M, Parisi MA, Doherty D, Tuchman M, Tsilou E, Kleiner DE, Huizing M, Turkbey B, Choyke P, Guay-Woodford L, Heller T, Szymanska K, Johnson CA, Glass I, Gahl WA (2009) MKS3-related ciliopathy with features of autosomal recessive polycystic kidney disease, nephronophthisis, and Joubert Syndrome. J Pediatr 155:386–392

Han YG, Spassky N, Romaguera-Ros M, Garcia-Verdugo JM, Aguilar A, Schneider-Maunoury S, Alvarez-Buylla A (2008) Hedgehog signaling and primary cilia are required for the formation of adult neural stem cells. Nat Neurosci 11:277–284

Hills CB, Kochilas L, Schimmenti LA, Moller JH (2011) Ellis-van Creveld syndrome and congenital heart defects: presentation of an additional 32 cases. Pediatr Cardiol 32:977–982

Hodgkins PR, Harris CM, Shawkat FS, Thompson DA, Chong K, Timms C, Russell-Eggitt I, Taylor DS, Kriss A (2004) Joubert syndrome: long-term follow-up. Dev Med Child Neurol 46:694–699

Holroyd S, Reiss AL, Bryan RN (1991) Autistic features in Joubert syndrome: a genetic disorder with agenesis of the cerebellar vermis. Biol Psychiatry 29:287–294

Hopp K, Heyer CM, Hommerding CJ, Henke SA, Sundsbak JL, Patel S, Patel P, Consugar MB, Czarnecki PG, Gliem TJ, Torres VE, Rossetti S, Harris PC (2011) B9D1 is revealed as a novel Meckel syndrome (MKS) gene by targeted exon-enriched next-generation sequencing and deletion analysis. Hum Mol Genet 20:2524–2534

Iannaccone A, Mykytyn K, Persico AM, Searby CC, Baldi A, Jablonski MM, Sheffield VC (2005) Clinical evidence of decreased olfaction in Bardet-Biedl syndrome caused by a deletion in the BBS4 gene. Am J Med Genet A 132:343–346

Iannicelli M, Brancati F, Mougou-Zerelli S, Mazzotta A, Thomas S, Elkhartoufi N, Travaglini L, Gomes C, Ardissino GL, Bertini E, Boltshauser E, Castorina P, D'Arrigo S, Fischetto R, Leroy B, Loget P, Bonniere M, Starck L, Tantau J, Gentilin B, Majore S, Swistun D, Flori E, Lalatta F, Pantaleoni C, Penzien J, Grammatico P, Dallapiccola B, Gleeson JG, Attie-Bitach T, Valente EM (2010) Novel TMEM67 mutations and genotype-phenotype correlates in meckelin-related ciliopathies. Hum Mutat 31:E1319–E1331

Inglis PN (2006) Piecing together a ciliome. Trends Genet 22:491–500

Jagger D, Collin G, Kelly J, Towers E, Nevill G, Longo-Guess C, Benson J, Halsey K, Dolan D, Marshall J, Naggert J, Forge A (2011) Alstrom Syndrome protein ALMS1 localizes to basal bodies of cochlear hair cells and regulates cilium-dependent planar cell polarity. Hum Mol Genet 20:466–481

Janssen S, Ramaswami G, Davis EE, Hurd T, Airik R, Kasanuki JM, Van Der KL, Allen SJ, Beales PL, Katsanis N, Otto EA, Hildebrandt F (2011) Mutation analysis in Bardet-Biedl syndrome by DNA pooling and massively parallel resequencing in 105 individuals. Hum Genet 129:79–90

Juric-Sekhar G, Adkins J, Doherty D, Hevner RF (2012) Joubert syndrome: brain and spinal cord malformations in genotyped cases and implications for neurodevelopmental functions of primary cilia. Acta Neuropathol 123:695–709

Kamiya A, Tan PL, Kubo K, Engelhard C, Ishizuka K, Kubo A, Tsukita S, Pulver AE, Nakajima K, Cascella NG, Katsanis N, Sawa A (2008) Recruitment of PCM1 to the centrosome by the cooperative action of DISC1 and BBS4: a candidate for psychiatric illnesses. Arch Gen Psychiatry 65:996–1006

Kana RK, Wadsworth HM, Travers BG (2011) A systems level analysis of the mirror neuron hypothesis and imitation impairments in autism spectrum disorders. Neurosci Biobehav Rev 35:894–902

Katsanis N, Ansley SJ, Badano JL, Eichers ER, Lewis RA, Hoskins BE, Scambler PJ, Davidson WS, Beales PL, Lupski JR (2001) Triallelic inheritance in Bardet-Biedl syndrome, a Mendelian recessive disorder. Science 293:2256–2259

Keppler-Noreuil KM, Adam MP, Welch J, Muilenburg A, Willing MC (2011a) Clinical insights gained from eight new cases and review of reported cases with Jeune syndrome (asphyxiating thoracic dystrophy). Am J Med Genet A 155A:1021–1032

Keppler-Noreuil KM, Blumhorst C, Sapp JC, Brinckman D, Johnston J, Nopoulos PC, Biesecker LG (2011b) Brain tissue- and region-specific abnormalities on volumetric MRI scans in 21 patients with Bardet-Biedl syndrome (BBS). BMC Med Genet 12:101

Khaddour R, Smith U, Baala L, Martinovic J, Clavering D, Shaffiq R, Ozilou C, Cullinane A, Kyttala M, Shalev S, Audollent S, d'Humieres C, Kadhom N, Esculpavit C, Viot G, Boone C, Oien C, Encha-Razavi F, Batman PA, Bennett CP, Woods CG, Roume J, Lyonnet S, Genin E, Le MM, Munnich A, Gubler MC, Cox P, MacDonald F, Vekemans M, Johnson CA, Attie-Bitach T (2007) Spectrum of MKS1 and MKS3 mutations in Meckel syndrome: a genotype-phenotype correlation. Online. Hum Mutat 28:523–524 (Mutation in brief #960)

Khan AO, Oystreck DT, Seidahmed MZ, AlDrees A, Elmalik SA, Alorainy IA, Salih MA (2008) Ophthalmic features of Joubert syndrome. Ophthalmology 115:2286–2289

Khanna H, Davis EE, Murga-Zamalloa CA, Estrada-Cuzcano A, Lopez I, den Hollander AI, Zonneveld MN, Othman MI, Waseem N, Chakarova CF, Maubaret C, Diaz-Font A, MacDonald I, Muzny DM, Wheeler DA, Morgan M, Lewis LR, Logan CV, Tan PL, Beer MA, Inglehearn CF, Lewis RA, Jacobson SG, Bergmann C, Beales PL, Attie-Bitach T, Johnson CA, Otto EA, Bhattacharya SS, Hildebrandt F, Gibbs RA, Koenekoop RK, Swaroop A, Katsanis N (2009) A common allele in RPGRIP1L is a modifier of retinal degeneration in ciliopathies. Nat Genet 41:739–745

Kulaga HM, Leitch CC, Eichers ER, Badano JL, Lesemann A, Hoskins BE, Lupski JR, Beales PL, Reed RR, Katsanis N (2004) Loss of BBS proteins causes anosmia in humans and defects in olfactory cilia structure and function in the mouse. Nat Genet 36:994–998

Lancaster MA, Gopal DJ, Kim J, Saleem SN, Silhavy JL, Louie CM, Thacker BE, Williams Y, Zaki MS, Gleeson JG (2011) Defective Wnt-dependent cerebellar midline fusion in a mouse model of Joubert syndrome. Nat Med 17:726–731

Lee SK, Kim DI, Kim J, Kim DJ, Kim HD, Kim DS, Mori S (2005) Diffusion-tensor MR imaging and fiber tractography: a new method of describing aberrant fiber connections in developmental CNS anomalies. Radiographics 25:53–65

Leitch CC, Zaghloul NA, Davis EE, Stoetzel C, Diaz-Font A, Rix S, Alfadhel M, Lewis RA, Eyaid W, Banin E, Dollfus H, Beales PL, Badano JL, Katsanis N (2008) Hypomorphic mutations in syndromic encephalocele genes are associated with Bardet-Biedl syndrome. Nat Genet 40:443–448

Leung GK, Fan YW (2005) Chronic subdural haematoma and arachnoid cyst in autosomal dominant polycystic kidney disease (ADPKD). J Clin Neurosci 12:817–819

Logan CV, Abdel-Hamed Z, Johnson CA (2011) Molecular genetics and pathogenic mechanisms for the severe ciliopathies: insights into neurodevelopment and pathogenesis of neural tube defects. Mol Neurobiol 43:12–26

Louie CM, Caridi G, Lopes VS, Brancati F, Kispert A, Lancaster MA, Schlossman AM, Otto EA, Leitges M, Grone HJ, Lopez I, Gudiseva HV, O'Toole JF, Vallespin E, Ayyagari R, Ayuso C, Cremers FP, den Hollander AI, Koenekoop RK, Dallapiccola B, Ghiggeri GM, Hildebrandt F, Valente EM, Williams DS, Gleeson JG (2010) AHI1 is required for photoreceptor outer segment development and is a modifier for retinal degeneration in nephronophthisis. Nat Genet 42:175–180

Maria BL, Boltshauser E, Palmer SC, Tran TX (1999) Clinical features and revised diagnostic criteria in Joubert syndrome. J Child Neurol 14:583–590

Marshall JD, Bronson RT, Collin GB, Nordstrom AD, Maffei P, Paisey RB, Carey C, Macdermott S, Russell-Eggitt I, Shea SE, Davis J, Beck S, Shatirishvili G, Mihai CM, Hoeltzenbein M, Pozzan GB, Hopkinson I, Sicolo N, Naggert JK, Nishina PM (2005) New Alstrom syndrome phenotypes based on the evaluation of 182 cases. Arch Intern Med 165:675–683

Marshall JD, Maffei P, Collin GB, Naggert JK (2011) Alstrom syndrome: genetics and clinical overview. Curr Genomics 12:225–235

McEwen DP, Koenekoop RK, Khanna H, Jenkins PM, Lopez I, Swaroop A, Martens JR (2007) Hypomorphic CEP290/NPHP6 mutations result in anosmia caused by the selective loss of G proteins in cilia of olfactory sensory neurons. Proc Natl Acad Sci U S A 104:15917–15922

McKusick-Nathans Institute of Genetic Medicine JHUSoM (2011) Online Mendelian inheritance in man

Moore SJ, Green JS, Fan Y, Bhogal AK, Dicks E, Fernandez BA, Stefanelli M, Murphy C, Cramer BC, Dean JC, Beales PL, Katsanis N, Bassett AS, Davidson WS, Parfrey PS (2005) Clinical and genetic epidemiology of Bardet-Biedl syndrome in Newfoundland: a 22-year prospective, population-based, cohort study. Am J Med Genet A 132:352–360

Nugent AA, Kolpak AL, Engle EC (2012) Human disorders of axon guidance. Curr Opin Neurobiol

Odent S, Le MB, Toutain A, David A, Vigneron J, Treguier C, Jouan H, Milon J, Fryns JP, Verloes A (1998) Central nervous system malformations and early end-stage renal disease in oro-facio-digital syndrome type I: a review. Am J Med Genet 75:389–394

Ozonoff S, Williams BJ, Gale S, Miller JN (1999) Autism and autistic behavior in Joubert syndrome. J Child Neurol 14:636–641

Paisey RB, Paisey RM, Thomson MP, Bower L, Maffei P, Shield JP, Barnett S, Marshall JD (2009) Protection from clinical peripheral sensory neuropathy in Alstrom syndrome in contrast to early-onset type 2 diabetes. Diabetes Care 32:462–464

Parisi MA (2009) Clinical and molecular features of Joubert syndrome and related disorders. Am J Med Genet C Semin Med Genet 151C:326–340

Parisi MA, Pinter JD, Glass IA, Field K, Maria BL, Chance PF, Mahurin RK, Cramer SC (2004) Cerebral and cerebellar motor activation abnormalities in a subject with Joubert syndrome: functional magnetic resonance imaging (MRI) study. J Child Neurol 19:214–218

Perrault I, Delphin N, Hanein S, Gerber S, Dufier JL, Roche O, Defoort-Dhellemmes S, Dollfus H, Fazzi E, Munnich A, Kaplan J, Rozet JM (2007) Spectrum of NPHP6/CEP290 mutations in Leber congenital amaurosis and delineation of the associated phenotype. Hum Mutat 28:416

Poretti A, Boltshauser E, Loenneker T, Valente EM, Brancati F, Il'yasov K, Huisman TA (2007) Diffusion tensor imaging in Joubert syndrome. AJNR Am J Neuroradiol 28:1929–1933

Poretti A, Huisman TA, Scheer I, Boltshauser E (2011) Joubert syndrome and related disorders: spectrum of neuroimaging findings in 75 patients. AJNR Am J Neuroradiol 32:1459–1463

Ptito M, Schneider FC, Paulson OB, Kupers R (2008) Alterations of the visual pathways in congenital blindness. Exp Brain Res 187:41–49

Putoux A, Thomas S, Coene KL, Davis EE, Alanay Y, Ogur G, Uz E, Buzas D, Gomes C, Patrier S, Bennett CL, Elkhartoufi N, Frison MH, Rigonnot L, Joye N, Pruvost S, Utine GE, Boduroglu K, Nitschke P, Fertitta L, Thauvin-Robinet C, Munnich A, Cormier-Daire V, Hennekam R, Colin E, Akarsu NA, Bole-Feysot C, Cagnard N, Schmitt A, Goudin N, Lyonnet S, Encha-Razavi F, Siffroi JP, Winey M, Katsanis N, Gonzales M, Vekemans M, Beales PL, Attie-Bitach T (2011) KIF7 mutations cause fetal hydrolethalus and acrocallosal syndromes. Nat Genet 43:601–606

Quisling RG, Barkovich AJ, Maria BL (1999) Magnetic resonance imaging features and classification of central nervous system malformations in Joubert syndrome. J Child Neurol 14:628–635

Rooryck C, Pelras S, Chateil JF, Cances C, Arveiler B, Verloes A, Lacombe D, Goizet C (2007) Bardet-Biedl syndrome and brain abnormalities. Neuropediatrics 38:5–9

Salomon R, Saunier S, Niaudet P (2009) Nephronophthisis. Pediatr Nephrol 24:2333–2344

Sang L, Miller JJ, Corbit KC, Giles RH, Brauer MJ, Otto EA, Baye LM, Wen X, Scales SJ, Kwong M, Huntzicker EG, Sfakianos MK, Sandoval W, Bazan JF, Kulkarni P, Garcia-Gonzalo FR, Seol AD, O'Toole JF, Held S, Reutter HM, Lane WS, Rafiq MA, Noor A, Ansar M, Devi AR, Sheffield VC, Slusarski DC, Vincent JB, Doherty DA, Hildebrandt F, Reiter JF, Jackson PK (2011) Mapping the NPHP-JBTS-MKS protein network reveals ciliopathy disease genes and pathways. Cell 145:513–528

Sartori S, Ludwig K, Fortuna M, Marzocchi C, Calderone M, Toldo I, Salviati L, Laverda AM, Tenconi R (2010) Dandy-Walker malformation masking the molar tooth sign: an illustrative case with magnetic resonance imaging follow-up. J Child Neurol 25:1419–1422

Sattar S, Gleeson JG (2011) The ciliopathies in neuronal development: a clinical approach to investigation of Joubert syndrome and Joubert syndrome-related disorders. Dev Med Child Neurol 53:793–798

Schievink WI, Huston J III, Torres VE, Marsh WR (1995) Intracranial cysts in autosomal dominant polycystic kidney disease. J Neurosurg 83:1004–1007

Schuil J, Meire FM, Delleman JW (1998) Mental retardation in amaurosis congenita of Leber. Neuropediatrics 29:294–297

Senocak EU, Oguz KK, Haliloglu G, Topcu M, Cila A (2010) Structural abnormalities of the brain other than molar tooth sign in Joubert syndrome-related disorders. Diagn Interv Radiol 16:3–6

Slavotinek AM, Biesecker LG (2000) Phenotypic overlap of McKusick-Kaufman syndrome with Bardet-Biedl syndrome: a literature review. Am J Med Genet 95:208–215

Spampinato MV, Kraas J, Maria BL, Walton ZJ, Rumboldt Z (2008) Absence of decussation of the superior cerebellar peduncles in patients with Joubert syndrome. Am J Med Genet A 146A:1389–1394

Spassky N, Han YG, Aguilar A, Strehl L, Besse L, Laclef C, Ros MR, Garcia-Verdugo JM, Alvarez-Buylla A. (2008) Primary cilia are required for cerebellar development and Shh-dependent expansion of progenitor pool. Dev Biol 317(1):246–259.

Stevens CA, Lachman RS (2010) New lethal skeletal dysplasia with Dandy-Walker malformation, congenital heart defects, abnormal thumbs, hypoplastic genitalia, and distinctive facies. Am J Med Genet A 152A:1915–1918

Tallila J, Salonen R, Kohlschmidt N, Peltonen L, Kestila M (2009) Mutation spectrum of Meckel syndrome genes: one group of syndromes or several distinct groups? Hum Mutat 30: E813–E830

Thauvin-Robinet C, Cossee M, Cormier-Daire V, Van ML, Toutain A, Alembik Y, Bieth E, Layet V, Parent P, David A, Goldenberg A, Mortier G, Heron D, Sagot P, Bouvier AM, Huet F, Cusin V, Donzel A, Devys D, Teyssier JR, Faivre L (2006) Clinical, molecular, and genotype-phenotype correlation studies from 25 cases of oral-facial-digital syndrome type 1: a French and Belgian collaborative study. J Med Genet 43:54–61

Thauvin-Robinet C, Lesca G, Aral B, Gigot N, Lambert S, Gueneau L, Macca M, Franco B, Huet F, Zabot MT, Attie-Bitach T, Attia-Sobol J, Faivre L (2011) Cerebral dysgenesis does not exclude OFD I syndrome. Am J Med Genet A 155A:455–457

Tobin JL, Di FM, Eichers E, May-Simera H, Garcia M, Yan J, Quinlan R, Justice MJ, Hennekam RC, Briscoe J, Tada M, Mayor R, Burns AJ, Lupski JR, Hammond P, Beales PL (2008) Inhibition of neural crest migration underlies craniofacial dysmorphology and Hirschsprung's disease in Bardet-Biedl syndrome. Proc Natl Acad Sci U S A 105:6714–6719

Tompson SW, Ruiz-Perez VL, Blair HJ, Barton S, Navarro V, Robson JL, Wright MJ, Goodship JA (2007) Sequencing EVC and EVC2 identifies mutations in two-thirds of Ellis-van Creveld syndrome patients. Hum Genet 120:663–670

Tory K, Lacoste T, Burglen L, Moriniere V, Boddaert N, Macher MA, Llanas B, Nivet H, Bensman A, Niaudet P, Antignac C, Salomon R, Saunier S (2007) High NPHP1 and NPHP6 mutation rate in patients with Joubert syndrome and nephronophthisis: potential epistatic effect of NPHP6 and AHI1 mutations in patients with NPHP1 mutations. J Am Soc Nephrol 18:1566–1575

Valente EM, Brancati F, Silhavy JL, Castori M, Marsh SE, Barrano G, Bertini E, Boltshauser E, Zaki MS, Abdel-Aleem A, Abdel-Salam GM, Bellacchio E, Battini R, Cruse RP, Dobyns WB, Krishnamoorthy KS, Lagier-Tourenne C, Magee A, Pascual-Castroviejo I, Salpietro CD, Sarco D, Dallapiccola B, Gleeson JG (2006) AHI1 gene mutations cause specific forms of Joubert syndrome-related disorders. Ann Neurol 59:527–534

Valente EM, Brancati F, Dallapiccola B (2008) Genotypes and phenotypes of Joubert syndrome and related disorders. Eur J Med Genet 51:1–23

Van Den AK, Craen M, Schuil J, Meire FM (2001) Ophthalmologic and systemic features of the Alstrom syndrome: report of 9 cases. Bull Soc Belge Ophtalmol 281:67–72

Vargha-Khadem F, Gadian DG, Copp A, Mishkin M (2005) FOXP2 and the neuroanatomy of speech and language. Nat Rev Neurosci 6:131–138

Walczak-Sztulpa J, Eggenschwiler J, Osborn D, Brown DA, Emma F, Klingenberg C, Hennekam RC, Torre G, Garshasbi M, Tzschach A, Szczepanska M, Krawczynski M, Zachwieja J, Zwolinska D, Beales PL, Ropers HH, Latos-Bielenska A, Kuss AW (2010) Cranioectodermal Dysplasia, Sensenbrenner syndrome, is a ciliopathy caused by mutations in the IFT122 gene. Am J Hum Genet 86:949–956

Weatherbee SD, Niswander LA, Anderson KV (2009) A mouse model for Meckel syndrome reveals Mks1 is required for ciliogenesis and Hedgehog signaling. Hum Mol Genet 18:4565–4575

Weiss AH, Doherty D, Parisi M, Shaw D, Glass I, Phillips JO (2009) Eye movement abnormalities in Joubert syndrome. Invest Ophthalmol Vis Sci 50:4669–4677

Wiszniewski W, Lewis RA, Stockton DW, Peng J, Mardon G, Chen R, Lupski JR (2011) Potential involvement of more than one locus in trait manifestation for individuals with Leber congenital amaurosis. Hum Genet 129:319–327

Yachnis AT, Rorke LB (1999) Neuropathology of Joubert syndrome. J Child Neurol 14:655–659

Yang HK, Hwang JM, Park SS, Yu YS (2010) Brain imaging studies in Leber's congenital amaurosis: new radiologic findings associated with the complex trait. Korean J Ophthalmol 24:360–363

Yzer S, Hollander AI, Lopez I, Pott JW, Faber JT, Cremers FP, Koenekoop RK, Born LI (2012) Ocular and extra-ocular features of patients with Leber congenital amaurosis and mutations in CEP290. Mol Vis 18:412–425

Zaki MS, Abdel-Aleem A, Abdel-Salam G, Marsh SE, Silhavy JL, Barkovich AJ, Ross ME, Saleem SN, Dobyns WB, Gleeson JG (2008) The molar tooth sign: a new Joubert syndrome and related cerebellar disorders classification system tested in Egyptian families. Neurology 70:556–565

Zaki MS, Sattar S, Massoudi RA, Gleeson JG (2011) Co-occurrence of distinct ciliopathy diseases in single families suggests genetic modifiers. Am J Med Genet A 155A:3042–3049

Index

K.L. Tucker and T. Caspary (eds.), *Cilia and Nervous System Development and Function*, DOI 10.1007/978-94-007-5808-7, © Springer Science+Business Media Dordrecht 2013

MIX
Papier aus verantwortungsvollen Quellen
Paper from responsible sources
FSC® C105338

If you have any concerns about our products,
you can contact us on
ProductSafety@springernature.com

In case Publisher is established outside the EU,
the EU authorized representative is:
Springer Nature Customer Service Center GmbH
Europaplatz 3, 69115 Heidelberg, Germany

Printed by Libri Plureos GmbH
in Hamburg, Germany